Friedrich Johannaber · Injection Molding Machines

Friedrich Johannaber

Injection Molding Machines

A User's Guide

3rd Edition

Hanser Publishers, Munich Vienna New York

Hanser/Gardner Publications, Inc., Cincinnati

Dr.-Ing. F. Johannaber is Manager, Technical Machine Laboratories, Plastics Division, Bayer AG, Leverkusen.

Translated by Rolf J. Kahl M. E., ret. Manager Technical Services, Polymers Division, Plastics, Miles Inc., Pittsburgh, PA

Distributed in the USA and in Canada by
Hanser/Gardner Publications, Inc.
6600 Clough Pike, Cincinnati, Ohio 45244-4090, USA
Fax: + 1 (513) 527-8950

Distributed in all ohter countries by
Carl Hanser Verlag
Postfach 860 420, 81631 München, Germany
Fax: + 49 (89) 98 12 64

Library of Congress Cataloging-in-Publication Data
Johannaber, Friedrich.
[Spritzgiessmaschinen. English]
Injection molding machines : a user's guide / Friedrich
Johannaber. -- 3rd ed.
p. cm.
Includes bibliographical references and index.
ISBN 1-56990-169-4 :
1. Injection molding of plastics. I. Title.
TP1150.J6413 1994 94-29974
668.4'12'028--dc20

Die Deutsche Bibliothek – CIP-Einheitsaufnahme
Johannaber, Friedrich:
Injection molding machines : a user's guide/Friedrich
Johannaber. [Transl. by Rolf J. Kahl]. – 3. ed. – Munich ;
Vienna ; New York : Hanser ; Cincinnati : Hanser/Gardner,
1994
Einheitssacht.: Spritzgiessmaschinen <engl.>
Aus: Kunststoffmaschinenführer
ISBN 3-446-17733-7 (Hanser)

Typesetting: Gruber, Regensburg, Germany
Printed in Germany by Schoder Druck GmbH & Co. KG, Gersthofen

Foreword

Most of the wide range of books available on plastics processing and injection molding in particular are aimed primarily at specialists with a scientific background who are often not concerned with the practical aspects of running a molding shop. Hence the need for an expert with that rare combination of both qualities.

That is still a fact 11 years after the first edition of this book. Altough we do not have changes in the basic of this technology, there are big progresses in some of the process applications where special necessities have to be fulfilled by the machinery and especially the control of machines. All this points are carefully selected and introduced into the new, the third issue.

Following a scientific training at the Institute for Plastics Processing in Aachen (IKV), the author spend several decades designing injection molding machines, managing molding plants and training operators.

The present work could only have been written by someone with such extensive experience. It joins the small number of standard reference books for the industry which have solid educational and factual foundations. It is easily understood yet exhaustive and is thus a "Guide" in the true sense.

The author has provided an elegant, succinct description of the injection molding process and, by concentrating on a few key parameters such as pressure, temperature, their rates, and their influence on the properties of moldings, gives the reader a clear insight into this technique. The subsequent comprehensive presentation of technical data relating to individual machine components and performance is unique and will be gratefully received by many readers and especially by machine builders.

And as mentioned before, all this represents the latest state of the art.

The book is a valuable tool both for trainess and students following courses in technical institutes and for specialists engaged in design and processing. It will almost certainly become required reading for everyone involved in the vast field of injection molding.

I am certain that the reader will be unable to put it down until he has reached the last page.

Aachen, March 1994 Prof. Dr.-Ing. G. Menges

Preface to the Third Edition

This third edition shall create an awareness of the particular variability of injection molding, which can be called a unique technique in this respect. Much of the information presented in 1982 is still correct, of course. It represents the fundamentals. However, a multitude of supplementations have increased the attraction of injection molding enormously. New procedures and solutions for difficult tasks have been added, which are the results from high-level engineering and often prove to be especially economical. Therefore it seemed necessary to present the demands on processing first and then the engineering solutions. This enables the reader to recognize the technical beauty and the economic relevance in each individual case.

This edition is published in a time of worldwide economic uncertainties. In spite of this, there should not arise any concern that the existence of injection molding is endangered. It is one of the technologies of the future and will be considerably instrumental in further technical progress. Injection-molded engineering products will displace those made of other materials or, merged with conventional materials, make more economical solutions possible and, above all, solutions which are kind to our environment. The technique of injection molding and especially of designing injection molding machines is not only an interesting field of activities but also one with great prospects. Today's almost unrestricted global borders will open up additional chances.

The present edition is not only meant to be a record of the state of engineering but it may also provide assistance in resolving actual problems and answering questions of how to correlate the most suitable machine with a specific objective.

Leverkusen, May 1994 F. Johannaber

Preface to the First Edition

The subject of this publication emerged from a more extensive description of all plastics processing machinery. I was happy to follow the suggestions of some of my colleagues and of the Carl Hanser Publishing Company and expand the chapter on injection molding machinery, for translation into English and publication as an independent volume.

I am very grateful to Rolf J. Kahl of Pittsburgh, who took on the very difficult task of translating and adapting this text to meet the requirements of a wider readership; to Bayer AG, Leverkusen, without whose great help the writing of this book would not have been possible; and to Axel J. Kaminski, who assisted in compiling the English version.

Leverkusen, November 1982 F. Johannaber

Notes

I am very grateful to Arburg Machinenfabrik Hehl Söhne, Bayer AG, Ferromatik Milacron, Krauss-Maffei AG, Mannesmann Demag Kunststofftechnik, Netstal AG and Paul Pleiger GmbH, who willingly placed atmy disposal some very detailed technical information, and to Battenfeld GmbH, Robert Bosch GmbH, Engel KG, Husky GmbH, Canada, Reiloy GmbH, all of whom supplied me with additional valuable documents. Without their assistance, this work would not have been possible.

I also want to thank H. Recker, Dipl.-Ing. and L. Spix, Dipl.-Ing., who allowed me to adopt substantial parts of their chapters dealing with measuring and control systems from the "Kunststoffmaschinenführer".

Special thanks are due to my wife Monica for her patience with me during the completion of this work.

Contents

The Economic Significance of Injection Molding

The future of injection molding is closely connected with the natural product petroleum. According to some estimates, the global reserves of this product, as far as known today, will last for another 40 years if the consumption remains the same [1]. Injection molding is in the peculiar situation to deal most economically with this exhaustible source of raw material.

Nowadays, the injection molding technique is the one among all significant plastics processing methods which has the best efficiency ratio between applied material and end product. One sees the future of injection molding too bright, however, if one assumes unlimited growth. With free trade and serious competition, injection molding is in constant conflict with classic methods of producing finished articles. However, a more optimistic outlook gives plastics a good chance to compete with classic materials such as sheet metals and replace them for car bodies [2, 3]. This kind of predictions does not remain unchallenged, though.

It is certain that injection molded plastic articles are only used where they offer an economic advantage and can be employed without harm to the environment. Most recent applications under the headline "plastics reinforce steel" promote a completely new engineering idea. The simple shaping of slightly curved sheet metal and the equally simple injection molding of reinforcing plastic ribs is utilized to obtain especially strong structural parts [4].

Additional questions of environmental compatibility and acceptability in the public opinion will become increasingly instrumental in the future. A greater effect even can be expected by legislative actions [5]. A new institutional scope of regulations have brought a dense and diverse network of laws to the plastics industry, too. This creates new obligations. To meet them, additional staff is needed for the resulting control, safety, informative, and administrative functions. Plastic parts will have to overcome the obstacles of official approval, evidence of safe production, and enhanced suitability for recycling before they can be marketed.

Injection molding, which takes place in a closed system, is particularly adapted to meet these requirements. Its characteristics provide for very good opportunities in the days to come. Injection molding contributes in more than one way to a careful treatment of resources by its inherent possibilities of using raw materials economically and producing lightweight but strong engineering objects [5].

Scrap and rejects generated during processing can be largely recycled, almost by 100% [9, 10, 12]. Restrictive requirements of vendors will, without doubt, be retracted rather quick since it is already well known today that a complete reuse of waste from production such as runners and sprues, short shots etc. is possible without any loss in quality. Recycling is of course not only the task of suppliers of regrind and compounders, it starts with the design. This means a tremendous chance for such engineers who are capable of having the idea of reuse and recycling already in mind during the design stage and who decide how economical recycling will turn out (Fig. 1).

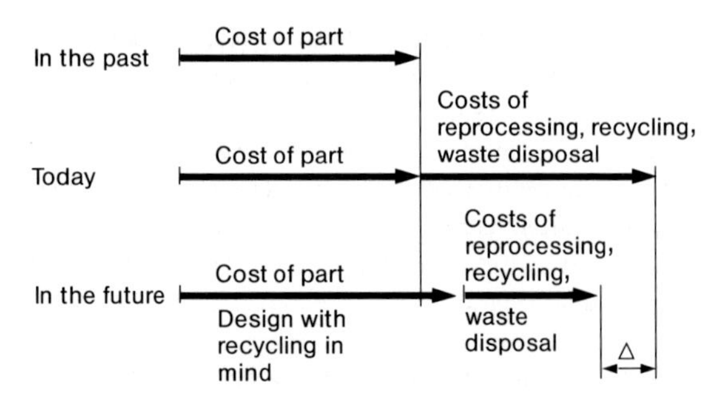

Fig. 1 In the future, the costs of an injection molded article will change because recycling and waste disposal must already be considered in the design stage

Re-employment of used, recycled plastic parts for the production of new ones will also be accepted very shortly at an increasing rate [7]. There are estimates that the amount of material used this way will increase globally from 2 million tons at present to 8 million in the year 2000 [8]. Even if this prognosis should be too optimistic, there will be a large market with new opportunities. There is no doubt that the bulk of these plastics will be processed by injection molding. Moldings, which have already been used, are well suited for materials of second choice as soon as the problem of separating the various grades is solved [12]. In contrast to materials like glass, metals, and even wood, where savings in raw material have to be made at the source, plastics can be used a second time as fuel to save precious petroleum because petroleum is the most important raw material source of the future. So far, this realization has remained concealed to the otherwise so active environmentalists; otherwise they would demonstrate against burning petroleum and call for employing used plastics as fuel.

In Europe the market for machinery developed parallel to that for plastic materials only to some extent. The reason for this is the large orientation of this industry towards export. At the top of the list of exporting countries is Germany [11]. Pressure from worldwide competition promotes economical production and application [14]. The author of this publication tends to provide a presentation in the sense of a progress report. New items are inevitably placed in the foreground. However machines with old design features are also new to the market with respect to the applied drive and control techniques. Such machines come from the Far East or South America. It is there, where the highly developed and computer-controlled automated European machines are considered uneconomic for a great deal of simple tasks because of their high prices. In spite of this, simple solutions e.g. a two-column machine of a German manufacturer can only remain the exception for highly industrialized countries. This does not mean that a value analysis in the sense of lean production and a consequent redesign are not important. This tendency became obvious at the K'92 in Duesseldorf [13]. Noticeable impressions of a technological progress over the past three years were better accessibility for part removal, less demand on floor space, lower noise level and less energy consumption, steps back to terminals close to the machine at the stationary platen without separate units and the provision of helpful data already present at standardized interfaces for the organization of subsequent work.

Conspicuous increases in efficiency occurred in the development of output capacity (screw speed) and reduction of idle times in the years after 1980.

However this so-called efficiency explosion has to be looked at in a relative manner because it partly resulted in a reduced quality of the melt (Sect. 4.3.1). There is a natural limit to idle times due to the problems of mass acceleration and their control. One has already arrived at these limits in earlier years. (Compare performance data in Chapter 8.2)

In the future, particular attention has to be paid to economic operation, cleanliness and noise generation of machines with electromechanical drive units. With this drive one reaches into a new dimension of an energy-saving and low-noise injection-molding technique [15, 16, 17, 18, 467].

It is significant that a considerable increase in available capacity and produced quantities could be achieved by recording and evaluating data of process and efficiency, and optimizing the process. Process models which make quality control possible by following up data relevant to the process automate quality control. Thus, electronic data processing results in a distinct improvement in the economics of injection molding.

1 General Design and Function

The beginning of injection molding and with this the origin of the injection molding machine can be dated rather precisely. The oldest known date should be 1872 when, in the US, J. W. Hyatt solved the problem of plasticating and shaping a mixture of nitrocellulose and camphor with his “packing machine” (US-patent 133229) [19].

The term injection molding was already used in typesetting in the early 19th century. The casting of types was called injection molding [20]. In his work “Injection Molding”, Uhlmann defined this procedure 1925 as a method to make formed bodies in “permanent molds under pressure”. The machine of J. W. Hyatt, however, was not yet called “injection molding machine”. He named it packing machine. It was preceded by the development of the world’s first synthetic plastic with the trade name “Celluloid” two years earlier.

A view at this early history of injection molding convincingly demonstrates the close association of plastics development, technology of injection molding machines, design, and process evolution [21, 22, 23, 24]. The unique success of these new materials had not taken place without this interplay, and it is still needed today.

The first machine which could be called injection molding machine was built by H. Buchholz in 1921. It was a plunger-type machine similar to the screw press as it was known in those days. This machine was, of course, actuated manually. The first injection molding machines produced in series were made by Eckert and Ziegler GmbH in 1926 [20, 25]. Although the mold was still clamped by hand, injection was done pneumatically. This machine was already built in a horizontal design, a principle that is still used today.

The road to prominence among plastics-processing machines began in 1956 with the introduction of the screw as plasticating and injection pressure generating component [26]. That machine already showed all the features of an injection molding machine as they are still typical today. H. Goller, (Ankerwerk Gebr. Goller), realized with it a pioneering idea of H. Beck. After all, it remains a surprising fact that the screw, obviously without the knowledge of H. Beck was subject of a patent application by Vorraber in 1905 but stayed undiscovered for more than 50 years [21, 29]. A very elaborate and informative description of this development until the year 1945 is provided by M. E. Laeis in his book “Injection Molding of Thermoplastic Materials” in 1959 [20]. It portrays the whole development with illustrations. A more abbreviated but likewise important description of this development has been written by R. Sonntag in 1985 [21, 29]. A retrospective view more directed at individuals and their engagement was recently published by W. Woebcken [22, 23]. It demonstrates how some inventors affected the history of injection molding to a high degree but did not always find recognition.

In the days of a G. Thilenius, H. Beck, H. Goller, and G. Friedrich, these inventors did not only invent the immaterial process but were capable of intellectually converting the process sequences and producing the material object likewise. With increasing complexity the generally common division of labor took place in the area of injection molding, too (Taylorism). Injection molding was divided into design, hydraulic, control, and application engineering, actual molding and much more. As far as German-speaking areas in Europe are concerned there is, however, an intense exchange of knowledge and experience at meetings of various organizations at a number of locations, and machine, mold-making and application techniques in this territory have reached a leading position in the world. Serious competition does not come from inexpensive third-world countries but from industrialized countries, which are stimulated to largely copy European techniques or even generate new and more efficient methods.

In the US, it is the Society of Plastics Engineers with its divisions covering almost every facet of plastics processing and application that represents about every fourth plastics engineer throughout the world. This organization takes care of a high level of continuing education through publications and many different educational conventions and meetings. Internationally well known and attended is the annual technical conference (ANTEC). Besides this, there are many regional technical conferences (RETEC) even abroad and monthly meetings of local SPE Sections which are attended by an estimated 90 000 members annually. All this enables the plastics professional to always familiarize himself with every development in plastics science and technology.

An injection molding machine can be defined as a "machine that produces formed objects in a discontinuous manner primarily from macromolecular materials. The forming is a primary forming under pressure. Part of the plasticated material in the machine is injected through a channel into the mold cavity. The essential components of an injection molding machine are the injection unit and the clamping unit.

As far as a differentiating view permits, drive and control unit are more in the foreground of the professional interest today than other machine components. However one can also consider these elements parts of the injection or clamping unit.

Today the prospective buyer of an injection molding machines has a choice among a large variety of products, sizes and models. The process of selection becomes less confusing as the equipment is thoroughly examined and the principles of operation are understood.

If the operator is placed in front of the machine, facing it, then the injection unit is on the right side of the support and the clamping unit on the left (Fig. 2). Although the majority of machines are horizontal, equipment with vertically operating clamping or injection units is frequently used, and some machines can be converted from horizontal to vertical operation.

Such a special design is gaining more and more significance with an increasing integration of necessary processing steps in one machine. The control unit is frequently housed in a separate, free-standing cabinet. This standard of many years, however, is reversed nowadays to a control unit attached to the stationary machine platen.

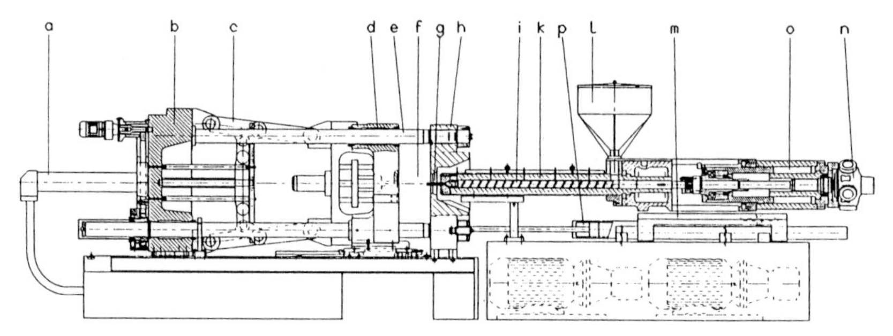

Fig. 2 Injection molding machine (sectional view), system Engel
a: Hydraulic cylinder for toggle clamp, b: Tail stock platen, c: Toggle, d: Moving platen, e: Tie bars, f: Space for mold, g: Nozzle, h: Stationary platen, i: Reciprocating screw, k: Barrel, l: Feed hopper, m: Carriage guide for injection unit, n: Rotary screw drive, o: Hydraulic injection cylinder, p: Pull-in cylinder for injection unit; in machine base drive unit with electric motors and hydraulic pumps (broken lines)

The hydraulic system of the machines, an extremely important component, does not appear as a unit, because its individual parts are located in various places in the machine.

The accessibility of a machine for the delivery and removal of parts play an important part today. Thus the concept of a German/Austrian manufacturer will certainly find its interested buyers. It divides the machine into three horizontal levels. The lowest level is kept open for part removal from the machine longitudinally or laterally. The following level contains the drive unit while injection and clamping unit are placed on the top level as usual [13].

The complete molding equipment consists of an injection molding machine, an injection mold, and a mold temperature control unit (heat exchanger) (Fig. 3). These three components exercise a direct influence on the fabricating process and determine its success or failure. They also interact with one another through pressure, temperature, and speed. Decisions on the effect from the plastic raw material to be used and the geometry of the molded part should have been made already in the planning stage. Development and design are essentially responsible for the costs of a molding (Fig. 4). Today these steps can be supported by proven CAE methods. The economical success is also affected by a good deal of peripheral equipment, which has to be synchronized with the machine. They take over tasks such as drying, transport, mixing, separation, removal, postoperation and more. The corresponding equipment belongs to an extended working station of an injection molding machine. Some of these systems will be treated in respective chapters of this book.

The raw material is supplied to the injection molding machine through the feed hopper, which is located on top of the injection unit. Thermoplastics usually are provided in the form of pellets; thermosets and, more recently, rubbers are used as pow-

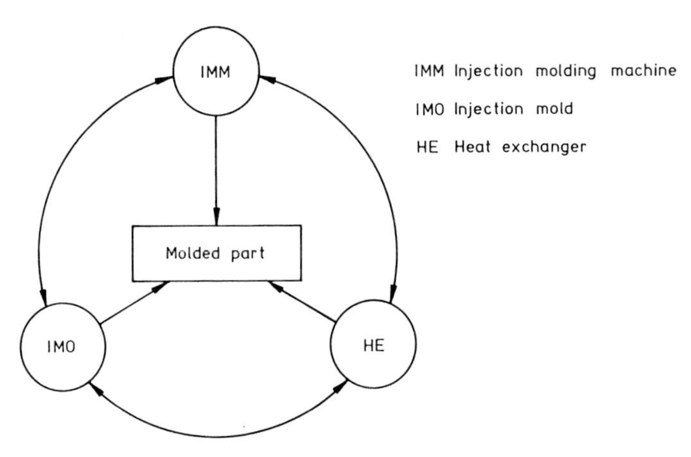

Fig. 3 Injection molding system - interactions [28]

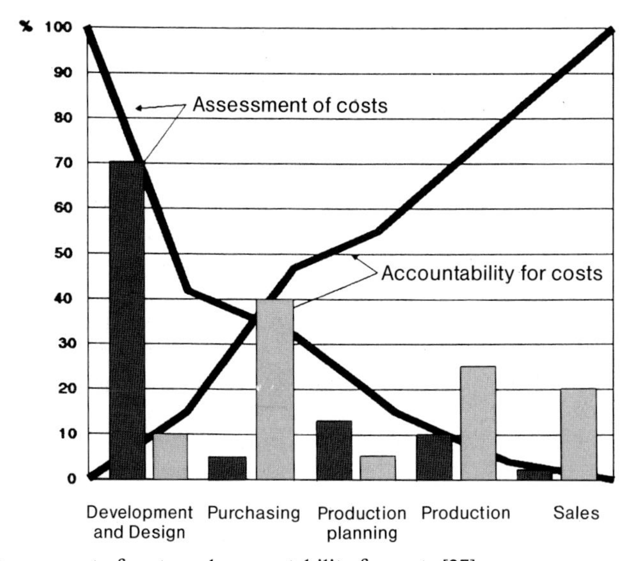

Fig. 4 Assessment of costs and accountability for costs [27]

ders. Good free flow is needed in all cases. Rubber frequently is supplied in ribbon form. Glass-reinforced polyester may be in ribbon form or used as a paste, which usually is fed to the screw under pressure through a cylinder with piston or an additional screw.

For processing thermoplastic pellets, the connecting channel between feed hopper and barrel, the feed chute, usually is of the same diameter or has the same width as the screw. An enlargement of the feed throat should not be considered or, if it is necessary, it should be done only in the axial direction to avoid the need for additional edge protection.

The screw takes in the fluid material generally directly from the feed hopper and coveys it to the screw tip (Fig. 5). On its way, the plastic passes through heated barrel zones, while the rotation of the screw results in a continuous rearrangement of the plastic particles in the flights of the screw. Shear and heat transfer from the barrel wall cause a largely homogeneous ($\Delta T \pm 5-20$ K) heating of the material. The conveying action of the screw builds up pressure in front of its tip. This pressure pushes back the screw. As soon as there is enough supply of melt in the space between tip and nozzle for one shot, the rotation of the screw stops. At that time the nozzle has been pushed against the sprue bushing of the mold and the mold is clamped, then a sudden controlled pressure surge in the hydraulic cylinder pushes the screw forward and pumps the melt into the mold cavity. There the plastic cools under temporary, usually decreasing pressure. A molding is generated. It can consist of one or several parts and the runner system (Fig. 6). When the molding is sufficiently solidified and cooled, the clamping unit opens. The mold is generally designed in such a way that, during opening, the molded part is kept in the mold half that is mounted on the movable platen. It is ejected by a mechanical or hydraulic ejector at an adjustable distance from the stationary platen.

The dynamics of the process are presented with Fig. 7. One can recognize that the pressure profile in the hydraulic cylinder (p_H) is similar to that in front of the screw tip (p_{SC}). The difference results from the friction between the hydraulic cylinder and its piston and between screw and barrel. The losses from friction may be as high as 10%. With high pressure, though, they are certainly considerably smaller (3–5%).

There is little correspondence between these pressure profiles and the pressure in the mold. The effect of the hydraulic pressure decreases with increasing distance from the nozzle. At some distance from the gate and more towards the end of the flow path, the pressure build-up during injection and holding pressure stage is not similar to either that in the runner system (p_{C1}) nor to that in front of the screw tip (p_{SC}). The pressure build-up is delayed; the maximum pressure is lower. There is a time lag in the pressure build-up in the mold, which grows with increasing distance from the sprue. The pressure continues to decrease during the holding pressure stage and reaches zero before the end of this stage. The pressure drop, in contrast to the pressure in front of the screw tip, begins as soon as the plastic starts to solidify in the runner system or part. This prevents any further pressure transmission.

This example demonstrates that the material solidifies under different local pressure conditions. Another inhomogeneity is caused by melt and mold temperature. The plastic leaves the nozzle with a temperature profile that varies with time and location. During injection and holding pressure stage, pressure and temperature variations are present in dependence on the system. They are even interdependent.

Surroundings and the process itself take effect. Machines exhibit a distinct particularity and hysteresis phenomenon e.g. of hydraulic elements. Every molder should have a mental picture of temperature and pressure differences so as not to misinterpret such terms as homogeneity of the melt, constant temperature, constant dimensions, and reproducibility.

One of the best possibilities of a direct control of such influences is provided by measuring the injection energy or the filling index [59-64]. The last method covers a range of the injection integral or the filling pressure which can be reliably recorded. This value correlates with the viscosity and therefore with the melt temperature and is a very good and representative quality criterion of the process. It is more than equal to the classic viscosity measurement with rheological equipment.

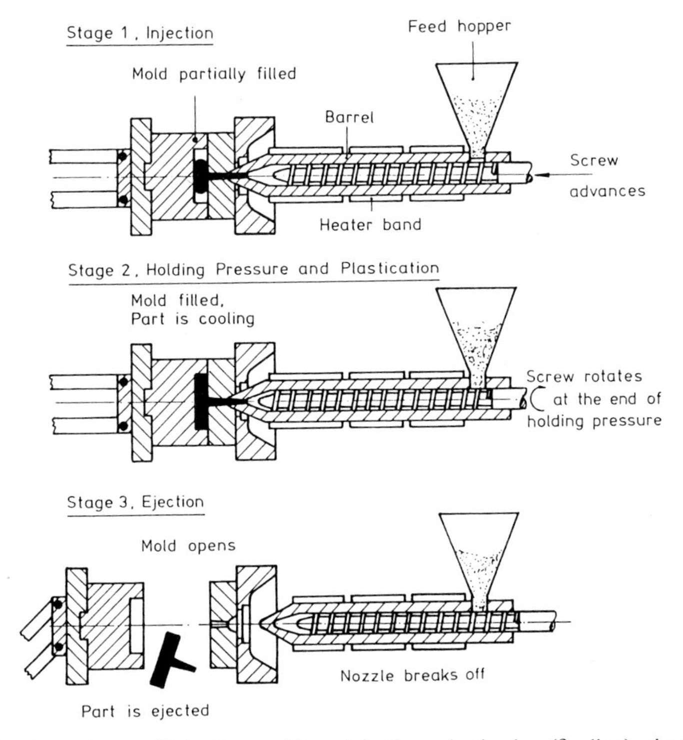

Fig. 5 The three stages of injection molding: injection, plastication (feeding), ejection

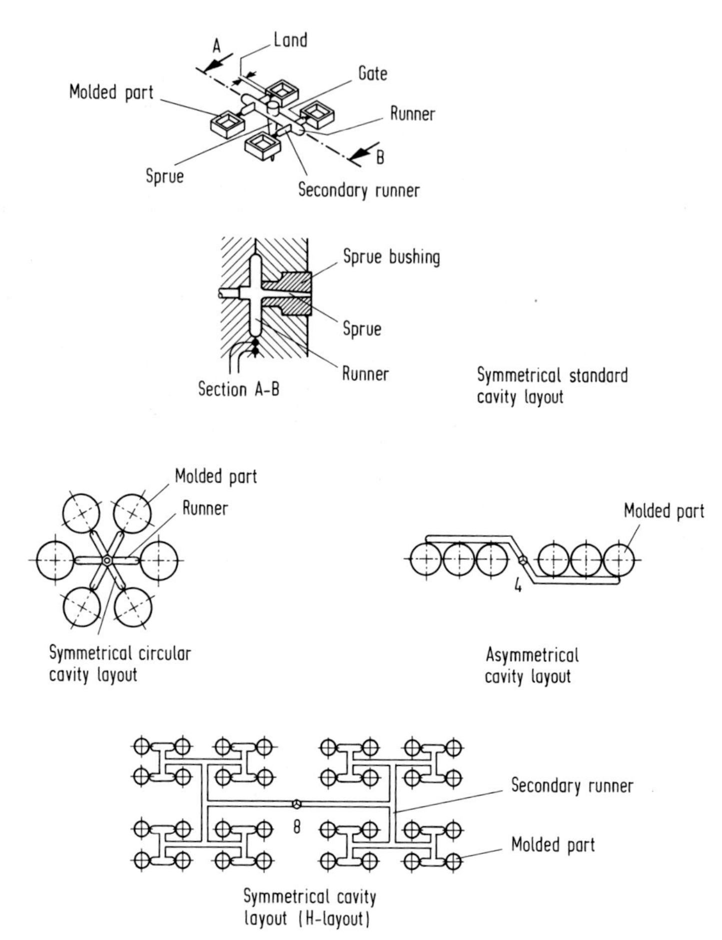

Fig. 6 Runner systems for injection molds [30, 31]

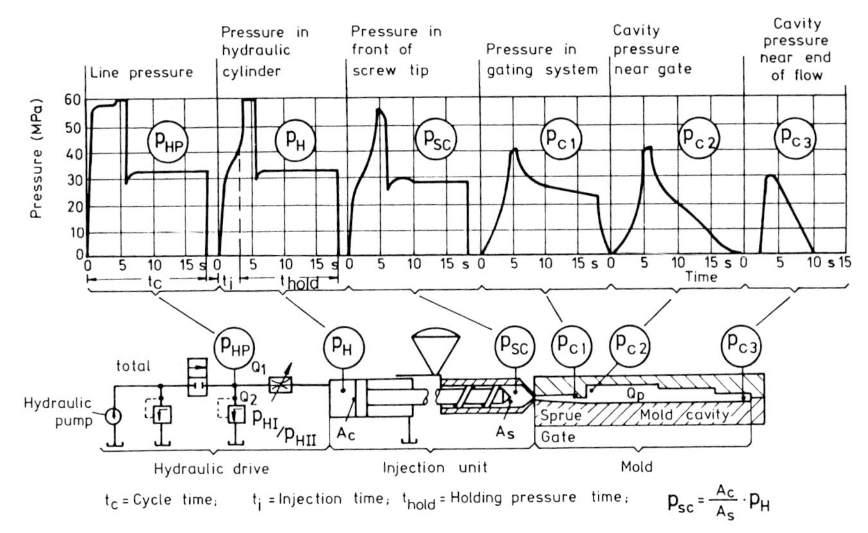

Fig. 7 Pressure development versus time at various locations in the injection molding system [32]

2 Characterization of the Injection Molding Process

A very general description of the molding process has been provided in Chapter 1 so that one can better understand the interaction of physical quantities and that of machine components, such as plasticating and clamping unit and mold. Above all, it should have become evident that an injection molding machine cannot function without a mold. In principle, the mold could be considered an essential machine part, although a highly variable one.

A number of process parameters affect the formation of a molded part. Pressure has been mentioned briefly. Temperature and time will be discussed in more detail now. For the most part, the information is provided in the form of a definition of the respective parameter. This should make sense, particularly in view of the widespread misconception about terms used in injection molding, as encountered in descriptions, discussions, or publications.

The injection pressure is frequently equated with the hydraulic pressure or confused with the holding pressure. Sometimes it is considered a constant value, and then again a variable, but incorrectly associated with time and location. In many cases it is assumed that injection pressure can be set at the machine. This is impracticable, of course, because it depends on the flow resistance of the runner system, the gate, and the cavity. Therefore it cannot be constant throughout the molding cycle, but increases from normal air pressure (about 100 kPa) to a maximum that cannot be predetermined. For a description of injection pressure to be accurate, it must include attributes such as minimum injection pressure, maximum injection pressure, or available injection pressure.

In recent years a collection of terms in four European languages has been generated to establish standards for the countries of the European Economic Community (EEC). This collection is called Euromap, and its terminology is used in this volume whenever it is applicable and not inconsistent with American practice.

2.1 Pressure

Various kinds of pressure are effective during the injection molding process. They are distinguished according to place and time of action. Local differences were already pointed out in Fig. 7.

2.1.1 Hydraulic Pressure on the Injection Side

Hydraulic pressure has to be supplied by the drive unit of the machine to overcome the resistance in nozzle, runner system, and cavity to the flow of material. Its characteristic is much the same as the pressure of the melt in front of the screw tip. It generally rises, within a short time, from barometric pressure (or the lowest possible pressure in the system) to a magnitude that corresponds to the flow resistance of the melt from nozzle to cavity.

High resistance results in a rapid build-up of high pressure, which scarcely permits observation of the onset of the compression stage after the volumetric filling of the cavity (Fig. 8, top). On the other hand, this onset normally can be recognized easily if the flow resistance is low (Fig. 8, center). The bottom graph in Fig. 8 demonstrates the pressure changes resulting from varying cross sections as they may occasionally become noticeable with extensive runner systems. Depending on their geometry, the pressure required to compensate their flow resistance may be considerably higher than the pressure drop in the cavity.

Other factors and their effects on the hydraulic pressure are shown in detail in Fig. 9. While the influence of the axial screw speed is commonly known, the effects of hydraulic oil and melt temperatures are frequently underrated.

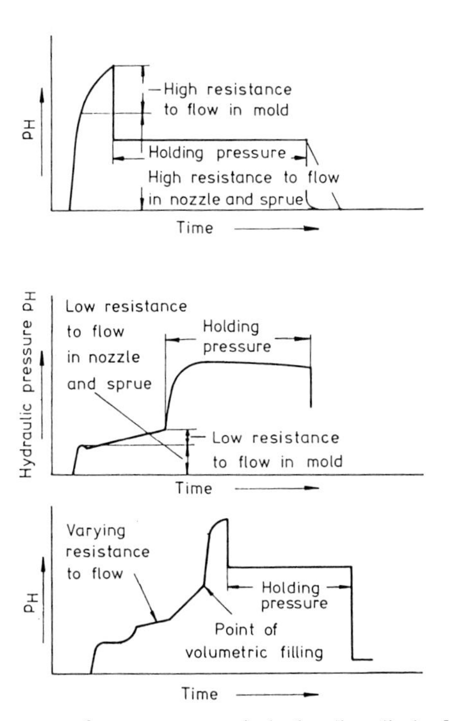

Fig. 8 Effects of resistance to flow on pressure in hydraulic cylinder [35]
(Top:) Point of volumetrie fill not noticeable owing to high resistance to flow; *(center and bottom)* Point of volumetrie fill visible because of low flow resistance in nozzle and runner system

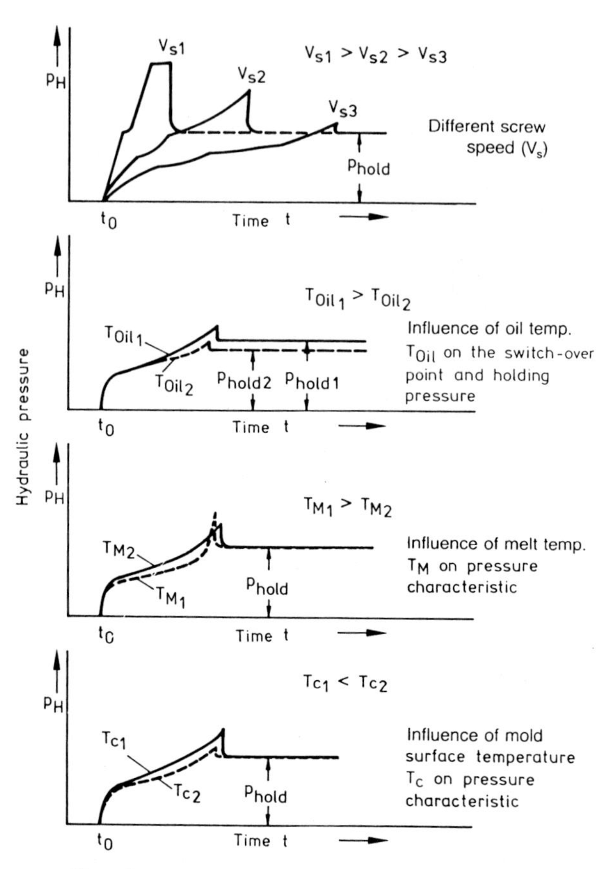

Fig. 9 Pressure profiles of hydraulic cylinder. Effect of various paramters on p_H [35]

The viscosity of the hydraulic fluid, which is temperature dependent, has an effect on the pressure. One is usually familiar with this correlation, which affects the constancy of production in many cases, particularly during start-up with cold oil. the effect of the mold temperature is more commonly neglected. Because the hydraulic pressure profile may be affected by the cooling process in the mold as shown in Fig. 9 (bottom), a distinct effect on injection time can be expected if the operation of the injection unit is viscosity dependent. This influences the production result.

Some other characteristic variations in the hydraulic pressure profile indicate, with relative reliability, irregularities during injection. Jetting or sticking to the cavity surface can result in irregular pulsation of the hydraulic pressure (Fig. 10, top). Less fre-

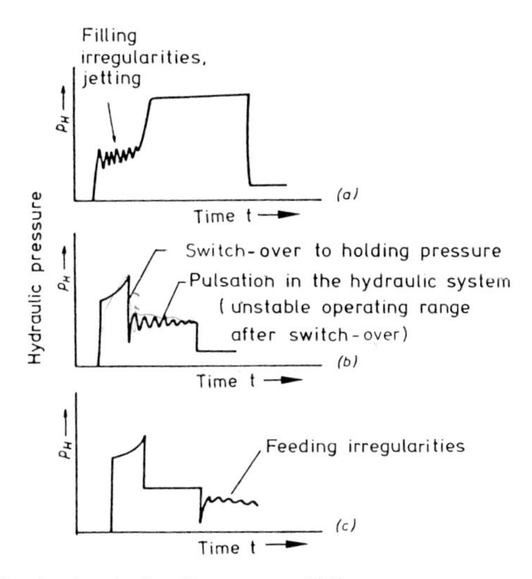

Fig. 10 Causes of pulsating hydraulic pressure [35]

quent is pulsating hydraulic pressure after abrupt pressure switches (Fig. 10, center). Such a pressure profile provides information about the reliability of operation of the machine's hydraulic system.

Variations of the hydraulic pressure during feeding are signs of feeding difficulties (Fig. 10, bottom). This would show up even more clearly in the recording of the pressure of a hydraulic screw drive or the amperage of an electric motor.

Measurements of hydraulic pressure are recommended, because they are simple to carry out and provide important information about the injection and feeding stages. The capability of reliably controlling the function of the plasticating unit during important process stages is the reward for a small effort.

The pressure in front of the screw tip shows qualitatively the same tendencies as the hydraulic pressure, although not during the holding-pressure stage. A conversion of the hydraulic pressure by means of the ratio between the cross section of the hydraulic piston and that of the screw is usually sufficiently exact. A frictional loss of about 5% in the operating range should be assumed. Only with very low pressure can this loss rise to 10%. Measuring the hydraulic pressure is preferably over pressure reading of the hot melt, which cannot reliably done during continuous operation. It also permits to draw conclusions on the rise in melt temperature which can be expected due to friction (Sect. 2.2).

Readings of the hydraulic pressure, however, provide no basis for conclusions regarding the holding-pressure stage and the pressure profile in the cavity. This is illustrated in principle by the curves of hydraulic and cavity pressure in Fig. 11.

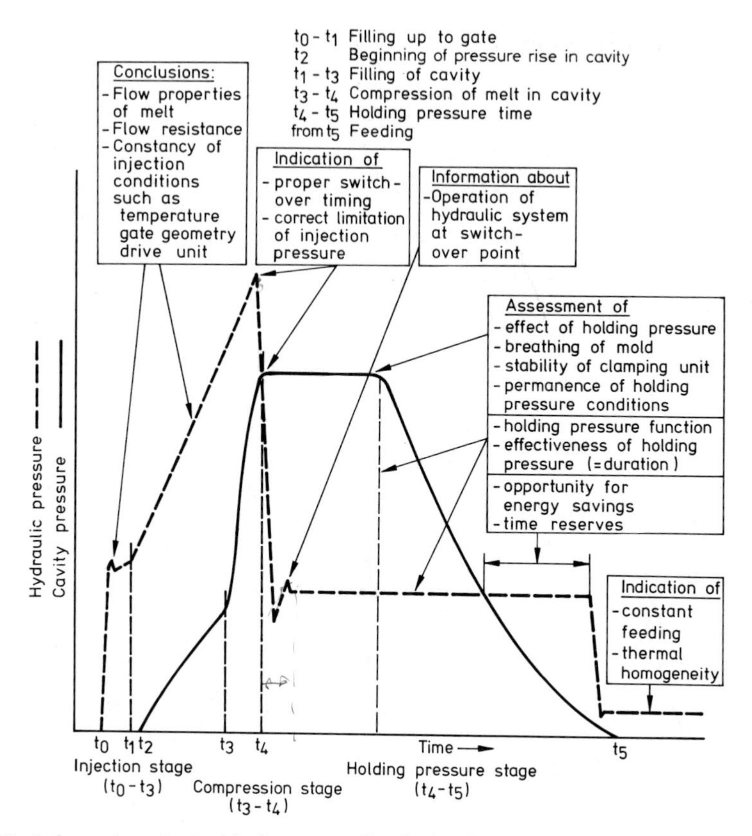

Fig. 11 Information obtainable from recording hydraulic pressure in injection cylinder and cavity pressure

One can recognize the influence of the flow resistance in nozzle and runner system (t_0-t_1), in some cases also the onset of the cavity filling (frequently by a small setback in pressure), and the moment of completed volumetric filling with a sudden and steep increase in pressure. This holds true only if the hydraulic pressure has not been switched over to holding pressure beforehand. Beginning with the compression stage (t_3), the hydraulic pressure provides only incomplete information about the process. Cavity pressure corresponds with hydraulic pressure only in very rare cases.

2.1.2 Cavity Pressure

Analysis of the injection molding process has substantially contributed to progress in process control. The cavity pressure plays a central role in this matter. The method of pressure recording has attained a high standard with the use of pressure transducers based on strain gages or piezoelectric crystals (Sect.6.3). Recording under harsh production conditions has been made possible if certain requirements are met. To avoid damage, all sensors in the mold must be installed in such a way that no connector projects over the outside contour of the mold.

The information obtainable from a cavity pressure curve is illustrated in Fig. 12 by a characteristic curve recorded during the molding of a technical part. It is possible to differentiate three fundamental stages: filling of the cavity (injection stage), compression of the melt (compression stage), and holding the solidifying material under pressure (holding-pressure stage). These three stages can be related to certain effects as well as to quality criteria. The injection stage primarily affects the appearance of the molded part, while the holding pressure controls, above all, the dimensions. The graph illustrates the relative significance of the injection pressure very well. One can see that this pressure has the function of overcoming the flow resistance from the nozzle to the cavity, but otherwise is mostly unimportant for the quality of the mold-

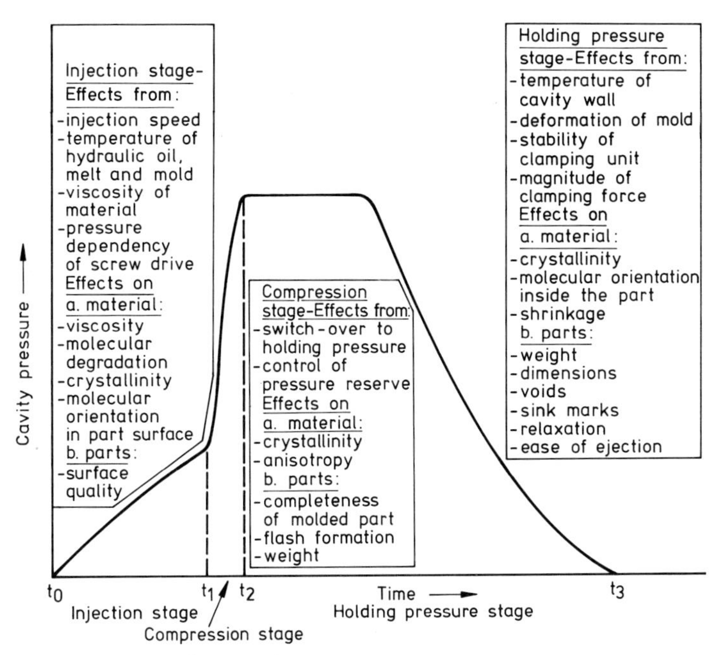

Fig. 12 Cavity pressure profile over time

ing. Compression and holding pressure usually are considerably more significant and effective.

The pressure profile also provides information about typical mistakes in the process technique. A high pressure peak in the compression stage can cause serious difficulty. It results from incorrectly set or unreliably functioning switch-over to lower holding pressure and produces flash or, worse, a packed mold. This leads to considerable differences in the weight of molded parts and in their dimensions, primarily in the direction of the mold opening. There is no dependable way to control the pressure peak; it must be avoided by proper selection of the switch-over point. More negative effects are treated in the following section, in which switch-over point and holding-pressure time will be discussed in more detail because of their particular importance.

Fig. 13 demonstrates typical effects of some processing parameters on the cavity pressure. Different axial screw speeds result in noticeable changes in the build-up of pressure during the compression stage (Fig. 13a). High mold temperature improves the pressure propagation in the mold (Fig. 13b). The gate design has a significant influence on the holding pressure during the cooling stage (Fig. 13c). Of course, there is also a pressure differential in the cavity between areas close to the gate and near the end of the flow path (Fig. 13d). Some of these effects will be detailed later. So far, Fig. 13 should provide the general information that the quality of molded parts can be affected by controlling the pressure in the mold during the molding process.

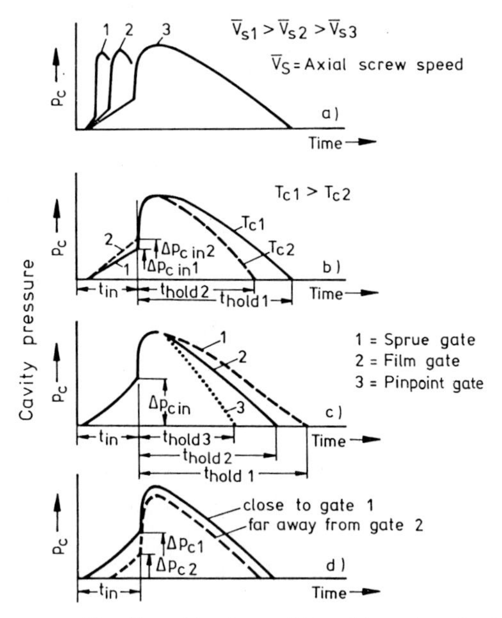

Fig. 13 Cavity pressure profile affected by the position of the point of switch-over from injection to holding pressure [35]

2.1.2.1 Switch-over from Injection to Holding Pressure

Since there is generally no reliable information about the pressure in the cavity, the choice of the point of switch-over to holding pressure is frequently incorrect. Fig. 14 demonstrates four basic possibilities:

(a) Injection without switch-over
(b) Injection with late switch-over
(c) Injection with premature switch-over
(d) Injection with optimum switch-over

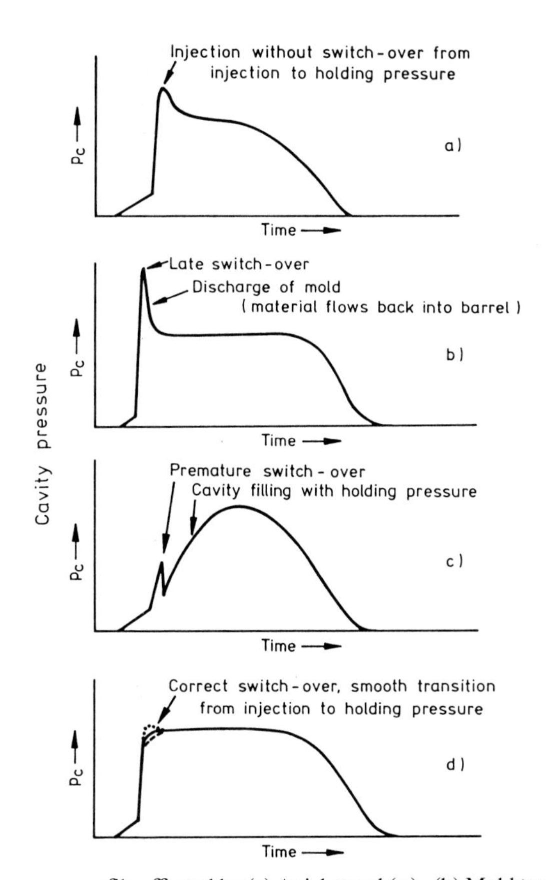

Fig. 14 Cavity pressure profile affected by (a) Axial speed (v_s) . (b) Mold temperature (T). (c) Gate geometry, and (d) distance of the pressure sensor from the gate [35]

Operations without switching over to holding pressure is practicable if the necessary final pressure is close to the filling pressure. this is mostly the case with small gates and with parts having a large length-to-thickness ratio.

With large gates, a relatively high injection speed is usually called for. Then the hazard of a late switch-over with subsequent packing is often great. Besides the adverse effect on dimensions and formation of flash, late switch-over is the principle cause of damage to the mold by deforming cavity edges and overloading the clamping unit. The latter can result in permanent damage to the tie bars and even failure by fracture in the long run.

Undesirable and unfavorable shear orientation in the part can be caused by back flow of the melt after late transition to lower holding pressure (Fig. 14b). The same occurs if the holding pressure is turned off before the gate is frozen.

Pressure setback (Fig. 14c) indicates a premature switch-over to holding pressure. The balance of the filling process takes place under holding pressure, which is too low for proper filling, and consequently with reduced injection speed. At the moment of switch-over, the flow may stagnate briefly, which produces troublesome surface marks.

The four primary methods of determining when to switch over from injection to holding pressure are dependent on (1) time, (2) travel, (3) cavity pressure, and (4) clamping force.

Switch-over Dependent on Time

This method calls for a signal to be released after a predetermined time has elapsed following the onset of injection. This method does not take into account compression of the melt in front of the screw tip and its viscosity, feeding accuracy, or variations in the hydraulic pressure, which can result in changes of the screw position (end of feeding, beginning of holding pressure) and consequently of the corresponding strokes (feeding stroke, injection stroke), as well as variations in the axial screw speed. The final outcome is a larger variability of quality criteria, particularly in the weight and dimensions of the molded part. For this reason, as a general principle, a time-dependent switch-over is not advisable. Without doubt, it is the worst of all options.

Switch-over Dependent on Travel

This method is the most often used one and has proved useful. The signal for switch-over is released by a limit switch of the holding-pressure position. Insofar as the injection stroke is largely constant, the switch-over can be assumed to take place each time at the same degree of volumetric cavity filling. This method becomes problematic, though, if the holding-pressure stroke is extremely short. There is the danger that small variations may prevent the switch from being actuated every time. In such cases it is better not to plan a switch-over at all. Variation in the feeding stroke, inaccuracies of limit switches, malfunction of the nonreturn valve, and differences in melt viscosity also cause problems with this method.

Switch-over Dependent on Cavity Pressure

In recent years, pressure-dependent switch-over has been successfully used, even under the tough conditions of practical production. Cavity pressure actuates the switch-over as soon as a preselected pressure has been reached. The advantage is based on the steady supply of a reliable signal (information) for an absolute magnitude of pressure at which the switch-over is most effective. A monitoring of the pressure takes placc. The influence of the screw stroke and the function of the nonreturn valve are eliminated. This method, however, like the previous two methods, cannot compensate for temperature variations of the hydraulic fluid, the melt, and the mold, or for changes in the injection speed.

The more rapid the pressure increases during the compression stage, the more effective this method is, because in such a case an especially precise and timely switch-over is mandatory to avoid the undesirable pressure peak. Among the large variety of moldings, those particularly suited for this method have their principle extension in the plane of the parting line and cannot tolerate flashing, such as flat covers of equipment housings, which have little depth.

Switch-over Dependent on Clamping Force

Recent developments indicate that the pressure build-up in the cavity during compression and holding-pressure stage can also be measured indirectly. The method is based on determining the reactive forces in the machine platens or the tie bars of the clamping unit beginning with the onset of the injection stage. The locking of the mold (hydraulic or with a toggle system) is assumed fully rigid. If the forces are considered which act on the mold, the machine platens, and the unit producing the clamping force, an equilibrium exists between the clamping force, the inertia of platens and mold and the reactive force and remaining clamping force:

$$F_C + F_{PL} + F_M = F_B + F_{CR}$$

The inertia is generally so small that it can be neglected. The clamping force can be measured in the tie bars or the tie bar nuts inductively or with strain gages. If these measuring devices are installed permanently, a cost-effective method has been found to determine the precise switch-over point. The frequent exchange of pressure transducers from one mold to another one, can be eliminated, and the chances of damaging them are largely reduced.

Another arrangement using a load cell between mold and machine platen has been evaluated in fully hydraulic clamping units, toggle systems, or a combination of both. The signal from the deformation of tie bars or tie bar nuts, however, proved to be superior. The characteristic of this signal reflects the typical pressure pattern in a mold near the gate much better.

2.1.3 Holding Pressure

The holding pressure is the pressure exerted on a molded part during a secondary pressure stage. As already demonstrated, there is rarely any correspondence between the set hydraulic pressure and the effective pressure in the mold (Fig. 7).

The magnitude and duration of the holding pressure are of major importance for dimensional accuracy and cosmetic quality of a part. They determine how well the cavity surface is duplicated. While the optimum pressure level can easily be established by controlling dimensions or observing the disappearance of sink marks, the duration of the holding pressure is frequently guesswork. The cavity pressure will provide reliable information, though, if it can be measured. As long as sprue, gate, or any other narrow passage is not yet frozen, changes in the magnitude and duration of the holding pressure will have an effect on cavity pressure. After the gate is sealed (frozen off), no further influence can be exercised. Fig. 15 depicts profiles of hydraulic and cavity pressures for different holding-pressure times. One can see that a time span of more than 8 seconds no longer causes a cavity pressure change near the gate. The gate

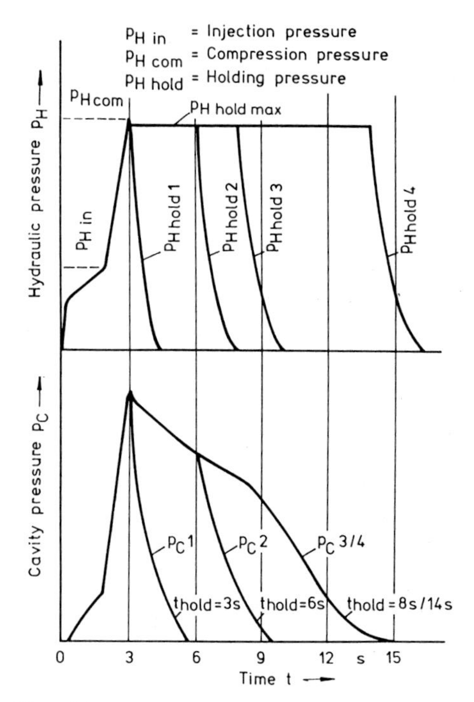

Fig. 15 Pressure profiles of hydraulic system and cavity resulting from varying duration of holding pressure stage [35]

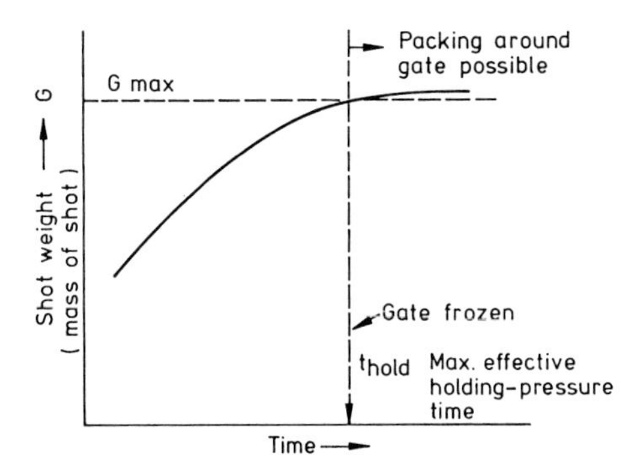

Fig. 16 Shot weight (mass of shot) dependence on holding pressure time

was sealed after about that time. Extending the holding-pressure time beyond 8 seconds does not affect part quality and would be nothing but a waste of energy.

Without the capability of measuring cavity pressure, there is another way to determine the maximum effective holding-pressure time by systematically monitoring part weight and holding-pressure time. Fig. 16 reveals that no increase in weight occurs after the maximum effective time. The impact of a longer time period is infinitely small. This corresponds with the case of 8 second holding-pressure time in Fig. 15. The holding pressure in the mold differs, of course, from place to place, because pressure losses occur along the flow path (Fig. 7).

2.2 Temperature

The temperature of the melt, the hydraulic fluid, and, above all, the mold are of great importance for maintaining constant properties and dimensions of molded parts. This will be briefly illustrated here in detail.

2.2.1 Temperature of the Hydraulic Oil

Energy losses in valves as well as pump efficiency depend on the viscosity of the hydraulic fluid. Therefore, all motions of the injection molding machine, which are not pressure compensated, are influenced by the oil temperature. Consequently, it is necessary to have start-up control that prevents start-up before the operation temperature has been attained, and an oil-temperature control if high-quality parts are to be molded.

2.2.2 Melt Temperature

The thermodynamic properties of the molten plastic, such as viscosity, enthalpy, and specific volume, change simultaneously with melt temperature. Let us look primarily at what happens in the mold. Fig. 17 shows that cavity pressure decreases with melt temperature. Sealing of the gate shifts to shorter time periods. This means that the time during which the formation of the part can be influenced is shortened, but it also means a reduction in cycle time. Dependent on melt viscosity, however, an increase in injection time may also be experienced. If this should be avoided, an automatic (valve characteristic) or controlled rise of the hydraulic pressure could be used.

This example should support the need for a constant melt temperature, even if only a smaller effect can be found with crystallizing plastics.

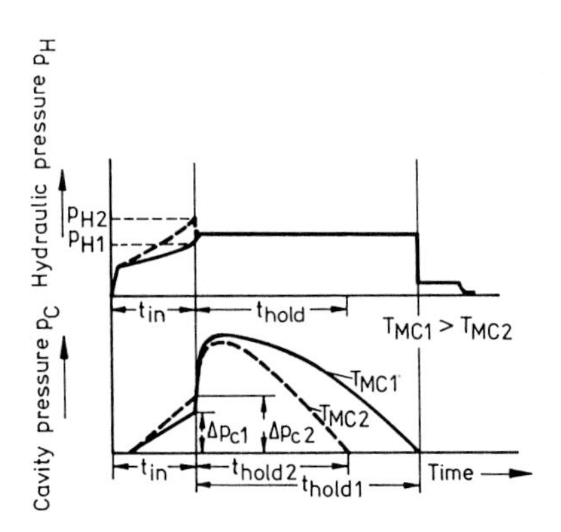

Fig. 17 Pressure profiles of hydraulic system and cavity affected by melt temperature T_M [35]

2.2.3 Mold Temperature

The temperature of the cavity wall is of major importance for part quality, economy of the process, exact dimensions, and accurate duplication. It is this temperature which, besides the thermal characteristics of the material, determines the cooling time. It should be stipulated here that the temperature of the cavity-wall surface is meant when mold temperature is mentioned. Local temperature variations will not be discussed here. With thin parts of less than 2.5 mm wall thickness, a clear increase in hydraulic pressure can be noticed very early during the injection stage. This can be attributed to an increase in viscosity from the cooling effect of the cavity wall and the decreasing thickness of the hot molten core.

The magnitude of the maximum cavity pressure in the vicinity of the gate is hardly affected by the mold temperature, but holding pressure is as a result of changes in the cooling process. This can be clearly seen in Fig. 18.

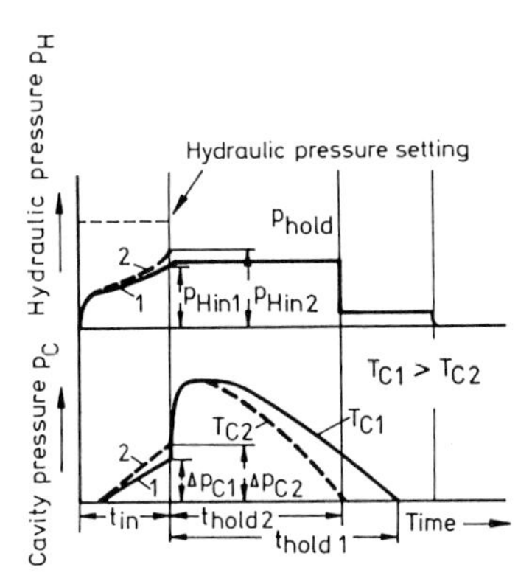

Fig. 18 Effect of cavity temperature T_c on pressure profiles [35]

2.3 Velocity

The only velocity that is important for the molding process is the velocity of the advancing screw, the axial screw speed. It is only effective during the injection stage. The different reactions of hydraulic and cavity pressure make it clear again how little information about the process is provided by the hydraulic pressure. Fig. 19 presents curves for hydraulic and cavity pressure for three different screw speeds. Injection times vary inversely with screw speed, and Fig. 19 (top) shows that hydraulic pressure rises faster with increasing injection speed. This is due to growing flow resistance in nozzle and gate. The pressure loss (Δp) from filling the cavity, measured near the gate, on the other hand, increases with decreasing injection speed, reflecting the effect of the superimposed cooling process. The latter causes an increase in melt viscosity in the cavity during injection and a more rapid formation of a growing solid surface layer, which narrows the available flow-channel. This, in turn, impedes the pressure transmission, which is reflected in the graph by the different levels of maximum cavity pressure. If this affects the duplication of the cavity surface, then the holding pressure must be raised considerably to compensate for slow injection.

These results are summarized in Fig. 20, which illustrates pressure drop and cavity pressure dependence on flow front velocity. For reasons of economy (energy savings) and part quality, injection speed should be selected as high as practical.

Other important parameters, which are not discussed here, are the geometry of nozzle and gate, cooling time, feeding time, operation of nonreturn valve, rotation and injection characteristics of the screw drive, and the rigidity of mold and machine platens, just to name the most essential ones.

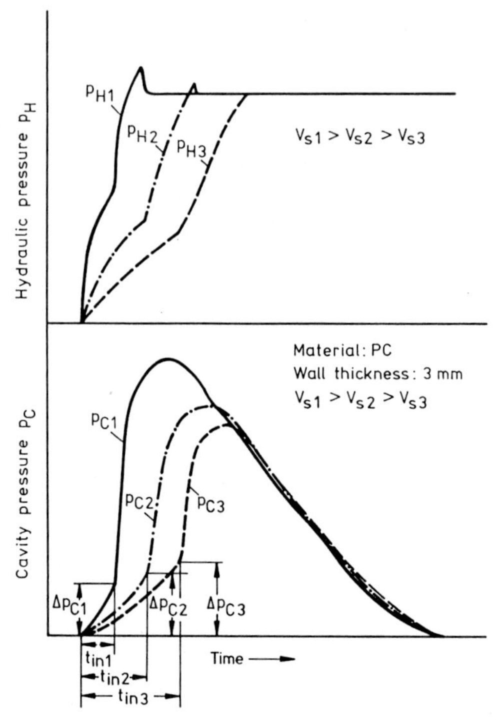

Fig. 19 Pressure profiles of hydraulic system and cavity, affected by axial screw speed v_s [35]

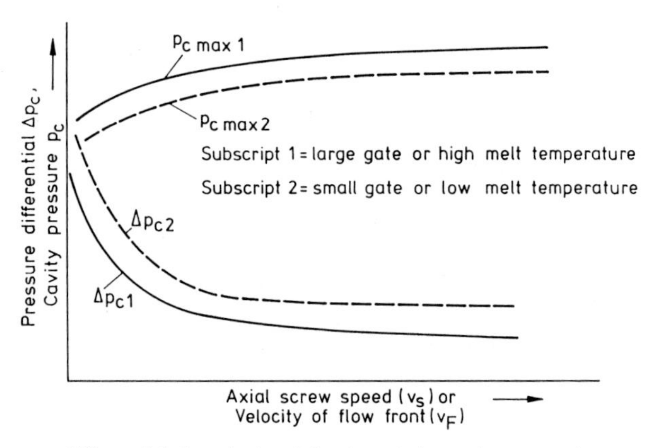

Fig. 20 Pressure differential Δp_c during injection and maximum cavity pressure $p_{c\ max}$ as a function of axial screw speed or velocity of flow front

2.4 Effect of Molding Technique on Properties

2.4.1 Dimensional Stability

All the parameters mentioned so far have a direct or indirect effect on the dimensions of a part (Table 1). Their general influence has been indicated to some degree. Therefore a discussion of the so-called P-V-T diagram is in order now. It depicts the dependency of dimensions on pressure and temperature. To begin with, there is an almost linear correlation between an arbitrary dimension of a part and the specific volume

Table 1 Effect of Processing Parameters on Part Dimensions

Processing parameter	Effect	Part dimension resulting from	
		an invariable mold dimension	a mold dimension affected by breathing of the mold
Increasing melt temperature	(a) Increase in volume (b) Better pressure transmission	(a) ↓ (b) –	(a) – (b) ↑
Increasing mold temperature	Higher release temperature	↓	↓
Increasing injection speed	Better pressure transmission during holding pressure stage	↑	↑
Increasing holding pressure	Better compensation of volumentric shrinkage during cooling stage; better compression of the melt	↑	↑
Increasing holding pressure time	Same as above, before gate is frozen After gate is frozen	↑ –	↑ –
Increasing mold deformation from cavity pressure	Packing in areas of large deformation affecting dimensions by breathing	↑	↑
Gate geometry: increasing cross section of gate	Indirect effects, (refer to injection speed, holding pressure, melt temperature)	↑	↑

(Fig. 21). This correlation can be easily established with a few experiments. As soon as the specific volume $v_p = v_4$ at 10^5 Pa (1 bar), which corresponds to the desired dimension, is known, the pressure has to be controlled in such a manner that this value is attained on the 10^5 Pa curve (point 4 in Fig. 23). The part continues to shrink along this curve until room temperature is reached at point 6 or temperature of use at point 5. Thus, constant dimensions can be maintained only if pressure and temperature are controlled and directed through the P-V-T diagram in such a way that the 10^5 curve, which presents the pressureless state of the cavity at the moment of mold opening, is always reached at point 4.

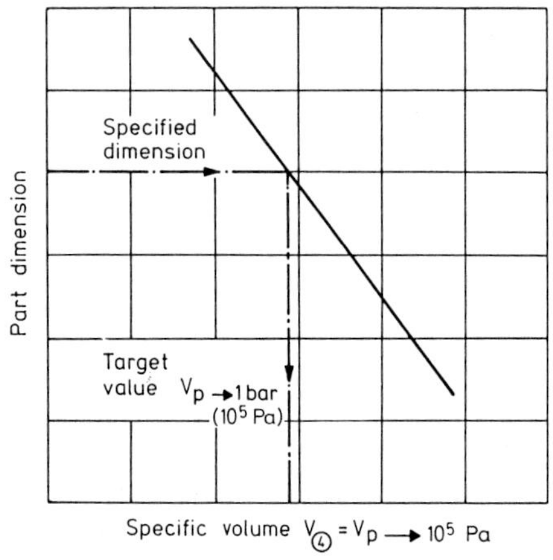

Fig. 21 Relationship of part dimension to specific volume [32]

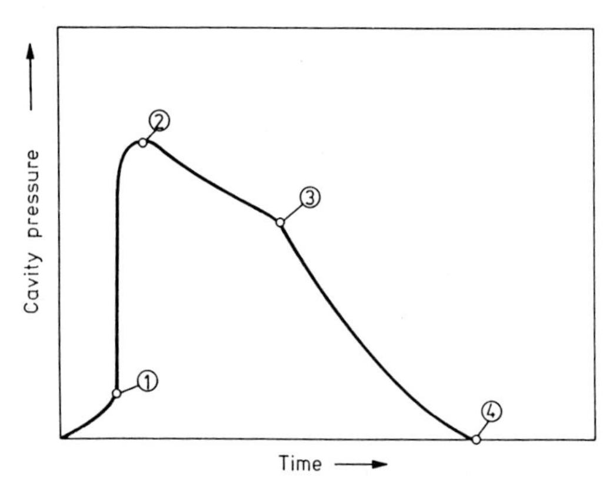

Fig. 22 Cavity pressure versus time, affecting parameters and effected part properties [35]

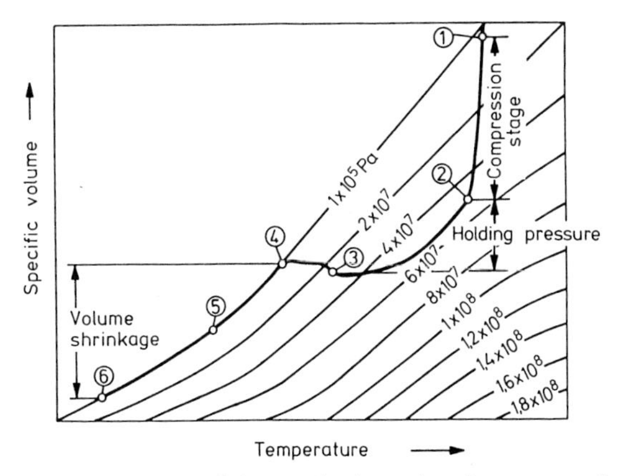

Fig. 23 *P V T* diagram (schematic) of thermoplastics and cavity pressure drop [35]

Process techniques which lead to the points 1, 2, 3, 4 in Fig. 22 are shown as characteristic examples for different strategies.

2.4.2 Mechanical and Physical Properties

The mechanical and physical properties of molded parts, particularly of those made of thermoplastics, do not only depend on the chemical constitution of the material and its corresponding properties. The processing conditions exercise a considerable influence, either as an inevitable or a selected effect. Properties of use such as strength, toughness, hardness, heat distortion, dimensional stability, and tendency to stress cracking may vary, more or less, with one and the same material, or can be selectively varied depending on the processing technique. Those factors which determine the part quality are frequently not apparent externally, but are reflected by the internal structure of the molding.

The most important structural characteristics of thermoplastics dependent on processing conditions are:

Molecular orientation
Residual stresses
Crystalline structure and degree of crystallinity (of crystalline materials)
Orientation of fillers (of filled or glass-reinforced materials)

Possible changes in the molecular structure through reduction of the chain length or degradation are not discussed here. The factors responsible for this result from residence time in the barrel, melt temperature, and an intense shear effect in the runner system during injection.

2.4.2.1 Molecular Orientation

Molecular orientation is the alignment of molecular chains in one particular direction. In a plastic melt at rest, individual molecular segments are, or rather move about, in a random, tangled state of maximum irregularity. As they flow during the forming process, e.g. during injection, the molecular chains are forced into a preferred direction; they take on a particular orientation. Orientation, as schematically shown in Fig. 24, occurs during flow in channels, either in injection molds or extruder dies.

The shear velocity in molten plastics is especially high in narrow channels and in areas close to the cavity wall, where the melt is also stretched (Fig. 25). The melt tends to adhere to the cavity wall while it flows rapidly in the center of a part. When the oriented melt comes to a standstill, the molecular chains regain their irregular, random state after some time due to their thermal motion (Brownian motion). This process is called relaxation. The relaxation rate depends on the structure of the molecules and the properties of the eventually used additives (internal lubricants), as well as on temperature and pressure. Low molecular weight, high temperature, and low pressure promote higher relaxation rates.

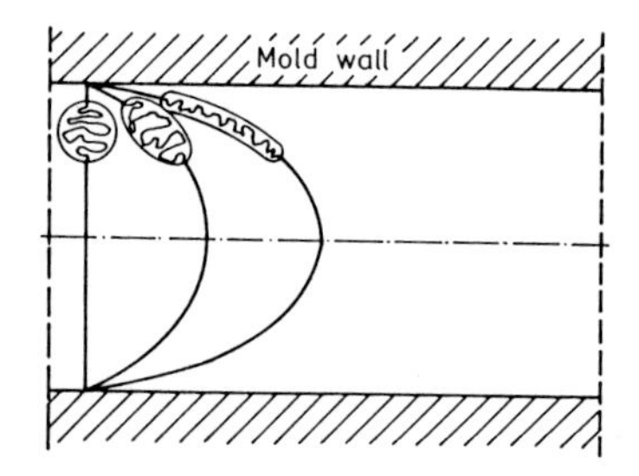

Fig. 24 Formation of molecular orientation [36]. An originally circular segment is deformed by the flow of melt, which causes the molecular chains to become oriented in the direction of flow.

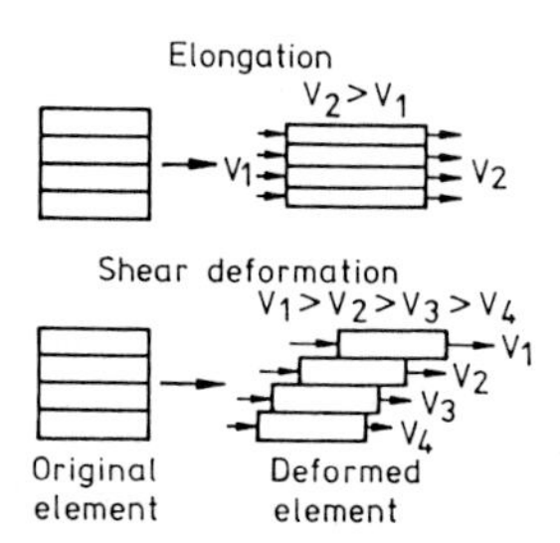

Fig. 25 Elongation and shear deformation of segments of melt caused by velocity differential [36]

In any case, the plastic melt relaxes within a few seconds at temperatures customary in injection molding. On the other hand, it usually takes only a fraction of a second, especially in the surface layers, for the material to solidify, and a considerable part of the orientation produced during molding is always "frozen in". With decreasing temperature, the relaxation rate slows down very rapidly. Below the glass-transition temperature, practically no relaxation occurs, provided the parts are not exposed to higher temperatures again later on.

Because the relaxation process of oriented plastic parts is always accompanied by shrinkage, the extent of shrinkage serves as a measure of the degree of orientation. Usually orientation from injection molding is a very complex superimposition of shear deformation and elongation. Around a pinpoint gate, the melt experiences shearing action radially and elongation tangentially to the flow front (Fig. 26). The radial direction is generally called the direction of flow. This is not identical with the actual path of a melt particle, but describes the principal direction of the material flow. One can conclude from the flow lines that biaxial orientation may be assumed near the gate. At some distance from the gate, there is no tangential elongation in the case illustrated in the figure, and the melt particles are oriented primarily by shear deformation in the direction of flow.

The complexity of orientation becomes even more obvious if a cross section is examined. The maximum shear velocity and orientation are just underneath a thin layer adhering to the cavity wall. With increasing growth of the solid layer, the orientation maximum moves towards the center.

The conditions in the flow-front region cause a biaxial elongation and orientation (Fig. 27). The melt flows transversely to the axial direction [41,42], and the relatively highly viscous plastic is biaxially stretched like a skin. This causes an orientation which is immediately frozen in when the material comes in contact with the cavity wall and

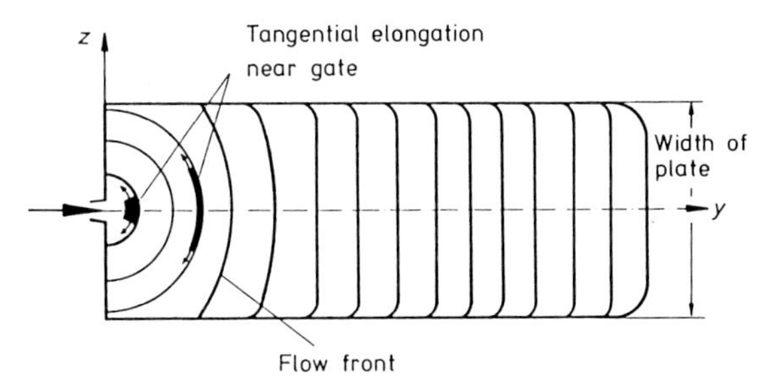

Fig. 26 Form of the flow front at various moments during filling of a plate mold [37]

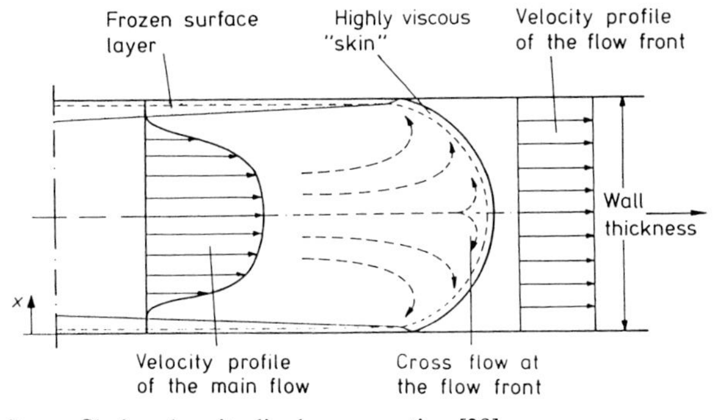

Fig. 27 Velocity profile in a longitudinal cross section [39]

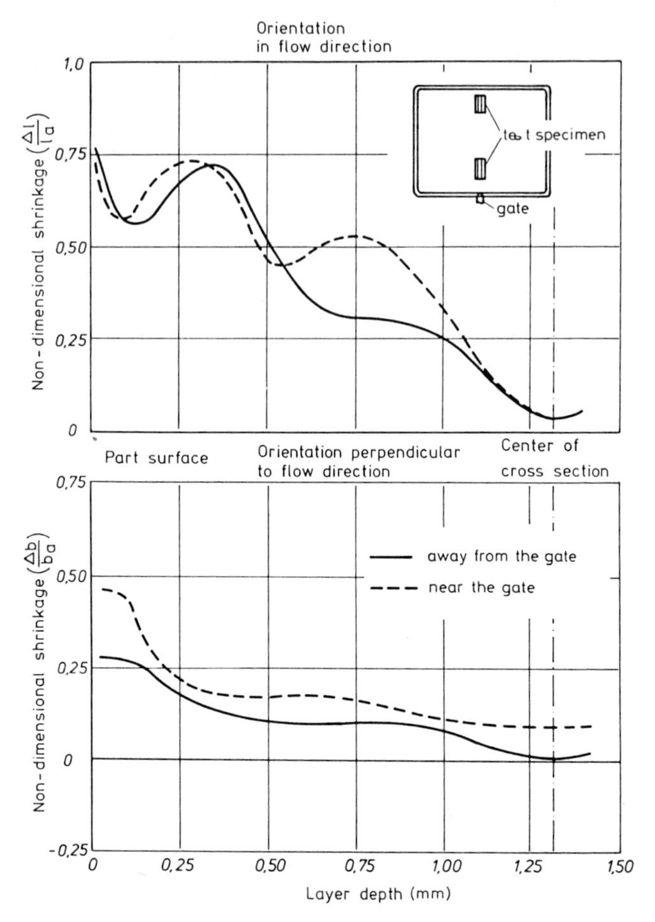

Fig. 28 Shrinkage variations in a polystyrene molding. Shrinkage close to and away from the gate [41]

results in an additional transverse orientation in the part surface. An orientation of this kind can be demonstrated by heating thin slices of the molding to a temperature above the glass-transition temperature and plotting the distribution of the observed shrinkage (Fig. 28).In the vicinity of the gate, orientation can be found in the cross section almost as deep as the center, which is attributed to the tangential orientation just discussed [41]. Longitudinal orientation frequently has a second maximum further inside the part that results from melt flow during the holding-pressure stage. Altogether, the degree of orientation is higher in the vicinity of the gate than away from it.

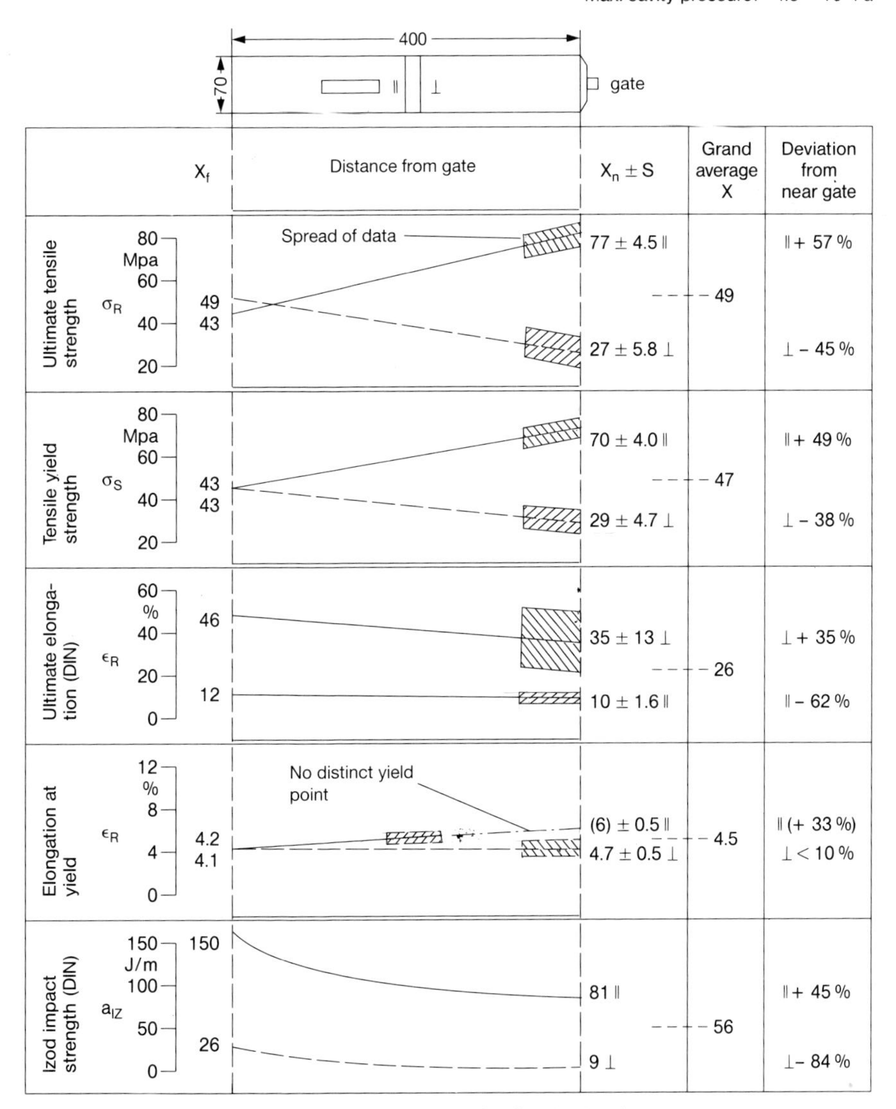

Fig. 29 Mechanical properties of a molded part in relation to distance from gate and direction of flow [44]

The molder has to understand the effect of orientation and should know that the orientation of molecular chains can often result in a marked anisotropy of the majority of part properties. This becomes especially noticeable in mechanical properties. The strength is always higher in the direction of orientation than perpendicular to it. Frequently such an effect may be desired, but every so often it is not because of reduced strength in the transverse direction. Fig. 29 presents an example of how mechanical properties can vary according to the distance from the gate and the flow direction. The effect on tensile and impact behavior is evident. Table 2 provides information about

Table 2 Effect of Processing Parameters on Orientation and Anisotropy

Processing parameter	Effect	Degree of orientation	Anisotropy
Incrreasing melt temperature	Increased relation time	↓	↓
Increasing mold temperature	Increased relaxation time	↓	↓
Increasing injection speed	Very thin surface layer Increased relaxation time in the core	↓ ↓	↓ ↓
Increasing holding pressure	Relaxation impeded, increased degree of cooling during injection of melt	↑ In part interior	↑
Geometry			
Increasing wall thickness	Smaller shear rate, long relaxation time	↓	↓
Extremely thin wall	Calls for high injection speed, orientation in surface layer predominant	↑↑	↑↑
Varying wall thickness	Same as above	↑↓	↑
Cross section of gate	Indirect effects, by way of injection speed, holding pressure, melt temperature	↑↓	↑↓

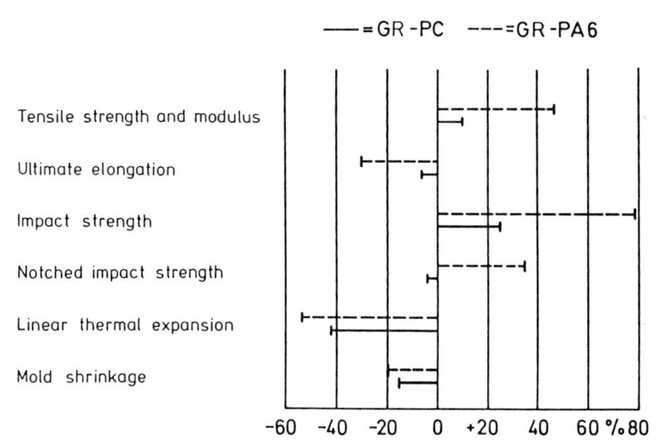

Fig. 30 Deviation of properties in the direction of flow as a percentage of those transverse to flow [47]

the relationship between processing parameters and degree of orientation and anisotropy.

Orientation does not produce autonomous (independent) residual stresses at temperatures of common use (Sect. 2.4.2.2).

The orientation of fillers occurs in the same way as molecular orientation. Especially fibrous fillers cause large-scale anisotropy of properties. The ranges of deviations from mean values are presented in Fig. 30.

2.4.2.2 Residual Stresses

Residual or internal stresses are mechanical stresses that are present in a part in absence of external loading. They are the result of changes from the position of equilibrium of atoms and the distortion of valence angles in the molecular chains, as well as from changes in the distance between segments of the molecule (secondary bonds). Deformations which cause residual stresses are of an energy-elastic nature. The strains associated with them are smaller than the yield strain if they do not result in stress cracking. Such strains are far below 10% for thermoplastics. Entropy-elastic deformation (orientation), however, can frequently be of a magnitude of several hundred percent in molded parts.

Energy-elastic stresses in a molding are usually in equilibrium. If there are compressive stresses in one place (e.g. in the surface layer), then there will be tensile stresses in another place (e.g. in the part center). If the residual stresses are not yet balanced after cooling in the mold, the molding will warp.

In contrast to orientation, residual stresses can cause failure (distortion, stress cracking) of a part without external loading. In addition, the value of a calculated permissible load has to be corrected by subtracting the residual stress acting in the same direction as the load.

Cooling Stresses

Residual stresses are generally caused by different cooling rates in various layers of the part. The rapidly cooling and solidifying surface layer forms a rigid shell, which restrains the still warm interior from contracting during the ensuing cooling process. This results in tensile stresses in the interior and compressive stresses in the external layer. The presence of such residual stresses can be demonstrated by removing layer by layer from bar-shaped specimens [45]. The resulting deformation is a measure of the existing stress.

A simpler method can be used for a number of plastics that are sensitive to stress cracking. Specimens are immersed in a medium that rapidly diffuses into the material and triggers visible cracking in regions of tensile stresses within a few seconds or minutes [47,48]. This test indicates that areas of high tensile stress near the surface are located primarily on the outside of corners, usually the result of nonuniform cavity-wall temperatures.

Residual Stresses from Mold Packing

A pressure peak during compression and excessive holding pressure can cause packing and overloading in the central section of a part, especially if a large gate permits this pressure to remain effective for a long time. Then the part is ejected under internal pressure. This causes high tensile stresses in the skin, which can result in stress cracking during ejection, although in most cases cracking will occur later, particularly under the influence of an adverse chemical medium. Similar effects are caused by yielding machine platens or molds as they are deformed during the holding pressure stage. During cooling, this pressure recedes, but the molded part remains under pressure. The normal shrinkage is fully compensated or even overcompensated. Again stress cracking can occur.

In some cases one can observe a tendency to stress cracking in a thin surface layer [49, 50]. It is difficult to distinguish whether this effect is produced by cooling under high internal pressure or is a frozen-in energy-elastic deformation from another mechanism. In the gate area, last material already near the glass-transition temperature is moved to compensate for volume shrinkage (last filling with cold melt) and can be expected to generate high strains. This results frequently in concentric cracks around the gate.

2.4.3 Crystallinity and Structure

Crystalline plastics possess the feature of attaining an orderly structure of molecular chains in the micro range. Such groups of molecules are called crystallites [51]. They cannot be observed under a light microscope, but easily form superstructures by spherulitic arrangement, which can be recognized with adequate magnification. A greater degree of crystallinity of a plastic results in greater hardness, strength, and brittleness. The degree of crystallinity depends primarily upon the molecular structure, but can

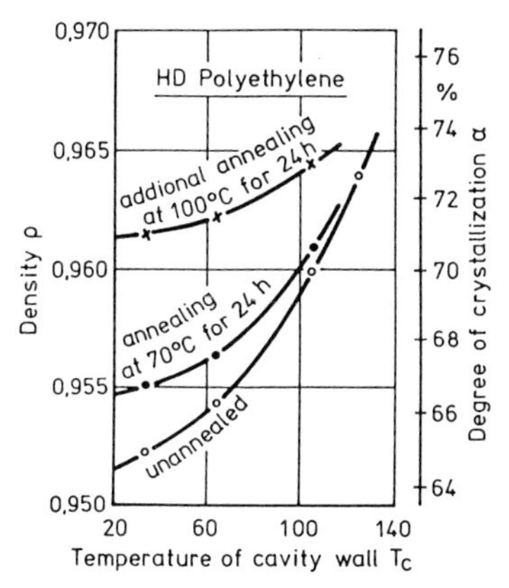

Fig. 31 Degree of crystalinity and density of parts molded with different mold temperatures [41]

also be influenced by processing and post-treatment [52, 53]. The cavity-wall temperature has the greatest effect (Fig. 31). Higher cooling rates result in less crystallinity. Rapidly cooled parts have a more or less transparent skin. For such moldings a post-crystallization can be expected, which causes a change in properties and dimensions for a long time to come. Annealing at elevated temperatures can accelerate post- crystallization. Shearing the melt particles at low temperatures also affects crystallinity. The resulting orientation promotes the formation of spherulites. Consequently, a low mold temperature on the one hand and shearing the melt at low temperature on the other can result in a very heterogeneous structure. This differences are related to areas close to and far from the gate.

Such adverse effects of processing are often the cause of failure of parts. It can be taken as a general rule that the more homogeneous the structure is, the more favorable the mechanical behavior will be.

3 The Injection Unit

The injection unit has to carry out the functions of accepting free-flowing pellets, heating and plasticating them, injecting the melt into a shape-providing cavity, and keeping it there under pressure (holding pressure).

One distinguishes single-stage plunger units, two-stage screw-plunger units with screw preplastication, and in-line reciprocating screw units. Single-stage plunger units are of importance only for mini injection molding machines when the desired shot weight can be provided with a plunger diameter of 10–20 mm. Today, most injection molding machines are of the reciprocating screw type. Its essential components are shown with Fig. 32.

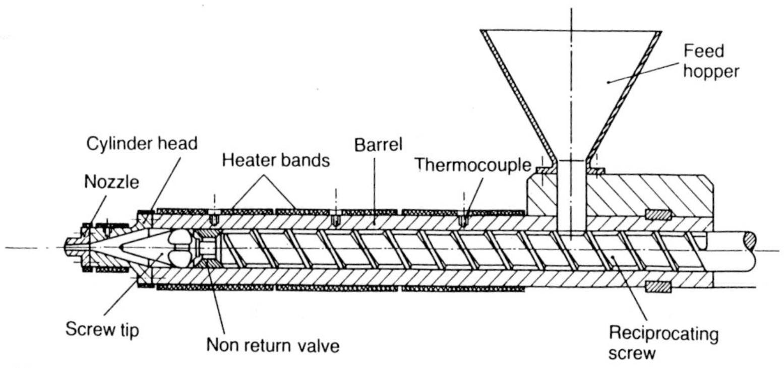

Fig. 32 Plasticating unit

The most important functions are:

- movement on its support to bring the nozzle into contact with the sprue bushing of the mold (in the clamping unit) or to retract it,
- generation of contact pressure between nozzle and sprue bushing,
- rotation of the screw during the feeding stage,
- axial motion of the screw during the injection stage,
- build-up of holding pressure.

Travel distances and positions associated with these functions are diagrammed in Fig. 33. Besides these indispensable operations, many modern machines also provide the following ones:

- a mechanism for hydraulic retraction of the nonrotating screw,
- the simultaneous axial and rotational motion of the screw for injection with a rotating screw,

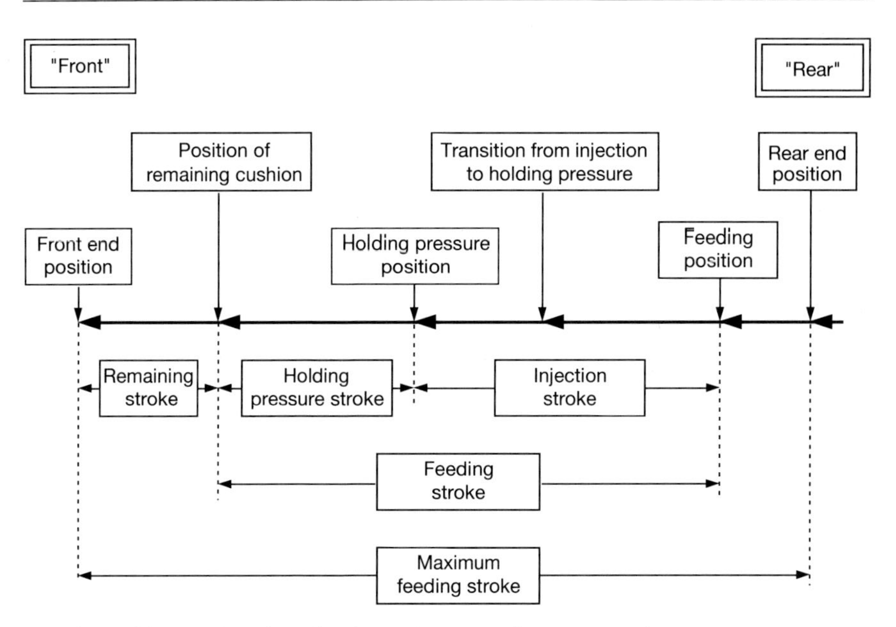

Fig. 33 Positions and strokes of reciprocating screw during one cycle

- control of variable strokes, velocities, rotational speed, and pressure,
- easy assembly or disassembly with swivel mount for time-saving exchange of screws,
- automatic exchange of whole plasticating units for a change of material (usually a special design).

3.1 Injection Pressure

The injection pressure is one of the most important performance parameters of an injection unit. This is the pressure exerted on the melt in front of the screw tip during the injection stage, with the screw acting like a plunger. It affects both the speed of the advancing screw and the process of filling the mold cavity with material. There are no reliable informations about the true magnitude of this pressure during injection molding. Therefore one usually refers to the corresponding pressure in the hydraulic cylinder.

The injection pressure generally rises within a short time from barometric pressure (or the lowest possible pressure in the system) to a magnitude which corresponds to the flow resistance of the melt in nozzle, runner system, and cavity. High resistance in nozzle and runner results in a high pressure build-up, making it difficult to recognize the onset of the compression stage after the volumetric filling of the cavity (Fig. 8a). On the other hand, this onset can normally be easily observed if the flow resistance in nozzle and runner is low and a sudden rise from injection to a much higher compression pressure takes place (Fig. 8b). A pressure profile with a varying flow resistance is pictured in Fig. 8c. Fig. 9 illustrates some more effects on the pressure development.

This demonstrates that the constancy of theses parameters is of great significance for the reproducibility of the molding process. Neither can optimizing programs for controlling the process [56, 57] be done without data from pressure profiles.

The disclosure from hydraulic pressure is especially consequential. In spite of this, pressure is often measured directly in the mold today. It can be used as an input variable in the framework of a process control of the molding process. The maximum injection pressures needed for some characteristic thermoplastics are listed in Table 3. Because processing thermoplastics is the standard task for an injection molding machine, the force on the hydraulic piston and the ratio between piston and screw cross sections are designed for a minimum injection pressure of 150 MPa required for thermoplastics without exceeding the power maximum of the drive system (necessary maximum pressure of the system ca. 180–200 MPa). A pressure between 15 and 80 MPa is effective in the mold most of the time. The total package of an injection unit mostly includes three plasticating sets with different screw diameters. The screws should be selected that the pressure requirements of Table 3 can be met (Fig. 34). The listed data have been determined under the assumption that a highly constant injection speed is required even at high pressure (high-quality molding). The maximum pressure of the standard barrel is calculated according to:

$$p_{max} \sim 1.25\ p_{req} \tag{1}$$

Example: $p_{req} = 150\ MPa$,
$p_{max} = 187.5\ MPa$.

The data in Table 3 make allowance for a power drop if about 80% of the pressure maximum is reached, and 5% losses from friction. In Sect. 5.8.1 Fig. 129 reveals that a hydraulic drive system only operates with adequately constant injection performance (= constant velocity of screw advancement during injection) up to a certain pressure level (mostly about 80% of the maximum pressure). Distinct losses in the injection performance at higher pressure can also be expected with direct electric drives. They can even be larger than from injection units with hydraulic drives.

Furthermore, one has to consider that 5 to 10% of the available energy

$$E_I\ (E_I = A_{Pist} \cdot \int p_H \cdot ds_{Scr})$$

for injection is consumed by friction in the hydraulic cylinder. It is certain that this energy consumption differs from one machine to another. This distinct particularity is responsible that, for high quality requirements, a change of machines is not possible without adjustment of the process parameters.

Several CAD programs permit an adequately exact predetermination of pressure losses in nozzle, hot runner, runner system, and along the critical flow path. With such data and the use of equation 1, a correctly dimensioned machine can be utilized or selected when it is purchased.

Table 3 Injection Pressure Required for Various Plastics

Material	Necessary injection pressure (MPa)[a]		
	Easy flow material,[b] heavy sections	Medium flow material,[b] standard sections	High viscosity material,[b] thin sections, small gates
ABS	80 - 110	100 - 130	130 - 150
POM	85 - 100	100 - 120	120 - 150
PE	70 - 100	100 - 120	120 - 150
PA	90 - 110	110 - 140	> 140
PC	100 - 120	120 - 150	> 150
PMMA	100 - 120	120 - 150	> 150
PS	80 - 100	100 - 120	120 - 150
Rigid PVC	100 - 120	120 - 150	> 150
Thermosets	100 - 140	140 - 175	175 - 230
Elastomers	80 - 100	100 - 120	120 - 150

[a] Not identical with listed maximum injection pressure.
[b] Definition depends on flow properties of the material, temperature, and restistance to flow.

Employed for processing of:	Injection pressure (MPa)	Diameter of screw and barrel bore	Hydraulic unit	Pressure ratio
Special cases	Highest pressure	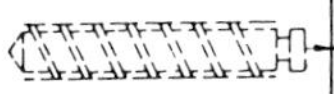	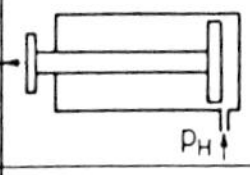 p_H	10-12
Thermosets, PC, PMMA and rigid PVC under difficult conditions	200-250		same as above	9-11
Standard processing of thermoplastics and elastomers	170-180 a)	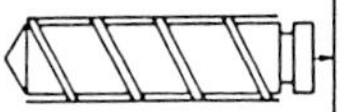	same as above	7- 9
Easy flow materials and cases with low resistance to flow	120-140 a)		same as above	6- 8
Special cases	100-120 a)	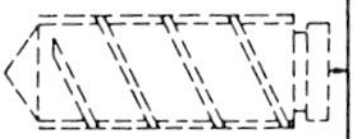	same as above	6- 7

Fig. 34 Injection pressure ranges for modular system of injection units
[a] Increase necessary if pressure loss in hydraulic servo and control system is larger than 20% of maximum pressure

3.2 Carriage Guide for the Injection Unit

In a standard design, the injection unit is supported by the right-hand section of the substructure or the stationary platen respectively. Small and medium-sized machines frequently support the carriage on bars located parallel to the main axis. The axis of the barrel is also parallel to these bars and usually in the same axial plane. The bars can be furnished with pistons over which hydraulic cylinders move. The cylinders are connected to the carriage and can provide the full-length stroke for carriage travel. Oil can be supplied through the bars. This leaves two alternative for the attachment of the hydraulic injection cylinder. With the design just described, the cylinder can be mounted coaxially with the screw or the injection cylinders are mounted coaxially over the traverse cylinder (Fig. 35).

Bars or slide ways on the machine base are preferred for machines between 3000 and 10 000 kN clamp force. Pull-in and retraction strokes can be provided by a piston-cylinder design in combination with the guide bars (Fig. 35a). However, the hydraulic cylinder between stationary platen and carriage, either above or below the base surface, is preferred (Fig. 36).

More advantageous are two parallel cylinders in the same plane with the barrel because they do not produce any lever action and cause a concentric positioning of nozzle and sprue bushing (Fig. 36b). Slide ways must be selected as carriage guide for

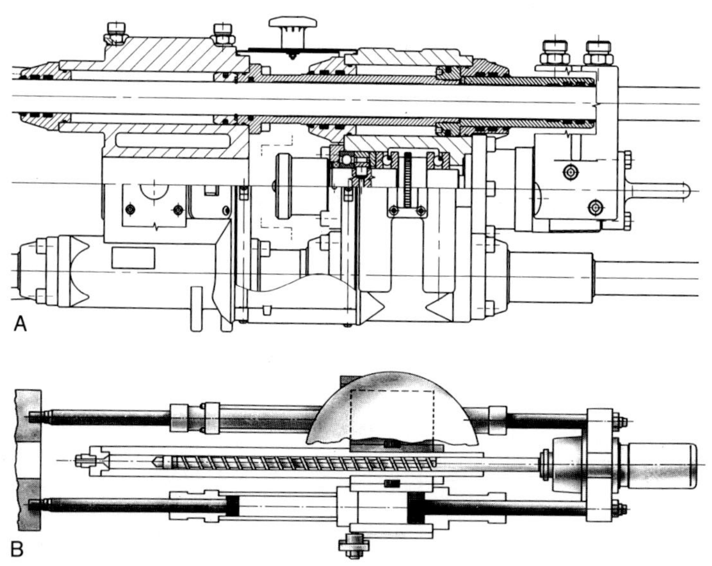

Fig. 35 Carriage guide of the injection unit with double-acting cylinders in coaxial design with the injection cylinders
A: System Arburg, B: System Battenfeld

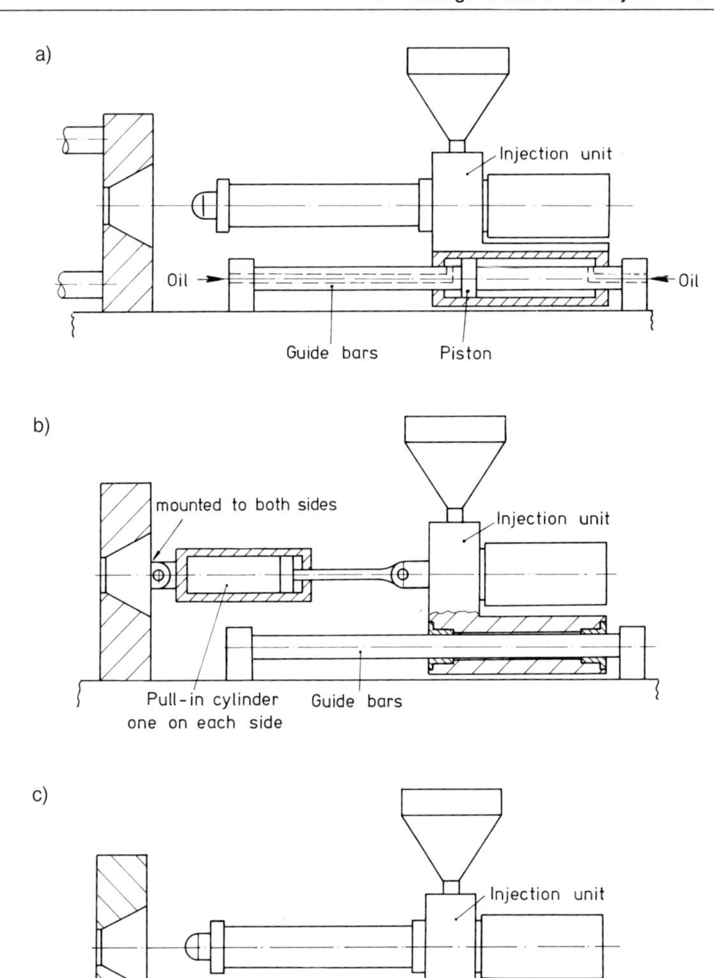

Fig. 36 Guide for heavy injection units.
a: Guide bars combined with piston and cylinder. b: Guide bars with separate pull-in-cylinders. c: Carriage slide

injection units of big machines because of their weight (Fig. 35c). The sliding carriage is generally guided in two planes on horizontal slide ways. It is backed up by adjustable guide shoes. For medium-sized and bigger machines it becomes necessary to support the barrel in the head region to secure it against tilting.

Table 4 Travel Velocity of the Injection Unit

Clamping force (kN)	Maximum velocity (mm/s)	Minimum velocity (mm/s)
< 500	300 - 400	20 - 40
501 - 2 000	250 - 300	30 - 50
2 001 - 10 000	200 - 250	40 - 60
> 10 000	200	50 - 100

Table 5 Contact Force between Nozzle and Sprue Bushing

Clamping force (kN)	Contact force (kN)
500	50 - 80
1 000	60 - 90
5 000	170 - 220
10 000	220 - 280
20 000	250 - 350

The last-mentioned versions generally have the hydraulic cylinder for axial screw movement mounted coaxially in the rear. In all cases, it must be possible to precisely align the injection unit with the center of the sprue bushing. This normally coincides with the machine axis. The motion has to be smooth and steady, although frequently at a high speed. During carriage movement, there must be no tilting, twisting, or other deviation from the center axis. Speed variations of modern machines can be programmed without sudden transitions. The average travel velocities of units are presented in Table 4. Table 5 lists the common forces with which the nozzle is in contact with the sprue bushing. This contact pressure prevents melt from leaking into the open at the interface between nozzle and sprue bushing.

3.3 Rotatory Screw Drive Systems

A considerable portion of the energy necessary for plastication (up to 60% for thermoplastics at low processing temperatures, 90% for thermosets) is the heat of friction provided to the material by the screw drive through the rotating screw. This results in a relatively high energy consumption during the feeding stage. The drive must be appropriately powerful because it also has to develop a high starting torque.

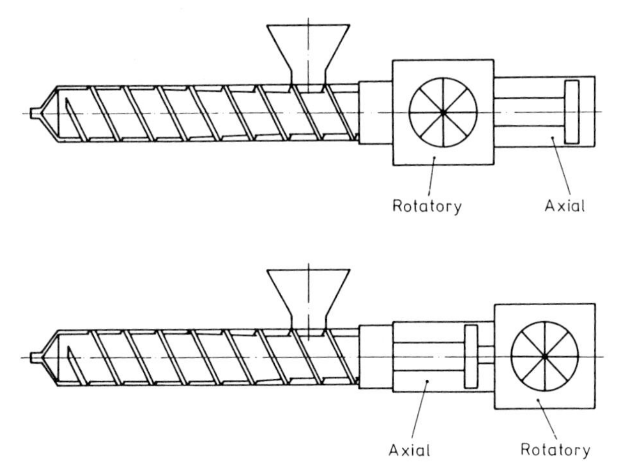

Fig. 37 The two possible locations of a rotatory and axial screw drive [65]

The rotatory drive is characterized by the the method and the position of the drive:

- electric motor with reduction or worm gear,
- hydraulic motor with reduction or worm gear,
- direct hydraulic drive.

There are two options for positioning the drive. A location between hydraulic piston and screw, or at the far end of the extended screw shank (Fig. 37).

3.3.1 Electromotive Screw Drive

While the electromotive drive is seldom used for small injection molding machines, it is the rule for machines with a clamping force of more than 15 000 kN. Three-phase shunt motors are generally employed; in certain cases they are equipped with an electrically actuated mechanical brake. As a direct screw drive, speed-controlled three-phase motors with frequency converter are employed in increasing numbers [66, 467] (Sect. 1.3.2).

When feeding begins, either the electric motor is started by an appropriate signal or a torque transmission between the already running motor and the screw gear is produced by actuating an electromechanical coupling. The latter version is preferable for the largest injection molding machines. The assembly of the rotatory drive in the injection unit is shown in Fig. 38. Electric motors have a high starting torque. Therefore screws with small to medium diameters must be especially safeguarded against being sheared off. Driving the screw in steps of constant speed is a particular advantage besides high dependability. This results in very good reproducibility even with varying loads. Late-model drives with controllable three-phase motors permit an almost exact adjustment at any speed.

a)

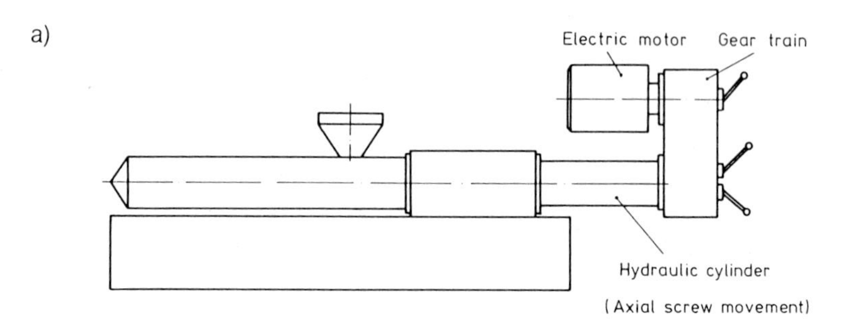

b)

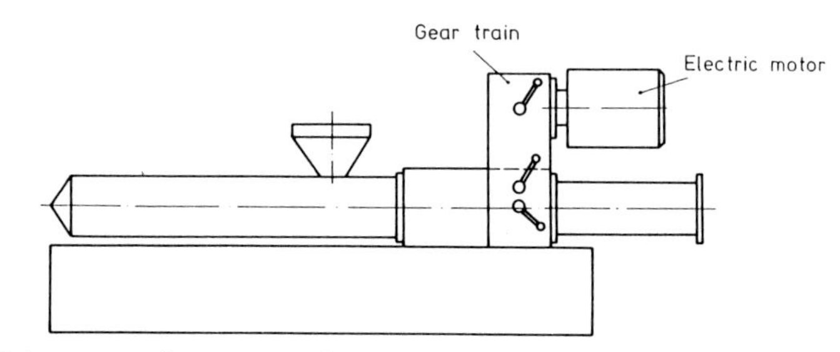

Fig. 38 Drive systems for screw rotation with electric motor. a: Gear box in the rear without movement during injection or recovery. b: Gear box between hydraulic cylinder and screw, moves with the screw

Electric motors with an electrically actuated mechanical brake stop precisely at the end of the feeding stage and prevent the screw from reversing during injection. This is especially important if no nonreturn valve is used. In cases where the braking effect of the electric motor cannot be utilized, a nonreverse lock is employed, which acts on the screw shank.

3.3.2 Hydraulic Screw Drive

Instead of electric motors, hydraulic motors are frequently used. They convert the supplied hydraulic into mechanical power ($p_H \times V_H = T \times n$). Hydraulic motors have a design similar to that of hydraulic pumps and can operate like them in many cases (Sect. 5.3.2).

One can distinguish (for the rotatory screw drive): Gear motors, internal gear motors, vane motors, radial and axial piston motors. The last three types are utilized by preference.

For driving screws with about 50 to 200 mm diameter mostly slow- running radial piston motors are employed. They are characterized by their smooth operation.

The exact positioning of the screw during injection is very important for the quality of the moldings. This can be achieved best if the mass of the movable parts is small.

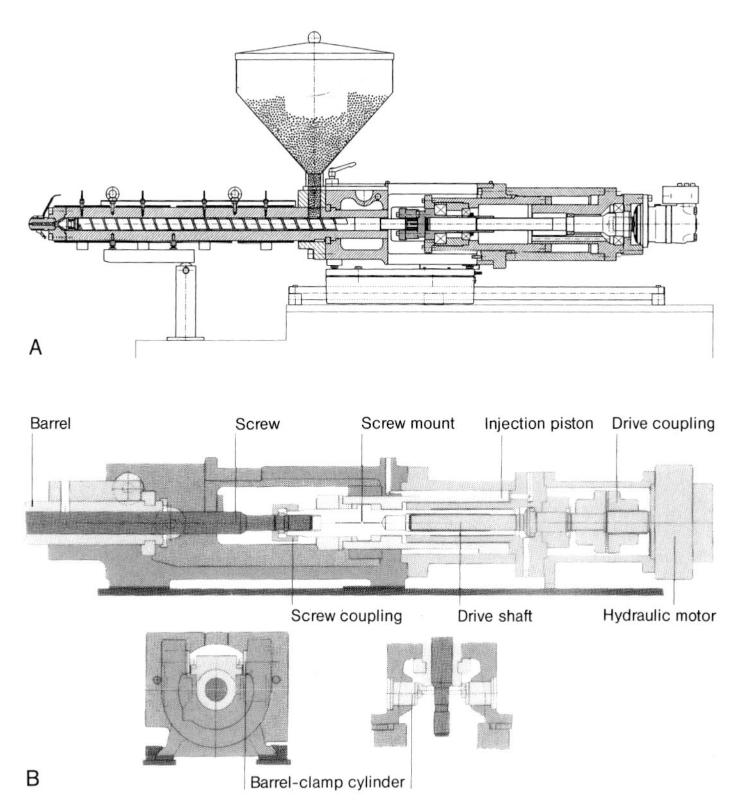

Fig. 39 Hydraulic screw drives
A: Conventional mounting as coaxial slide-in unit (system Engel),
B: With quick-clamping device for barrel (system Krauss Maffei)

Thus, a design that keeps the rotatory drive stationary and connects it with the screw coupling by means of a hollow shaft is favored (Fig. 39a and b). The direct hydraulic drive, however, is so lightweight that it can be directly mounted to the screw shank without any disadvantage.

The advantage of light weight is all but lost if a reduction gear is employed to provide for interchangeability between electric and hydraulic motor (Fig. 38).

3.3.3 Torque

The rotatory drive of the screw is of great importance for the process. Torque and speed of the rotating screw cause the conveyance of the material and the desired shear and homogenization. A good screw drive is expected to function without flaws up to the largest screw sizes and even with materials such as rigid PVC, PC, and Acrylics, which are difficult to process. A screw drive is particularly well designed if it can also supply the torque required for processing thermosets.

The torque which is needed for thermoplastics and thermosets can be taken from Fig. 40 [68–71]. The torque demand for the processing of elastomers is approximately equal to that for thermoplastics. Curve 1 pictures roughly the torque dependence on the screw diameter that can generally be realized with commercial hydraulic or electric drive units. The torque required for thermosets is somewhere between curve 1 and 2. For other materials to be processed, the torque demand may lie along curve 2, curve 3, or in between. Rigid PVC, PC, and Acrylics need high torque (curve 2), PE and PS a lower one (curve 3) [71]. Curve 2 serves as a basis for rating injection molding machines for processing thermoplastics.

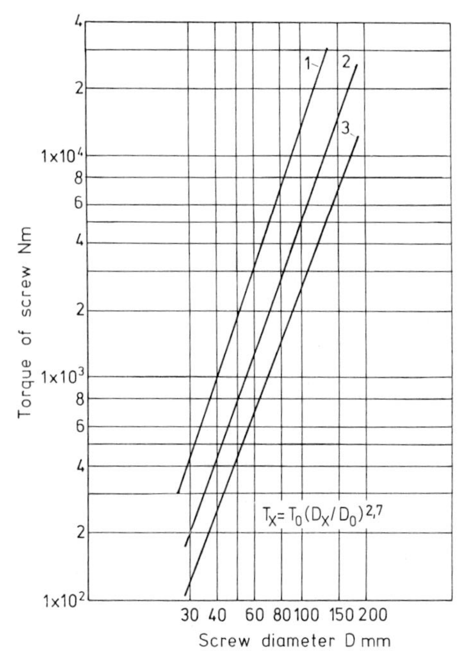

Fig. 40 Torque dependent on screw diameter for length 20 L/D
1: Thermosets, 2: Engineering thermoplastics, 3: PS and PE [68, 71]

In accordance with this presentation, the torque T_x required for a screw of diameter D_x can be calculated by employing the principles of similarity [72, 73, 74, 75]:

$$T_x = T_0 \times (D_x/D_0)^{2.7}$$

where T_0 is the known torque value of a given screw diameter D_0. This corresponds with the law of transformation, which is valid for extrusion.

The necessary power input at the desired speed (material output) can be computed with the torque taken from Fig. 40 and the correlation between torque and power input:

$$N_S = C \times n_S \times T_S$$

where N_S = power input (kW), C = conversion factor (0.001), T_S = torque of the screw, n_S = speed in revolutions per second (s^{-1}). N_S is the power that must be provided by the screw drive motor.

3.3.4 Screw Speed

It is still common to consider the speed of the screw an essential machine-related process parameter and to use it as input variable for establishing the operating point. It is the circumferential screw speed, however, which is relevant to the quality of the melt. The molder should get used to deal with circumferential speeds. With equation (2), the rpm (rotations per minute) can be computed from the circumferential speed.

$$n_S = \frac{v_c \cdot 60}{D \cdot \pi} \quad (min^{-1}) \tag{2}$$

D = diameter (m)

The speed ranges should conform preferably to one of the following three demands (Fig. 41), compare [75]:

- high flow rate at maximum speed for molding thin-walled parts for packaging, plastics such as PS and PE, eventually PP,
- medium flow rate for technical moldings, so-called engineering plastics,
- low flow rate at minimum speed for processing thermosets and elastomers.

To accommodate these requirements, most machine producers offer the following ranges of circumferential speeds for screws:

- high speed for high melt-flow rate, speed v_c = 0.4 to 1.5 m/s,
- standard speed, v_c = 0.2 to 0.4 m/s,
- low speed, v_c = 0.05 to 0.2 m/s.

The circumferential speed or alternatively the revolutions per minute are preselected at the display. It is necessary to verify the selected data very precisely or, at least, repro-

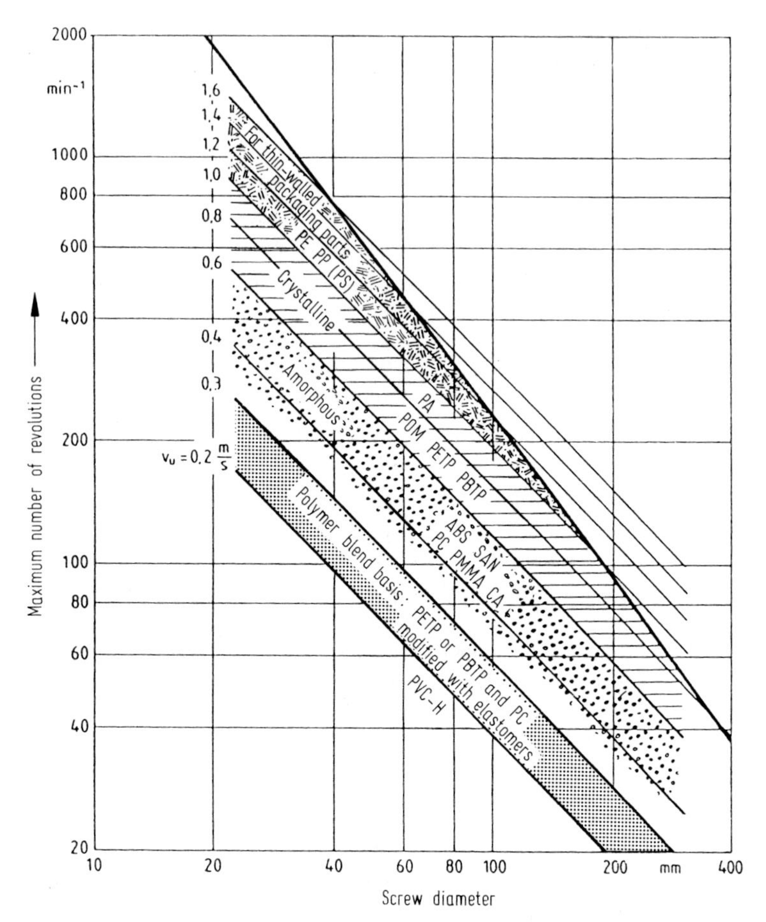

Fig. 41 Optimum screw speed dependent on screw diameter and critical values, v_u = circumferential speed [75]

duce them exactly. Today most machines do not measure rotational or circumferential speed and, accordingly, do not record the true values. Such a display, however, would help to increase the quality of injection molding, especially, if, for a "just-in-time" production, exchange of machines was programmed. Investigations [78, 209, 474] have demonstrated that reproducing the speed and with it the feeding stage is not one of the best properties of an injection molding machine. Therefore measuring the data and their actual output would be desirable.

At present, the control of the screw-driving hydraulic motor is almost exclusively done with position-controlled proportional valves or servo valves. A very precise speed control is also available with a three-phase electric servo motor, which is controlled by a resolver. These solutions, which are not inexpensive, are offered by most machine producers today because one can do without the otherwise required reduction gear.

3.4 Screws

In-line screw injection molding machines were employed for the first time in 1956 [76]. The screw provides for a comparatively rapid heat exchange between the "hot" barrel and the relatively "colder" material by continuous exchange and rearrangement of material in the screw channel. Compared with a single-stage plunger machine, which is built today only for small barrel diameters up to about 20 mm, a considerable increase in the melting rate of a material is achieved. The screw itself has also experienced a certain development although today's standard screw is the same in its essential outline as in 1956. It usually is a screw with three zones and has a length-to-diameter ratio (L/D) of 20:1 ± 10%.* Shorter screws do not provide an adequate quality of the melt. With longer screws, 24:1 as the utmost, one has to anticipate degradation of a number of engineering plastics from too long a residence time [79, 80, 81].

This is the reason for using screws with L/D ratios of 22:1 to 26:1 only in fast-running molding machines e.g. in the packaging sector. Then they are called "packaging screws". They are often equipped with mixing sections or with a combination of shear and mixing sections and eventually with an efficient feeding zone [75, 80, 82].

Recently, screws again undergo an intense development at which output, mixing and homogenizing effects are in the foreground. Standard plastics often call for high output (circumferential speed of 0.4 to 1.5 m/s). For blends, however, homogenizing at a low temperature level is of importance (0.1 to 0.3 m/s). Promising programs are being developed to compute even more difficult injection molding screws. This work should come to a conclusion during the next two years [83, 84, 85].

Another screw design known from extrusion is now introduced to the injection molding technique as barrier screw. Melted material is separated from solid pellets beginning at the point of melting and both are carried in two parallel channels.

The information provided in the following sections is restricted to so-called thrust screws and excludes preplasticating screws.

---

*According to European standard, the length of the screw is the distance from the forward edge of the feeding throat to the end of the last flight of the screw (Fig. 42), with the screw in its foremost position. It is related to the active length. In contrast to this, the American standard (SPI) determines the total length as the distance from the outset of the first flight next to the shank to the front end of the screw. Thus the L/D ratio of 20:1 is a larger ratio in the United States and may well be 22:1 or more, depending on design details.

3.4.1 Standard Screws for Thermoplastics

Modern screws for thermoplastics are generally designed like the one depicted in Fig. 42. The terminology connected with screw design is presented and explained with Fig. 43. Essential dimensions are flight depth and the channel-depth ratios. These data are compiled for several screw diameters in Table 6. If a range of variation of ± 10% is accepted as a basis for all dimensions the whole scope of commercially offered variations is considered almost completely. Fig. 44 demonstrates the relationship graphically. Smaller flight depths are applicable for crystalline thermoplastics. They are not suited for rigid PVC. An increase in the length of the standard screw cannot be expected. Such a development would be faced with distinct disadvantages such as hazard of degradation from extended residence time and intense exposure to shear and heat. Modern screws with an L/D ratio of 20:1 provide an adequate output in almost all cases (Refer to footnote on page 51).

The standard three-zone screw is not designed for efficient mixing. It is only conditionally qualified for compounding or dry coloring of plastics. Should such duties be desired in addition to plasticating and injection, then shear and mixing elements must be employed, which cause a significant improvement in dispersing additives. A shear element should always be placed near the end of the compression section, and the mixing element within the range of the metering section [80, 81, 82].

Usually, geometry, energy demand, and capacity of a screw can be computed rather easily with the following principles of similarity if the adequate function of a screw with length L_0, the flight depth h_0, the pitch p_0, and the diameter D_0 is known.

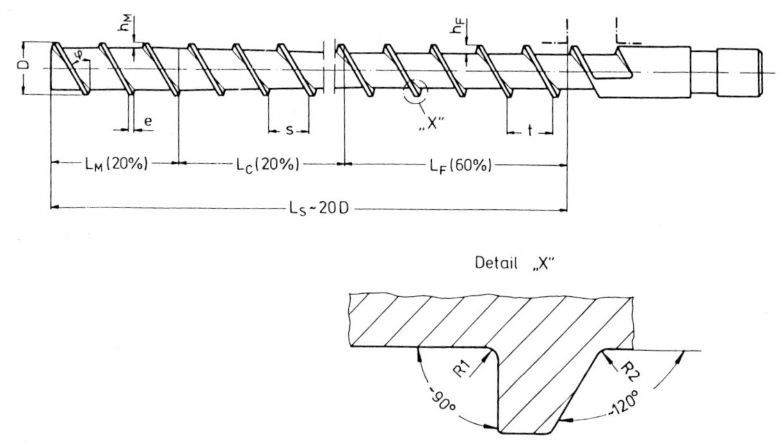

Fig. 42 Screw for processing thermoplastics

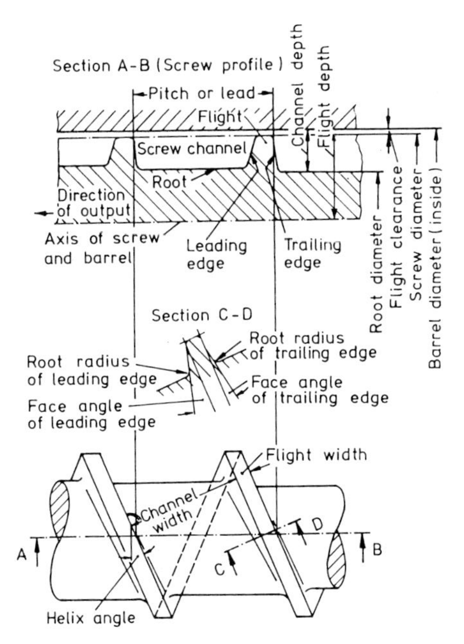

Fig. 43 Design terminology for screw of Fig. 42

Table 6 Significant Dimensions (Averages) of Screws for Processing Thermoplastics. Deviation of ± 10 % is common. L_s = effective screw length according to European standard (Euromap 1). Model law: $h_x = h_0\,(D_x/D_0)^{0.74}$ [77]

Diameter (mm)	Flight depth (feed) h_F (mm)	Fligth depth (metering) h_M (mm)	Fligth depth ratio h_F/h_M	Radial flight clearence (mm)	Comments
30	4.3	2.1	2.0 : 1	0.15	Peak-to-valley height of surface: 2 - 4 μ m
40	5.4	2.6	2.1 : 1	0.15	$R_1 \sim 1$ - 4 mm
60	7.5	3.4	2.2 : 1	0.15	$R_2 \sim 5$ mm (30 - 60 mm dia)
80	9.1	3.8	2.4 : 1	0.20	$R_2 \sim 10$ mm (61-150 mm dia)
100	10.7	4.3	2.5 : 1	0.20	Pitch $t = D$ (to 0.7 D)
120	12	4.8	2.5 : 1	0.25	L_s/t always about 20
> 120	max. 14	max 5.6	max 3 : 1	0.25	Flight width 0.1 D
					Maximum feed travel D

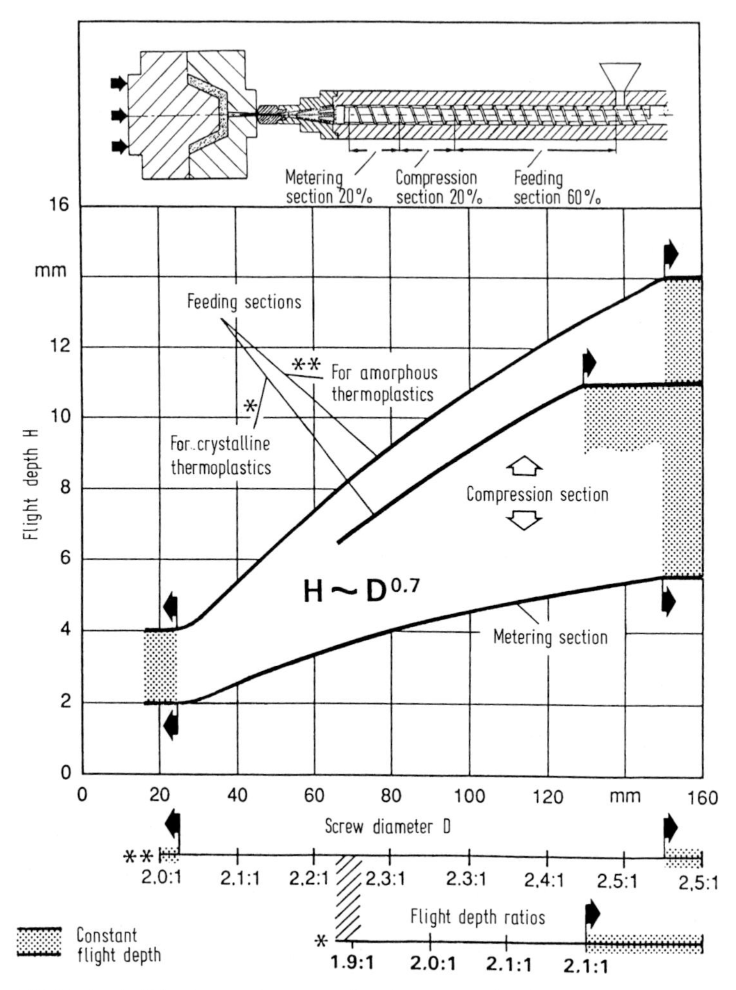

Fig. 44 Flight depth (H) of screws for processing thermoplastics dependent on screw diameter (D) [87]

Assuming that

$$L/D = L_0/D_0 \text{ and } p/D = p_0/D_0 \tag{3}$$

and melt temperature, dynamic pressure, and heat-flux density at the inside barrel wall are constant, then

flight-depth ratio and residence-time ratio are

$$h/h_0 = (D/D_0)^{0.74} = t_r/t_{r0} \tag{4}$$

the circumferential-speed ratio is

$$n/n_0 = (D/D_0)^{0.74} \tag{5}$$

output ratio and heat consumption are

$$V/V_0 = (D/D_0)^2 = E_H/E_{H0} \tag{6}$$

torque ratio is

$$T/T_n = (D/D_0)^{2.7} \quad (Sect.\ 3.3.3) \tag{7}$$

The ratio of the shear velocities is

$$\gamma/\gamma_0 = (D/D_0)^{0.48} \tag{8}$$

To begin with, the exponents can be used as a good approximation to plasticating data and with an accuracy generally sufficient in injection molding without considering the material. If this should be insufficient, then the exponents have to be determined by experiment [72, 73, 74, 75, 83].

Until a few years ago, one could rely on the assumption that the cooling time (t_c) affects the molding cycle substantially. In the meantime, molding with large screws and consequently the time needed for feeding and plastication have created a frequent bottleneck. This is a result of the steady development of plastic materials and processing technique. Presently, the simplest and most practical solution is the use of a larger screw with a sufficient diameter. The nomograph of Fig. 45 can assist in the selection of such a screw. Very questionable, however, is a tendency during the last 10 years to achieve higher shot weights by increasing the feeding stroke of the screw. Italian and Japanese manufacturers led the way with this method. However, this is a way into the wrong direction [88, 89]. With a feeding travel of more than three times the diameter, the homogeneity of the melt is incompatible with high quality demands. There is the additional hazard of getting air into the melt, which results in unavoidable surface blemishes.

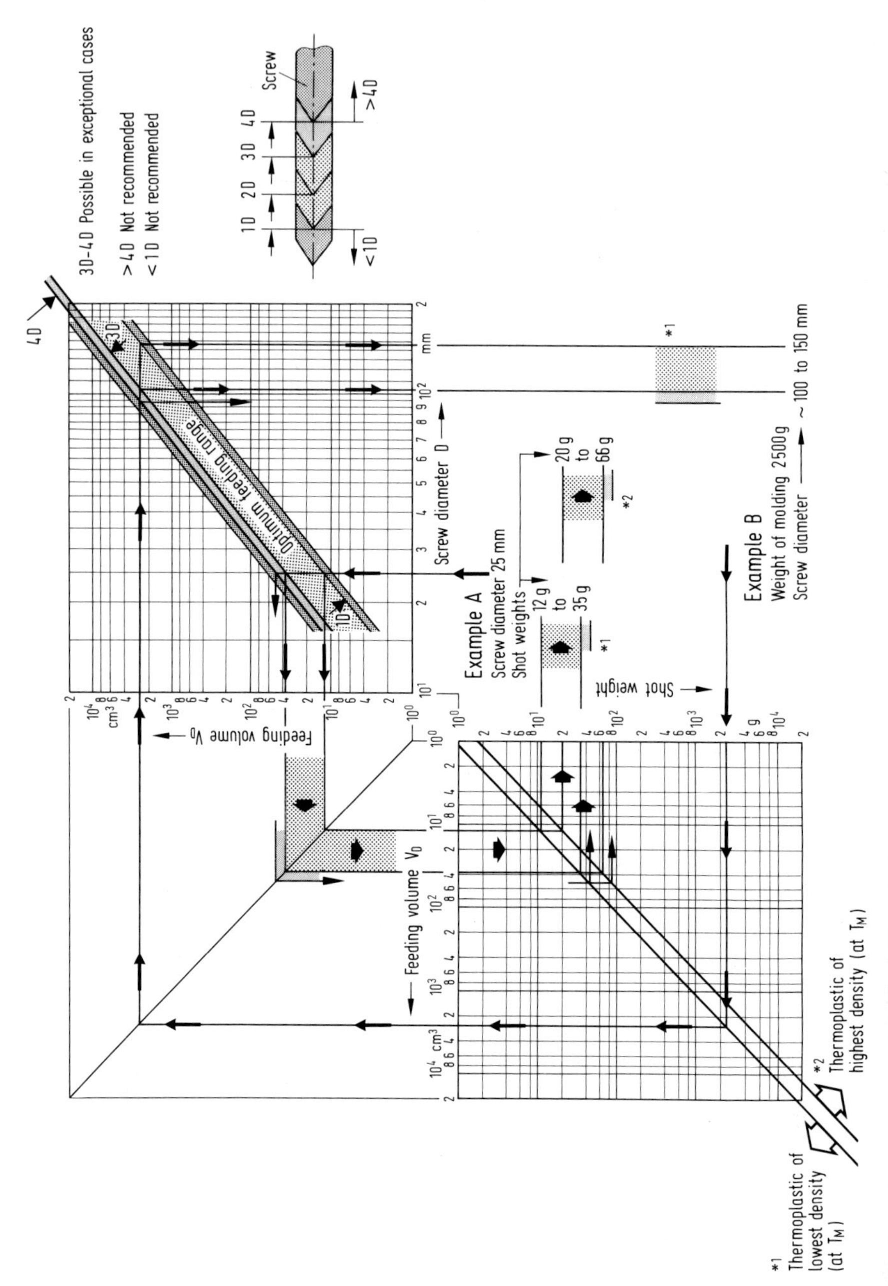

Fig. 45 Correlations among screw diameter, feeding volume, density and shot weight [87, 95]

3.4.2 Special Screws for Thermoplastics

A special screw geometry is common only in cases where a screw is used exclusively for processing a particular plastic material [75, 77, 80, 82, 83, 84]. Thus, screws with a relatively shallow flight depth are utilized for processing Nylon, PBT, PET, and POM (flight depth from Fig. 44), while screws for CA and CAB can eventually be modified in the opposite manner [91]. Heavy-duty screws for processing especially PS, PE, and PP for packaging parts are extremely long screws with a ratio L/D ~ 25:1 [75, 80, 82].

3.4.3 Barrier Screws

The barrier screw is a design known from extrusion, which separates pellets and melt and conveys them in two parallel channels beginning at the point of melting. It is introduced now to the injection molding technique [90–94]. Starting at the theoretical point of melting a second, at first narrow channel is added. The melted material flows through a small gap between flight and barrel wall into this channel and is collected there. After the whole materiel has been melted the channel that has conveyed the solid pellets disappears. After this, the melt is intensely mixed by a shear and mixer combination, which can be consist of a so-called "Maddock" section and a diamond-shaped pin mixer. A complete screw is shown with Fig. 46.

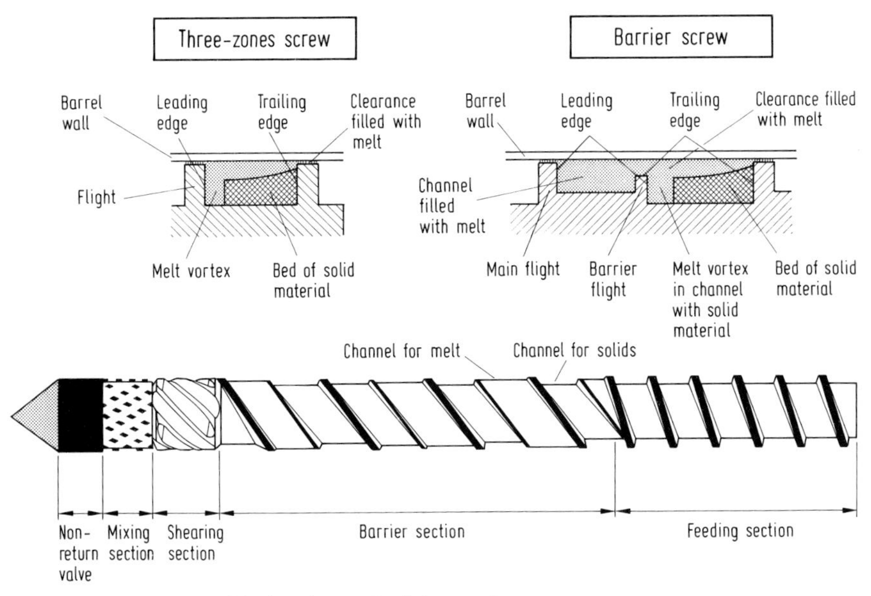

Fig. 46 Barrier screw with shearing and mixing section

If this screw design is compared with that of a three-zone screw, there is not much difference between them as far as melt temperature and output are concerned. There are indications, though, that the barrier screw displays a greater dependence of the feeding capacity on the screw stroke. This somewhat greater sensitivity to a variety of plastics may be eliminated with screws a little longer than L/D = 20:1.

Striking, however, is the gained homogeneity of the melt. It exceeds by far that of the standard three-zone screw. This opens up new prospects for the molder concerning quality and economics. Especially if recycled material is added, the compounding can immediately be done with the molding machine, and the intermediate step of separately compounding reused material is saved. Thus, not only costs are cut but the material is preserved from an additional processing step [94].

Concurrent with the barrier screw, double-flighted screws equipped with shear and mixing components are being tested at present. Similar positive results can be expected.

Besides immediate use of recycled material at the machine, such screws with intensive mixing capability should also permit coloring tasks.

3.4.4 Screws for Rigid PVC

While plasticized PVC is commonly processed with standard screws, special screws are generally used for rigid PVC. The sections along the screw remain the same as in Fig. 42 but the total length and the flight depths are different (Table 7). For processing PVC it is customary to plate the screw and screw tip – occasionally also the barrel and the barrel head if no bimetallic lining is preferred – as protection against corrosion. Suitable materials are chromium and nickel. Steel grades have to be selected which

Table 7 Significant Dimensions (Averages) of Screws for Processing Rigid PVC. Deviation of ± 10% is common. L_s = effective screw length according to European standard (Euromap 1). Model law: $h_x = h_0 (D_x/D_0)^{0.74}$

Diameter (mm)	Flight depth (feed) h_F (mm)	Fligth depth (metering) h_M (mm)	Fligth depth ratio h_F/h_M	Radial flight clearence (mm)	Comments
30	4.5	2.5	1.8 : 1	~0.15	Peak-to-valley height of surface: 2 μm
40	5.5	3	1.8 : 1	~0.15	R_1~3 mm
60	7.3	4	1.8 : 1	~0.2	R_2~5–10 mm (30–50 mm dia)
80	9	5	1.8 : 1	~0.2	R_2~10 mm (60–100 mm dia)
100	10.5	5.5	1.9 : 1	~0.2	Pitch $t = D$ (to 0.8 D)
					L_s/t 18 – 20
					Flight width 0.1 D
					Maximum feed travel D

provide good adhesion of the protective plating. Heating is accomplished with electronically controlled band heaters as it is with other thermoplastics. It is normally not necessary to heat the screw.

3.4.5 Screws for Vented Barrels

Vented barrels have been employed with Acrylics, CA, CAB, and ABS since about 1965 to extract water that is adsorbed or absorbed in the molecular structure [98–102]. Usually no vacuum is used and the water escapes as vapor through the vent into the open. The significance of this degassing technique during processing increased between 1976 and 1986 when PC and Nylon could be successfully degassed with appropriately designed units. A number of difficulties in practical operations led to a retreat from degassing. The problems are of the following kind:

- Degraded depositions in the pressureless degassing region leading to visible flaws in brightly colored moldings,
- difficult change of materials, products of degradation hard to remove,
- reliability of processing not as good as with standard screws
- availability of dryers operating independent from climate have become common in injection molding.

So far, degassing screws for vented barrels have been built with diameters between 25 and 170 mm.

There are three different concepts:

- two standard screws of medium length are arranged in tandem, $L_S/D = 2x(13 \text{ to } 16){:}1 = 26 \text{ to } 32{:}1$ [103–106];
- screws with starve feeding [107];
- short screws $L_S/D = 20{:}1$ (Fig. 47). (L_S = effective screw length)

Long degassing screws result in too long a residence time of the material in the plasticating unit and, with this, to a partial degradation of the plastic (e.g. Nylon, PC, PET, PBT). They are not suitable for these materials and have disappeared from the market. Degassing units with starve feeding make it possible to adjust the flow of material in the first screw to the one in the melt-conveying second one by an appropriate setting of the metering device e.g. a vibrator. It feeds the screw only with as much material as the second screw can handle. This ensures that no melt can escape through the vent. Because of their manufacturing costs this solution is not offered anymore on the market.

The short degassing screw as well as the screw with starve feeding can be installed in modern injection molding machines without significant constructional changes. They are designed in such a way that they can process all plastic materials considered for drying (Fig. 47) [100, 101, 108, 109]. The output decreases by 15 to 50%, though,

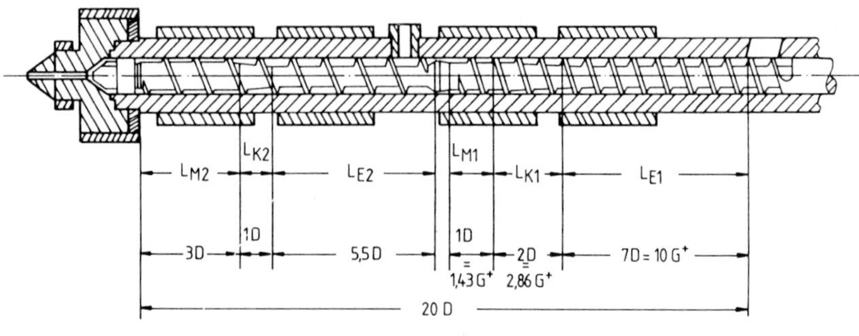

Fig. 47 Two-stage screw with vented barrel [108, 109, 110]

Table 8 Recommendations for Temperature Settings Along a Vented Barrel for Various Thermoplastics [109]

T Melt °C	Control setting °C				T Melt °C	Control setting °C		Material
230-245	230	230	230	230	225	225	240	ABS
260-270	265	265	260	250	255	235	250	ABS/PC
210-220	210	210	210	210	210	215	225	Cellulosics
220-245	220	230	230	210	230	220	230	PA 6
260-280	260	250	250	230	250	240	250	PA 66
250-260	245	245	240	240	250	235	250	PBTP
300-315	300	300	290	280	280	280	310	PC
240-255	240	240	240	240	230	220	240	PMMA
280-310	280	290	290	290	275	290	290	PPO
250-265	250	260	260	255	235	240	260	SAN

compared with the standard screw. This does generally not result in an increase in overall cycle time. The processing range, even in dependence on residence time and temperature, has become well known in the meantime (Table 8) [109, 110]. It is generally suggested to employ a pocket in the vent region. Such a suggestion is demonstrated with Fig. 48. Table 9 proposes flight depths for universally applicable degassing screws.

Degassing screws exhibit higher wear than standard screws because of their unfavorable design in this respect. In the open area, there is an additional strain on the

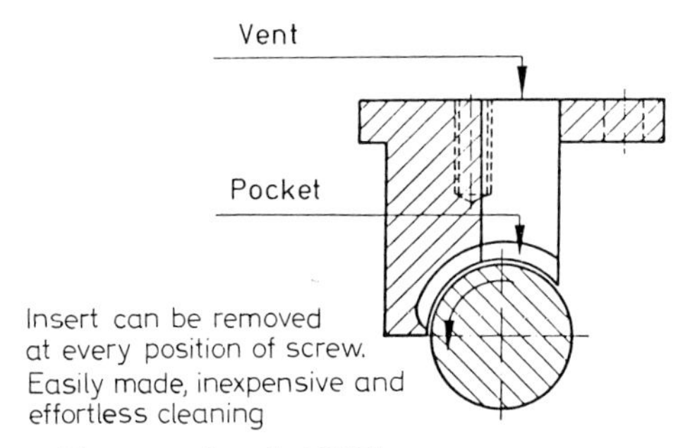

Fig. 48 Barrel insert with vent and pocket [109]

Table 9 Flight Depth and Clearance at Shear Element of Screw for Vented Barrel with L/D ratio of 20 : 1. Model law: $h_x = h_0(D_x/D_0)^{0.7}$

Diameter (mm)	Flight depth h_{11}	Flight depth h_{12}	Flight depth ratio	Flight depth h_{21}	Flight depth h_{22}	Flight depth ratio	Shear gap width s_1
30	4.0	2.0	2.0 : 1	6.3	2.2	2.85 : 1	0.5
50	5.4	2.7	2.0 : 1	9.3	3.2	2.90 : 1	0.8
70	7.0	3.2	2.2 : 1	11.7	3.9	3.00 : 1	1.0
100	9.0	4.1	2.2 : 1	15.1	5.0	3.00 : 1	1.3

$R_1 \sim 2-3$ mm, $R_2 \geq 10$ mm for up to 60 mm dia ≥ 15 for more than 60 mm dia

material from water vapor and volatile ingredients. For this reason, these screws should have a wear-resistant surface [108, 109]. In some cases a chromeplated surface has proved particularly suited (Sect. 3.4.10).

Color and material change can be considerably improved by pulling the insert (Fig. 48) and filling the opening with pellets. The necessary steps of this operation are described in detail in [109].

As a precaution, and in compliance with eventually existing safety regulations, the vapor that escapes from the vent should in any case be ventilated away from the working area with additional exhaustion equipment [109].

3.4.6 Screws for Thermosets

Screws for processing thermosets have less flight depth and a smaller channel-depth ratio than screws for thermoplastics. They are used without a nonreturn valve. All commercially available screws show similar features (Fig. 49). Their design should prevent heating the curable material unduly by shear to avoid a reaction in the flights of the screw. Therefore screws with a flight-depth ratio of 1:1 to 1:1.3 are employed most of the time. With a ratio of 1:1 the division into LF and LM shown in Fig. 49 no

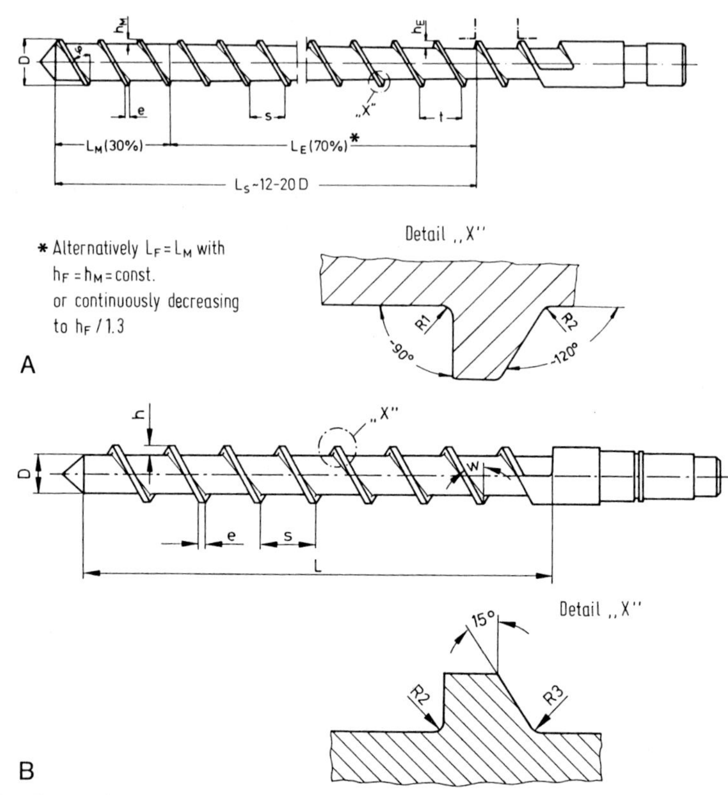

Fig. 49 Screws for processing thermosets
A: Screw with varying flight depth; B: Screw with uniform flight depth

longer holds [111]. According to [112] screws for thermosets are generally shorter than those for thermoplastics. The L/D ratios for free-flowing materials are 12:1 to 15:1. For already plastic polyesters (BMC, DMC, etc.) a ratio of about 10:1 is sufficient.

Some data concerning screws for processing thermosets are provided with Table 10. Because these screws must operate without nonreturn valves, there is a comparatively large back flow of material into the rear section during injection and holding pressure stage. Therefore flight depth and width are of particular importance. The flight is wider than that of screws for thermoplastics. It may be 0.15 to 0.2 times the screw diameter. The remaining smaller channel cross section impedes the back flow of the plastic, and the wear resistance of the thicker flight is relatively good. Generally, screws longer than L/D = 15:1 can also be successfully employed and without problems if a lower temperature profile is maintained in a range of three to four flights from the feeding throat. Screws with a slight increase in flight depth towards the tip exhibit little back flow and therefore less tendency to wear than other screws. Adjustments for special tasks are also common. For example, screws with slightly deeper flights than those in Table 10 are frequently used for processing polyesters.

Table 10 Significant Dimensions (Average) of Screws for Processing Thermosets. Deviations of± 10 %are common. L_s= effective screw length according to European standard (Euromap 1). Model law: $h_x = h_0(D_x/D_0)^{0.74}$

Diameter (mm)	Flight depth[b] (mm)	*L/D*-ratio	Fligth width (mm)
30	4	12 - 15 (∅ 14)	4
40	4/4.5	12 - 15 (∅ 14)	5
50	5/5.5	12 - 15 (∅ 14)	6
60	7	12 - 15 (∅ 14)	7
75	8.5	12 - 15 (∅ 14)	8.5
80	12	12 - 15 (∅ 14)	12

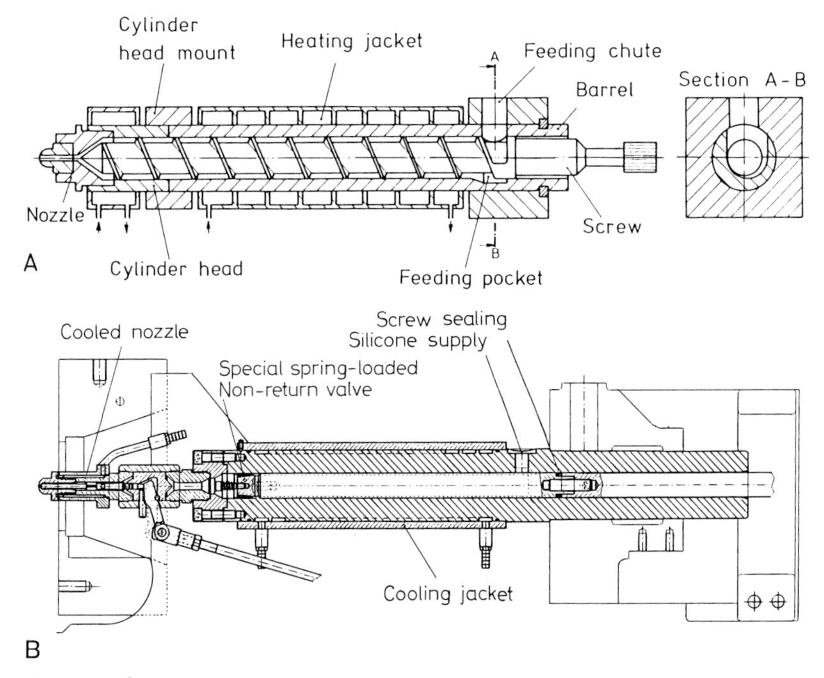

Fig. 50 Units for thermosets or elastomers
A: Standard design
B: Barrel for processing liquid silicones (System Engel)

Today most screws for thermosets are made of fully hardened steel of great toughness.

It is not common to heat the screw. On the other hand, care must be taken that only a small gap of 0.5 mm remains between screw tip and nozzle when the screw is in the foremost position. This prevents deposition and polymerization of the material in the nozzle. This adjustment should be checked with each machine. Occasionally, feeding pockets (Fig. 50) are provided [111], which improve output and compression. The barrel is primarily heated with water, which is circulated through a jacket that is tightly fitted onto the barrel. The temperatures are about between 60 and 90°C, increasing towards the front end of the barrel.

3.4.7 Screws for Elastomers

Elastomers have been processed on injection molding machines with reciprocating screws since the early 1960s. Whereas conveyance of the material presents no problems, temperature profile and residence time pose a difficulty. They must be selected in such a way that the rubber compound does not vulcanize prematurely in the flights, the space in front of the screw, or the nozzle. This is made possible by employing temperature-controlled barrels (Fig. 50A). The screw design is similar to that in Fig. 47. Some standard dimensions are given in Table 11. Screws with diameters of 30 to 120 mm have proven to be effective. Besides single-stage screw units, double stage units with screw preplastication and plunger injection are employed, too [113, 116].

A short length and a slight decrease in flight depth beginning with the feed section are typical. Both features assure that the rubber compound is treated with the necessary care during processing. If these provisions are inadequate, or if back pressure is used to achieve the necessary compression, the temperature of the screw must be controlled to avoid an undue rise in temperature. For this purpose, screws for elastomers usually have a hollow core, if the diameter permits, to provide for water cooling.

The screw must be able to take rubber ribbons. For this reason, the barrel has to be equipped with an enlargement, which is called entry or feeding pocket (Fig. 50A). The pocket prevents the entering ribbon from being sheared of between the screw flight and the edge of the throat. A special design is required for processing powdered rubber to produce a higher feeding rate and the necessary compression. If compounding is the task, the screw has to be equipped with mixing elements in the metering section. An especially good mixing effect was achieved with a screw which had a compression zone of L/D = 5.5:1 following the usual feeding zone of L/D = 4.5:1, a change in flight depth of 3:1, and was equipped with an ensuing "Maddock" mixing element and a mixing zone with shear flights with a length of L/D = 5:1 [117]. Elastomers can often be processed with a commercial nonreturn valve. Screws for thermosets are also employed in special cases.

The screws are exposed to considerable wear if rubber with inorganic fillers is processed. Fully hardened screws have proved to be particularly good in such cases.

Table 11 Significant Dimensions (Averages) of Screws for Processing Elasomers. Deviations of ± 10% are common. L_s = effective screw length according to European standard (Euromap 1). Model law: $h_x = h_0(D_x/D_0)^{0.74}$

Diameter (mm)	Flight depth (feed) (mm)	Flight depth (screw tip) (mm)	L/D-ratio	Flight clearance (mm)	Comments
30	5	4.2	12 – 15	~0.1	Pitch $t = D$
40	6.1	5.1	12 – 15	~0.15	Peak-to-valley height of
60	8.2	6.9	12 – 15	~0.15	surface: 2 – 5 µm
80	10	8.4	12 – 15	~0.2	R_1 ~ 2 mm
100	11.7	9.7	12 – 15	~0.25	R_1 ~ 5 mm (30 – 50 mm dia)
					R_2 ~ 10 mm (60 – 100 mm dia)
					Flight-depth ratio:
					1.1 : 1 to 1.3 : 1
					Maximum feed travel 4*D*

3.4.8 Screws for Processing Silicones

The behavior of silicones during processing is similar to that of rubber. Therefore the screws, which are used, differ only slightly from those for rubber processing. They should generally have no flight-depth differences over their whole length [115]. Their channel is fairly shallow, more so than listed in Table 11. A nonreturn valve is employed and an annular seal at the shank (Fig. 50B). This seal should yield, however, at a malfunction to prevent a back flow of the material to the feeding region. Now and then, screws carry mixing elements in the front region instead of flights, which can be a simple, smooth bolt.

3.4.9 Other Types of Screws

In the course of processing ceramics on screw injection molding machines, special screws have shown up which are not discussed here due to their particularity. Informations about them are not readily available because of strict secrecy.

3.4.10 Wear and Wear Protection

Causes of Wear

Problems with wear have been known since the start of processing thermoplastics reinforced with short glass fibers in the beginning of the 1960's. They are well investigated and extensively publicized [118]. More essential reading material is presented with [119 to 131]. In many cases the economics of injection molding became questionable through wear-causing effects. This theme occupied molders, machine manufacturers and plastic producers for many years [132, 133]. Initially, molders did not consider the high investment for wear-resistant plasticating units to be justified. When idle times because of difficult supply of spare parts and costs of producing rejects gained significance, the idea prevailed that expenses for wear protection had become necessary.

When injection molding of thermosets and rubber was introduced in the 60's, one was obviously well prepared for their wear behavior. Relatively wear-resistant, mostly fully hardened screws rapidly became the art of engineering [128].

When mass materials such as PE and PS were processed, initially no problems, or non that were recognized as wear-related, occurred. Plastics were mostly dark-colored, too, and discoloration did not become visible.

When PVC was processed, however, shortcomings in the service life of parts of the plasticating unit were observed. When the number of new plastics and those with an increasing variety of fillers grew, difficulties became clearly more severe. At the same time, the efficiency of injection molding machines was steadily increased by raising screw speed and torque. All effects combined resulted more and more in breakdowns of plasticating units. At first solely an adjustment of plastics with modifiers was tried. This, however, had only a limited effect on wear.

The speedy introduction of acceptable high-duty steels for screw and barrel was impeded by the thesis that "standard steels have to be suitable for processing standard plastics". There even was no definition of "standard plastics". The German definition of standard steel refers to a gas-nitriding steel, which was used, more by chance, for plunger and barrel of plunger-type molding machines to minimize friction and related wear. Using this steel was a success. When with the introduction of in-line reciprocating-screw molding machines in 1956 a similar wear between the rotating screw and the barrel occurred, this steel was obviously selected again. At that time it was not tested on its corrosive and abrasive resistance. Since some years, one has learned that it has distinct weaknesses although it has a service life of many years. The origin of this weakness is discussed in detail in [118]. This steel is, at best, conditionally suited for mass materials, preferably Polystyrene or Polyolefins [125].

Since about 1979 a steady increase in the use of wear-protected units can be noted (Fig. 51). A recognizable breakthrough was not achieved before the year 1983. In the meantime, most manufacturers deliver the machines with special wear protection for all components of the plasticating unit. This leads to the conclusion that meaningful wear protection is today's state of engineering and part of the standard equipment. The discussion about the necessity of wear protection has ceased after it had been sufficiently proved in everyday life that expenses for the consequences of wear and the loss in quality from worn out machine parts can easily reach a multiple of the costs for wear protection.

High demands on plastic material and plasticating and injection unit during the normal course of molding are caused by starting and brake torque, shear, contact pressure, molding defects, excessive pressures, high temperatures, depositions and additives.

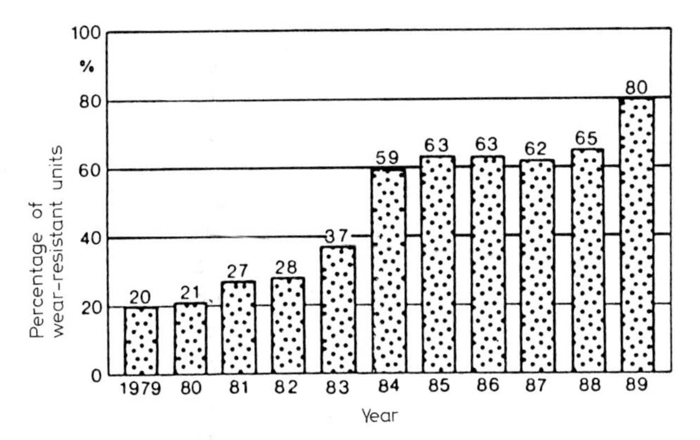

Fig. 51 Percentage of wear-resistant plasticating units in injection-molding machines made in Germany. After 1989 the figure should be between 80 and 90 %

Local weak points in the system are:

- nonreturn valve (primarily subject to wear),
- nozzle (subject to wear),
- connections between barrel, barrel head, and nozzle (sealing surfaces),
- turns in cavity,
- screw,
- barrel.

Wear Protection of Plasticating Units

The wear protection in an injection molding machine has two essential starting points:

- selection of suitable materials for plasticating components and their appropriate treatment,
- favorable design of components.

Temperature, injection speed, pressure, and residence time have to be chosen or optimized according to criteria different from those for wear resistance.

Therefore it is important to create a safety margin against wear as great as possible by means of inventive design and skillful material selection. Because injection for molding machines are universally employed for all kinds of thermoplastics, it can only be suggested to take care of a combined protection against abrasive as well as against corrosive wear.

Preventive wear protection by design is particularly related to screw geometry and configuration of barrel head, nonreturn valve, and nozzle. One can expect a comparably good wear resistance from the proposed geometry presented with the Tables 4, 5, and 8. Fig. 58 presents a design proposition in principle for the barrel-head region, nonreturn valve, and nozzle for processing thermoplastics. It corresponds to the demands of low wear and simple exchangeability [135].

The planeness of sealing faces at the head is an absolute must. The, also have to be without any damage. Contact pressure up to 400 MPa supports the sealing effect.

The nonreturn valve according to Fig. 54 has given the best results. Ball nonreturn valves with sealing center-ball valve do not as well.

It is important to pair materials, either by their nature or by their hardening treatment, in such a way that faces of different hardness are in friction against one another. This should be done regardless of the design of the nonreturn valve. Then, after a short run-in time, no abraded particles worth mentioning are generated, which otherwise would particles worth mentioning are generated, which otherwise would cause a gray to bluish discoloration of moldings.

Screws for thermosets without nonreturn valve are manufactured with angles at the tip of about 60–90°. This is a compromise between small back flow (obtuse tip) and sufficient flushing of dead space (long tip). The clearance with the nozzle should be selected as 0.5 mm maximum. Nonreturn valves for wet polyesters and, lately, phenolics are mostly made from hot-work steel.

Shut-off nozzle are particularly prone to wear. It would be beyond this scope to discuss this in detail. Often occurring problems are related to dead corners and leakage which lead to dark specks or silvery streaking on the molding surface. Needle shut-off valves are distinctly more sensitive than nozzles with rotating or sliding bolt.

Open nozzles for processing thermosets are made of alloy steel because of their wear. For larger nozzles, inserts of cemented carbides are even employed.

The most common solutions and combination of materials are listed in Table 12 according to their principal use by components of the plasticating unit. The list does not claim to be complete and is, of course, subject to change in accordance with future experience.

Nitrided through-hardening steels for barrels without bimetallic lining have to have a high amount of alloying constituents – 12 to 17% chromium as a rule – to provide adequate corrosion resistance besides hardness. Although higher carbon content increases hardness and abrasion resistance, formation of chromium carbide ties up the effective portion of free chromium needed for corrosion resistance. Merely a compromise is usually attainable. Such versions are employed only occasionally because of their tendency to fracture and their use is limited to small diameters.

Nitriding, or even better ion-nitriding improves the surface hardness of steel. It also reduces the corrosion resistance but experience proves that sufficient protection is maintained with a high content of chromium. However, ion-nitrided barrels of high-alloyed steels have only an effective depth of the protective layer of less than 0.4 mm, which may be easily worn away reducing the wear resistance to that of the steel substrate.

More recently, bimetallic barrels are gaining increased significance. They are a combination of an outer steel structure with a uniformly hard internal alloy lining, which is typically 1.5 to 2 mm thick. It contains carbides embedded in a tough matrix.

Table 12 Steel Selection for Wear-Resistant Components of Injection Molding Units [118]

Component	Materials
Barrels	a) Bimetallic lining by rotational casting with suitable alloys: tungsten carbide composites, chromium modified iron-boron alloys, nickel alloys on a substrate of nitriding steel with 12 - 17% Cr b) Shrink-fitting of cast liners c) Boriding, for small diameters only
Screws	a) High-alloyed, through-hardening chromium steels (diameter < 60 mm and length < 1500 mm) with 13 - 17% Cr, occasionally additional ion-nitriding b) Hard-facing of ion-nitrided alloy steels with Stellite or nickel-based alloys c) Hard-facing and chrome plating of roots and flanks d) Boriding, for small diameters only
Barrel head	a) Ion-nitrided high-alloyed chromium steels b) Chrome-plated nitriding steels
Nonreturn valve	A. Tip and seat: Hard-facing of fins with chromium-nickel-boron alloys containing carbides a) High-alloyed chromium steels, eventually ion-nitrided b) High-chromium alloy steels, through-hardened B. Sliding ring High-chromium alloy steels, with good toughness, through-hardened or heat-treated, ion-nitrided C. All other components of - hard alloys or - borided or - with CVD or PVD depositions

Depending on the composition, a sufficient protection against abrasion and corrosion is achieved.

There are three different materials employed for bimetallic barrels, tungsten carbide composites, chromium-modified iron-boron alloys and nickel alloys. Tungsten carbide provides the best overall resistance to abrasion and corrosion.

Screws are commonly made of medium-carbon alloy steel such as the ANSI 4140 steel or its foreign equivalent. It is hardened to about 30 Rc.

Screws with improved resistance to abrasion can be made of through-hardened vanadium-containing tool steels with about 56 Rc. Its use is limited to screws of less than 40 mm diameter because of costs and brittleness.

Through-hardening steels are only used up to a diameter of about 60 mm or a length of 1500 mm. Larger dimensions can cause insurmountable problems with distortion. Some chromium steels, which retain their hardness under heat, can, in addition, be ion-nitrided. This increases the hardness (ca. 69 Rc) in an admittedly thin surface layer, which is well supported by the through-hardened substrate, though.

An effective protection of screw flights is achieved by hard-facing with suitable alloys. Especially a nickel-based alloy, which achieves a hardness of 52 Rc is reported to have better durability than the often used cobalt-chrome-based Stellite. For highly corrosive materials, a nickel alloy that contains virtually no iron, is suggested.

Flights that are so protected are paired with chromium steel with 13 to 17% chromium (corrosion). The total screw surface is ion-nitrided, in particular to protect the roots of the flights (abrasion), whereas the steel is generally heat-treated. Screw for vented barrels should better not be ion-nitrided because the corrosion resistance of the steel will be reduced.

Chrome plating of the roots and the flanks of the flights offers moderate protection to those regions. If paired with hard-facing, it is somewhat sensitive to rough treatment and abrasion. Chipping is possible. Standard nitriding steels serve as substrate.

Recently, tungsten-carbide composites have become available as an overlay for all surfaces. This overlay is very resistant to fracture because it is metallurgically bonded to the steel substrate.

Other surface treatments such as electroless depositing nickel or nickel-silicon carbide or depositions from a gaseous phase (CVD or PVD process [118]) are without great significance yet.

Barrel heads are less exposed to abrasion than screws. Ion-nitrided high-alloy chromium steels or chrome plated nitriding steels offer sufficient protection.

Nonreturn valves, again, are more imperiled especially by the wear between sliding ring and the fins of the tip or by the axial motion of the sliding ring in the barrel. Chromium-alloy steels provide the necessary resistance to corrosion. An ion-nitriding is rarely needed. The sliding ring itself should have high hardness, wear resistance and still be sufficiently tough.

In increasing numbers, HIP-treated steels have gained interest. They are partially made by a powder-metallurgical process. Such steels and plasticating elements made of them have an especially high resistance to wear because of their extremely fine grain size. Meanwhile a sufficient number of screws, barrels, and nonreturn valves have been successfully employed, and their service lives have proved to be superior to those with conventional wear protection by a factor of about 5 to 10 at twice the price. Therefore, if one is faced with considerable wear, it is worthwhile using HIP-powder steels.

Wear Protection by Repair

Worn-out units of "standard" nitriding steel can be used again by providing them with a liner. At the same time, they are made more resistant to wear. Rotational-cast liners are shrink-fitted into turned-out old barrels. Good solutions are not necessarily cheaper than a new barrel, though. The old barrel part remains a weak point; so are gaps at the sealing surfaces, where plastic can degrade and corrosion take place.

Wear Protection by Design

Deposits of plastics in the barrel and friction of steel against steel or against plastic under high pressure should be avoided. Necessary steps are [121, 129]:

- producing contact pressure between barrel and barrel head by using a flanged ring with bolts that permit a contact pressure of 400 MPa after tightening with a torque wrench (Fig. 58);
- avoiding dead corners in the barrel head;
- designing an open cross section in the nonreturn valve of 80 to 120% of the free annular area of the front end of the screw;
- applying armored flights on the screw tip (but not the valve ring);
- allowing for adequate length of the valve ring (about one D for screws up to 70 mm diameter and 0.7 D for screws with larger diameter);
- providing a length of the feeding section preferably 10 to 12 times the diameter that no unmelted or few partially melted pellets are conveyed into sections with increasing pressure.

3.4.11 Screw Tips

The highest pressure occurs at the screw tip, which is the front end of the screw. Therefore it makes sense from the proceesing viewpoint to prevent the back flow of the material into the rear flights by means of a closing element. This is especially important during the injection and holding pressure stages. The solution with the simplest design is to use a screwed-on tip with a larger diameter than that of the screw root at the foremost end of the screw (Fig. 52). The narrow gap with the barrel causes a pressure rise that restrains the back flow, Such tips with an angle between 60 and 90° permit particular careful processing. A complete shutoff, however, cannot be achieved with such tips.

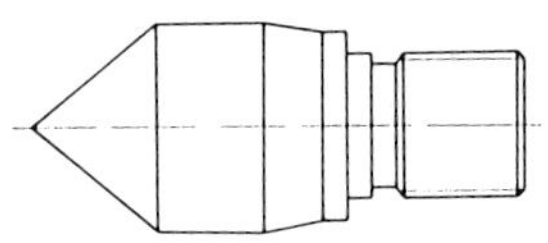

Fig. 52 Plain tapered screw tip

Rigid PVC without plasticizer calls for an open design of the screw tip. It should promote good flow of the melt and impede back flow during injection and holding pressure stages. Fig. 53 illustrates some tips that have proven effective in various circumstances. Output-promoting design such a helixes prevent deposits and restrain back flow during injection. Such screw tips should be protected against corrosion just like the screw (Sect. 3.4.10).

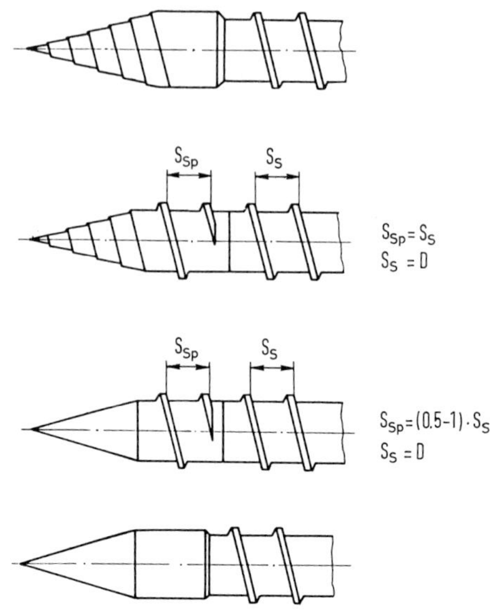

Fig. 53 Screw tips for processing rigid PVC

3.4.12 Nonreturn Valves

A nonreturn valve is a component at the foremost section of the screw that prevents the back flow of the plasticated material during the injection and holding pressure stages. It can accomplish this duty best if it produces a high pressure loss of small free cross sections which can be closed rapidly. A pressure loss also occurs during the feeding stage when the melt passes through these cross sections in the opposite direction. The screw must continue to convey the material in spite of this and eventually later applied back pressure. If the pressure rise is high in cross sections traversed by the melt, it interferes with the output. The pressure may become so high as to damage plastics.

Generally nonreturn valves should be dimensioned in such a way that the free cross section is not smaller than 80% (120% if possible) of the free annular area at the foremost end of the screw. Only a few plastics such as PS and PE do not cause difficulties with smaller cross sections. Fig. 54 illustrates the principle of ring-type nonreturn valve, which is most common. It consists of three parts: the tip, which can be attached to the screw by means of a threaded shaft, the seat, and the axially sliding ring. In the closed position the ring rests tightly on the seat, the conical contact surfaces forming an angle of 45 to 60 with the axis. During feeding, the ring is in the open position and rests against three or four ribs which are attached to the tip like fins.

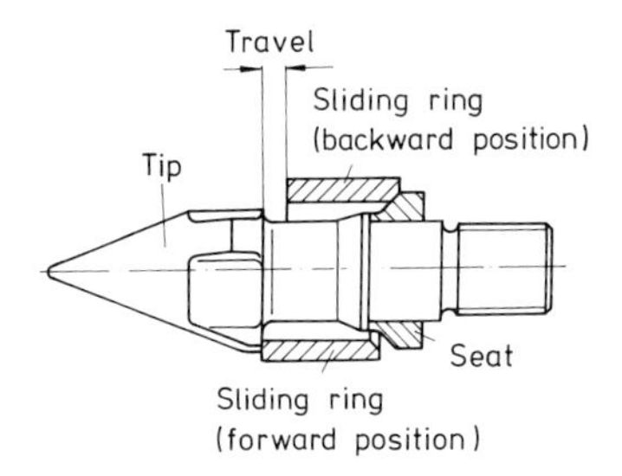

Fig. 54 Ring-type non-return valve for thermoplastics

The following criteria have to be considered for the design:

- refraining of sharp turns of the melt flow, sufficient length of the ring for its sealing function (about one D; 0.6 to 0.8 D for diameters more than 70 mm);
- rapid closing to avoid leakage flow;
- eventual hard facing the flights at the tip, which are in sliding contact with the ring.

The nonreturn valve with cam action (Fig. 55 D) is characterized by the recess (g) in an axially sliding closing sleeve (c). The recess has a helicoid surface into which a bolt (d), press-fitted into the tip, is engaged. Besides the axial force generated during injection, the turning screw creates a force on the closing sleeve with the bolt. With low output, the ring is pushed forward only by a little distance and the melt passes a narrow gap. Thus a balance is generated between gap width and output. Near the end of the feeding stage, when the output approaches zero, a force automatically acts in the direction of the screw and the sleeve is already in its closing position when injection starts [138]. This nonreturn valve is supposed to close faster and more uniform and, with this, a more uniform product is the result. A similar nonreturn valve is presented with Fig. 55 E.

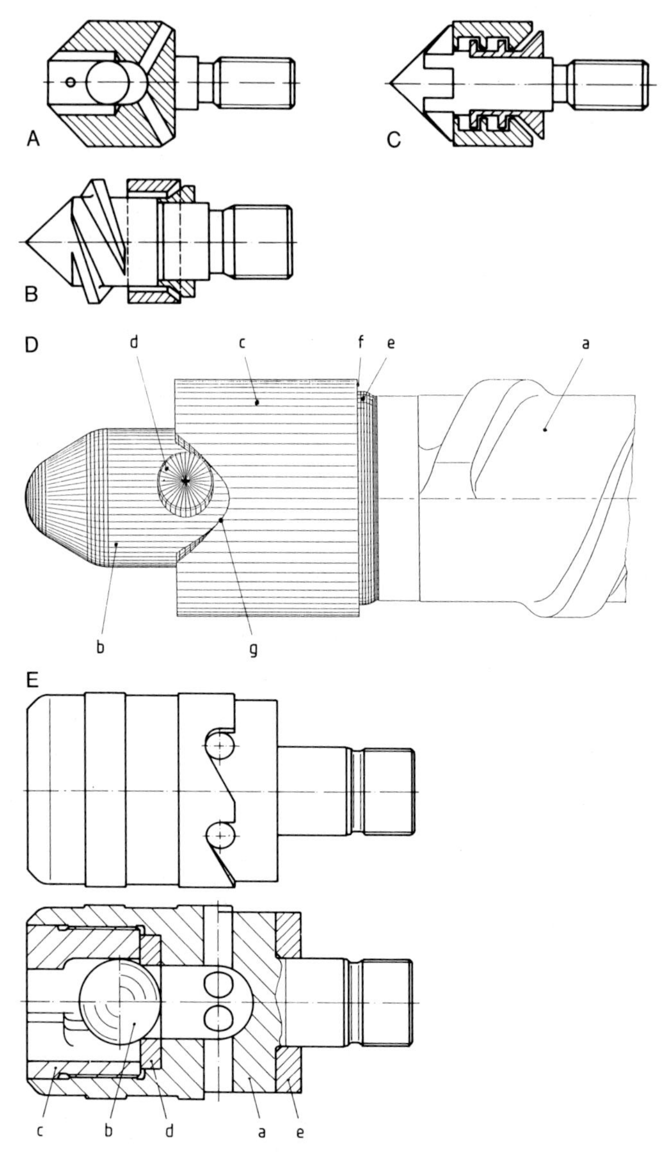

Fig. 55 Non-return valves of various designs (A, B, C)
D: Operation of a non-return valve with cam action
a: Screw, b: Tip of non-return valve, c: Blocking sleeve, d: Bolt, e: Blocking ring, f: Tapered sealing face between blocking sleeve and ring, g: recess with helical surface
E: Non-return valve with ring (top) and with ball (bottom)
a: Screw tip, b: Steel ball, c: Ball guide, d: Thrust washer, e: Spacer

3.5 The Barrel

The barrel is a tube that surrounds the screw. It forms the outer boundary of the screw channel as in an extruder. For injection molding one can assume that the major portion of the heat that must be applied to the plastic is supplied by the barrel. To do this, the barrel is equipped on the outside with electrical band heaters. The watt density of the commonly used ceramic heating elements is about 6 to 8 W/cm². In the nozzle and barrel head section, mica-insulated band heaters are employed to accommodate smaller diameters; they have a watt density of about 4 W/cm². Since it is desirable to avoid melting the plastic prematurely in the feeding throat and the first few flights of the screw, this section has to have provisions for cooling. Particular consideration has to be given to the control of the temperature of the barrel wall to achieve a high degree of precision of ± 2 K. This is done by using specified lead wires with a linearity up to the maximum temperature, specified junctions and connectors. The controllers should have automatic calibration, a control accuracy of 0.3 K or better, a temperature stability of ± 0.5%, a transient pulse suppression, and preferably a thermocouple break protection. Process temperature and set point can be read from a display.

The barrel should be easily disassembled for a rapid change of screws or cleaning procedures. It is important for assembly or disassembly that only few bolts are used. Systems for rapid barrel change are of serious interest today. They render a semiautomatic engagement or disengagement possible (Fig. 56). Proven solutions for a fully automatic change are already offered, too (Fig. 57).

On an injection molding machine, little importance is attributed to the feed throat of the barrel. This is justifiable only if no high output is required. It is known from the extrusion process that the geometry of the throat has a distinct effect on the output.

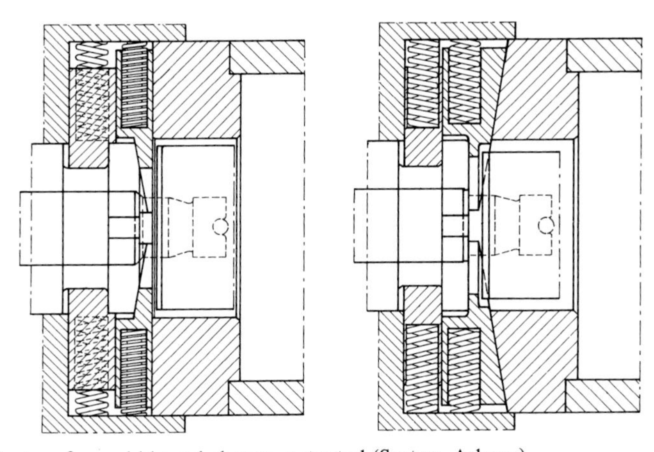

Fig. 56 System for rapid barrel change, patented (System Arburg)

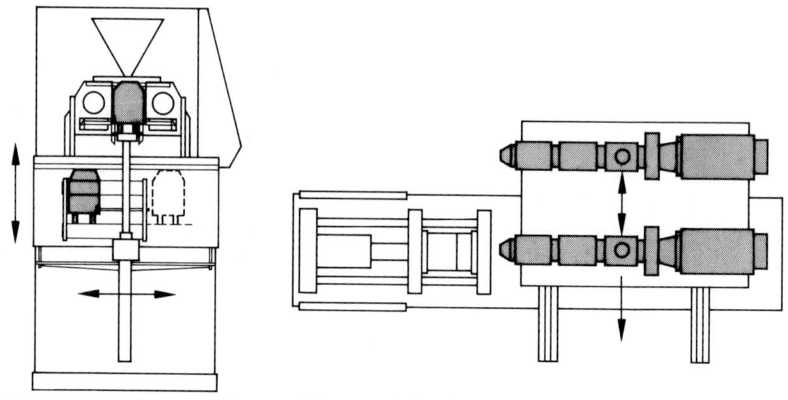

Fig. 57 System for automatic barrel change (System Battenfeld)

The opening should have a length of one to two diameters and, if possible, leave the screw partly covered opposite to the direction of rotation. This covering increases the feeding rate. Occasionally barrels, mostly with a rectangular throat, are supplied with a recess (feeding pocket) adjacent to the throat in the direction of rotation. It runs out into the lower cylindrical surface (Fig. 50). Barrel design that was supposed to promote feeding with longitudinal or spiral flutes, conical taper, and cooling over a length of three to five diameters have been tried experimentally [84, 137] but were not successful in injection molding, except for processing powder.

The barrel head is considered part of the barrel. The assembly of both is of some importance. The connection and transition contour to the nozzle must be streamlined to avoid hang-up of material in this section without using forced purging. Furthermore, the mating surfaces have to remain leakproof, even under maximum pressure. A common solution is shown in Fig. 58. The nozzle is either screwed into or, less often, flanged onto the cylinder head.

Metallic contaminants in the material can cause breakdowns of production and damage to screws, barrels, nonreturn valves, nozzles, runner systems, hot runners, mold surface, or directly to the molding. Metal arresters mounted at the transition from the feed hopper to the feeding throat offer a high degree of safety against getting the contamination into the screw [145].

3.6 The Nozzle

The nozzle, as a component of the plasticating unit, is forced against the sprue bushing of the mold prior to injection and produces a force-locking connection there. Large radii of the tip must be provided to accommodate the forces on the nozzle without wear (Table 5). In the United States, spherical radii which match the spherical radii of 1/2 or 3/4 in. of sprue bushings can be considered standard, whereas a proposed European standard (Euromap 2) suggests radii of 10, 15, 20, and 35 mm. A classification can be made with Table 13. There are, of course, also sprue bushings with a flat surface and without any radius, which call for a corresponding nozzle tip. Note that the O-dimensions of American-made nozzles, which are generally available in 1/32- in. increments, should be slightly smaller than the equivalent dimension of most standard sprue bushings of 5/32, 7/32, 9/32, and 11/32 in. Such standardization results in an effective simplification.

Table 13 Spherical Radii and O-Dimensions of Nozzles According to European Standards

Clamping force (kN)	Spherical radius (mm)			O-Dimension (mm)		
	Thermo- plastic processing	Thermo- set processing	Elasto- mer processing	Thermo- plastic processing	Thermo- set processing	Elasto- mer processing
< 500	10 (35)	10 (35)	10 (35)	3 - 5	5 - 8	4 - 6
500 - 1 000	10 (35)	15 (35)	15 (35)	4 - 6	6 - 8	5 - 7
1 000 - 5 000	15 (35)	20 (35)	20 (35)	5 - 8	8 - 10	6 - 8
5 000 - 10 000	35	35	35	6 - 10	8 - 10	8 - 10
More than 10 000	35	35	35	10 - 12	-	-

3.6.1 Open Nozzles

The nozzle usually has a channel that tapers in towards the tip and provides the connection between barrel head and the sprue bushing of the mold. The most common type is the open nozzle (Fig. 58). Technically, the open nozzle is the most favorable solution because it can have the shortest length. It should be employed whenever the process permits it.

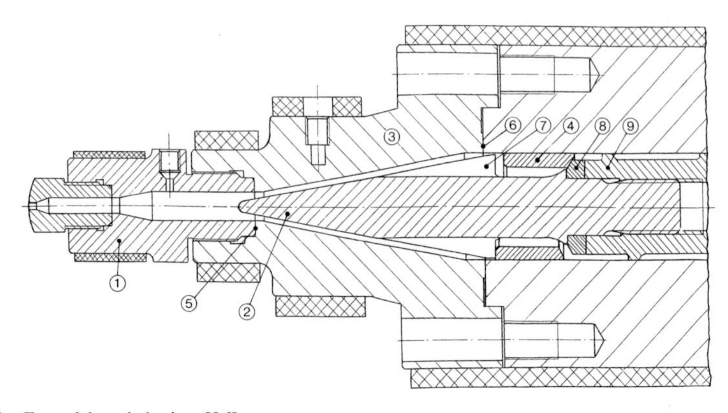

Fig. 58 Barrel-head design [96]
1: Nozzle with screw-in tip, 2: Screw tip, 3: Barrel head, 4: Sliding ring, 5: Sealing face of nozzle, 6: Sealing face of barrel head, 7: four (or three) armor-plated vanes, 8: Thrust washer, 9: Screw

3.6.2 Shut-off Nozzles

Shut-off nozzles are used to avoid drooling of the melt and stringing, or to be able to feed with the nozzle retracted. There are two major types, one that is positively actuated during the cycle and one that is controlled separately. Fig. 59 depicts some examples from the first group, nozzles with a sliding sealing element. They are opened when pulled in against the sprue bushing and closed by the pressure of the melt in the barrel. The nozzles shown in Fig. 60 are needle shut-off nozzles, which are kept closed by a spring. The injection pressure has to be as high as to overcome the force of the spring. The corresponding pressure loss means additional shear and consequently additional heat for the material, which may be a distinct disadvantage. If the shear effect should be avoided, a shut-off nozzle with externally actuated needle can be employed. Fig. 61 presents such a externally controlled shut-off nozzle with internal needle. The over-all length of such nozzles is mostly extensive. Cut-off nozzles with a sliding or rotating pin have also proved to be effective (Fig. 62).

Table 14 presents information about the suitability of various nozzle types and reflects the results of general experience.

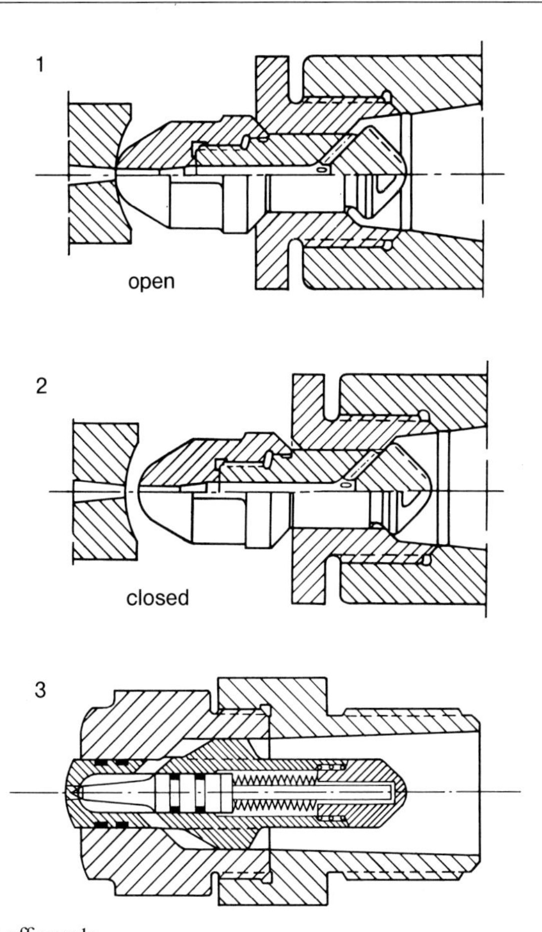

Fig. 59 Slide shut-off nozzle
1 and 2: Nozzle is opened by contact pressure. 3: Spring-loaded nozzle is opened by injection pressure acting against internal spring

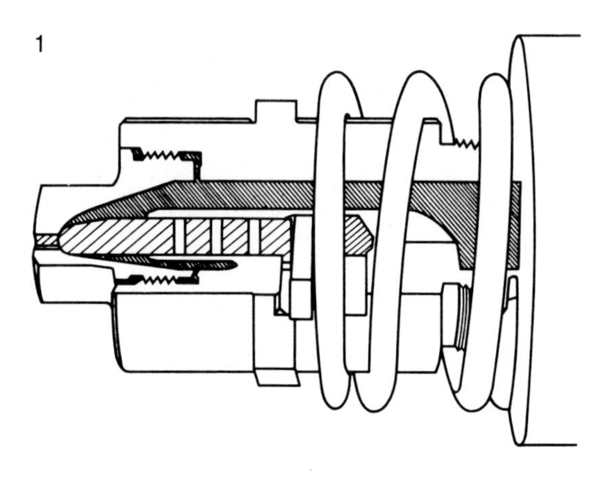

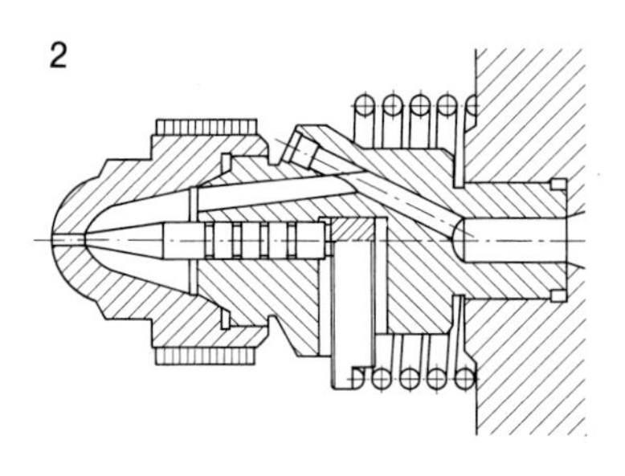

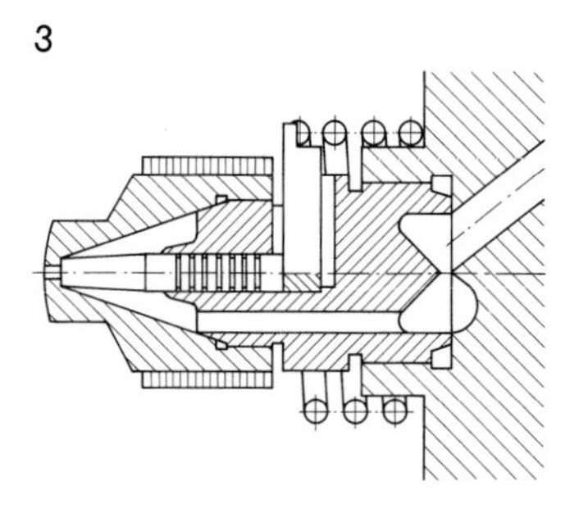

Fig. 60 Needle-type shut-off nozzles, spring loaded
1 to 4: With outside spring (1 System Fuchslocher, 2 System Bernex, 3 System Keil)
5 to 7: With internal spring (7 Belleville washer with ball)

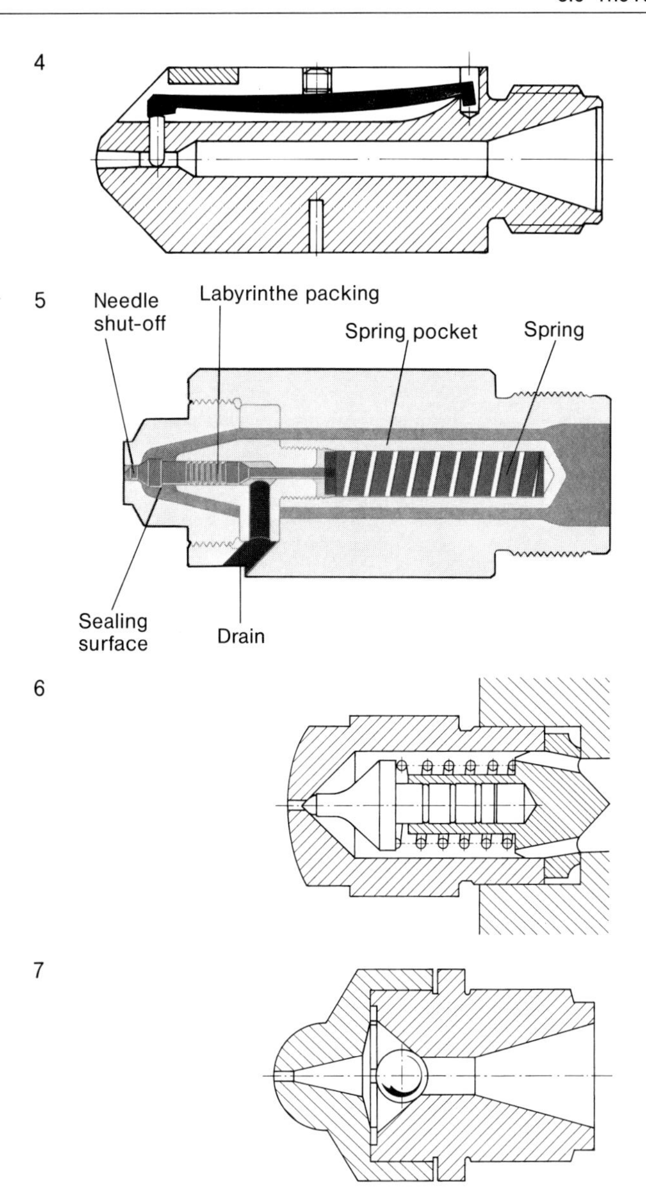

Fig. 60 (Continued)

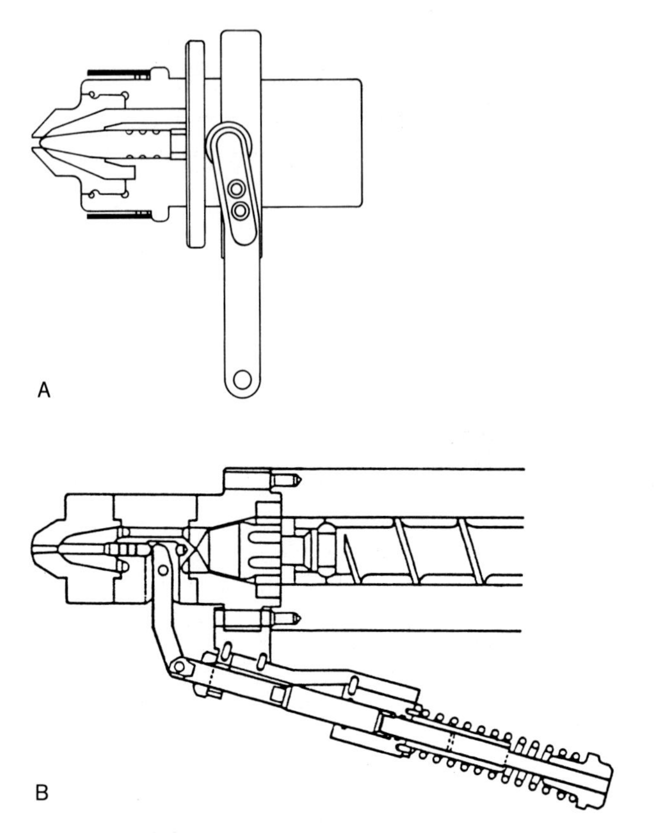

Fig. 61 Needle-type shut-off nozzles, separately controlledA: Needle directly actuated by lever [143], B: Hydraulically actuated needle

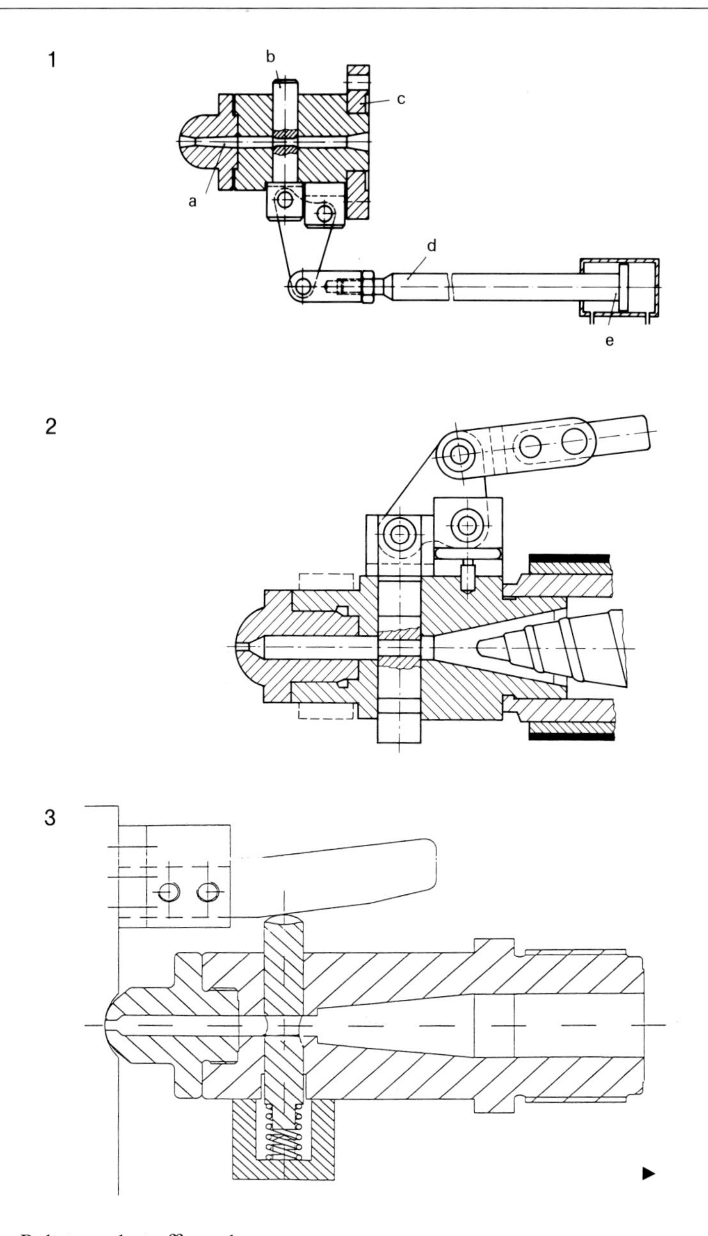

Fig. 62 Bolt-type shut-off nozzles
1 and 2: Bolt hydraulically actuated, a Nozzle, b Bolt, c Flange, d Linkage, e Hydraulic cylinder; 3: Sliding bolt mechanically actuated [150], 4 and 5: Pivot bolt mechanically actuated [150] 4: Cylindrical bolt, 5: Tapered bolt for complete sealing

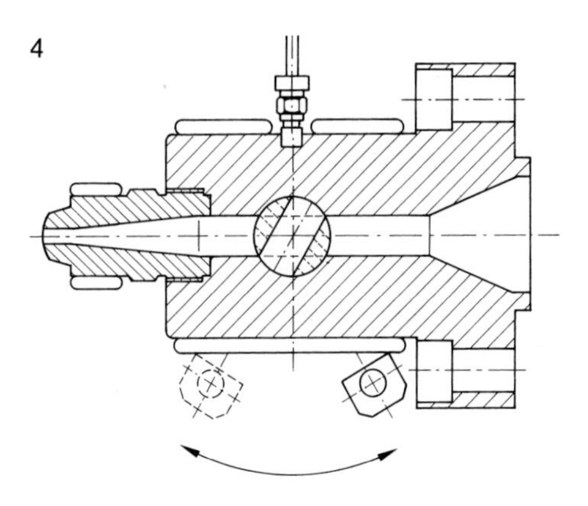

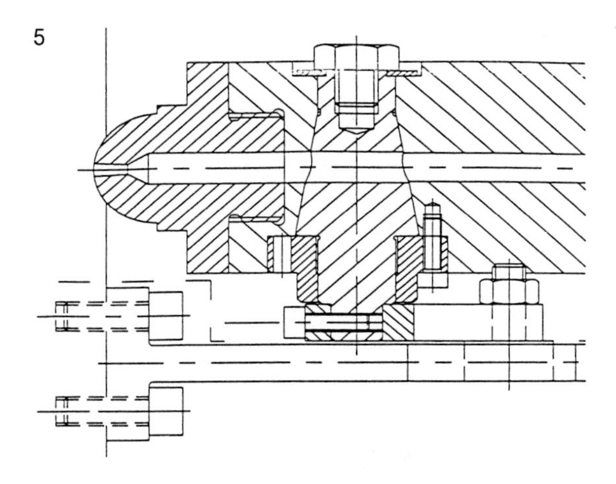

Fig. 62 (Continued)

Table 14 Recommended Nozzles for Plastics [143]

Plastic Material	Type of Nozzle				
	Open nozzle	Sliding shut-off nozzle[1]	Spring-loaded needle shut-off nozzle[2]	Hydraulic needle shut-off nozzle	Hydraulic bolt-type shut-off nozzle[3]
ABS	●	○	○	○	○
CA	●	○	○	○	○
CAB	●	○	○	○	○
PA	○	○	○	●	●
PAI	○	○	○	○	●
PBT	○	○	○	●	●
PET	○	○	○	●	●
PC	●	○	○	○	○
PE	●	●	○	○	○
PEEK	○	○	○	○	●
PMMA	●	○	○	○	○
POM	●	○	○	○	○
PP	●	●	○	○	○
PPO	●	○	○	○	○
PPS	○	○	○	○	●
PVC	●	-	-	-	-
SAN	●	○	○	○	○
TSG	-	-	○	●	●
Thermoset	●	-	-	-	-
Elastomer	●	-	-	-	-

● recommended
○ feasible
- nicht geeignet

[1] resistant to flow, poor temperature control
[2] throttle, high shear stresses
[3] channel cross section = nozzle-opening cross section

3.6.3 Nozzles with Material Filters

For molding parts with an extremely high degree of purity, the use of a nozzle is suggested, which retains all contaminants larger than the mesh size of the employed filter. This also prevents the contamination of hot runners and narrow gates. Fig. 63 presents a nozzle with a filtering core with up to 1062 holes 0.5 mm or, by choice, 1.0 mm diameter. Such a nozzle is only a mean to improve production as long as the total free cross section of all holes is sufficiently large as not to generate an undue pressure loss. Filters should only be used in connection with pressure control. The filtering core can be easily removed and cleaned after the front piece (actual nozzle) has been screwed off.

Material filters are also available as individual components to be mounted between barrel head and a common nozzle and therefore have standard thread sizes. Filtering is not only achieved by holes of appropriate size to retain larger particle but also by forcing the melt across ribs through a remaining small gap of 0.5 or 1.0 mm.

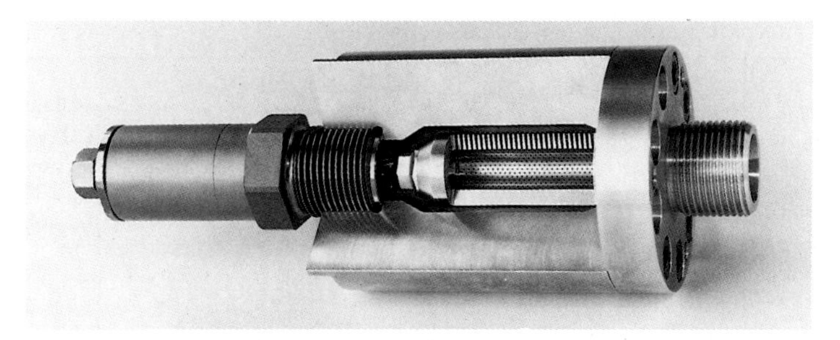

Fig. 63 Nozzles with material filter for particles larger than 0.5 mm or optional 1 mm (System Incoe)

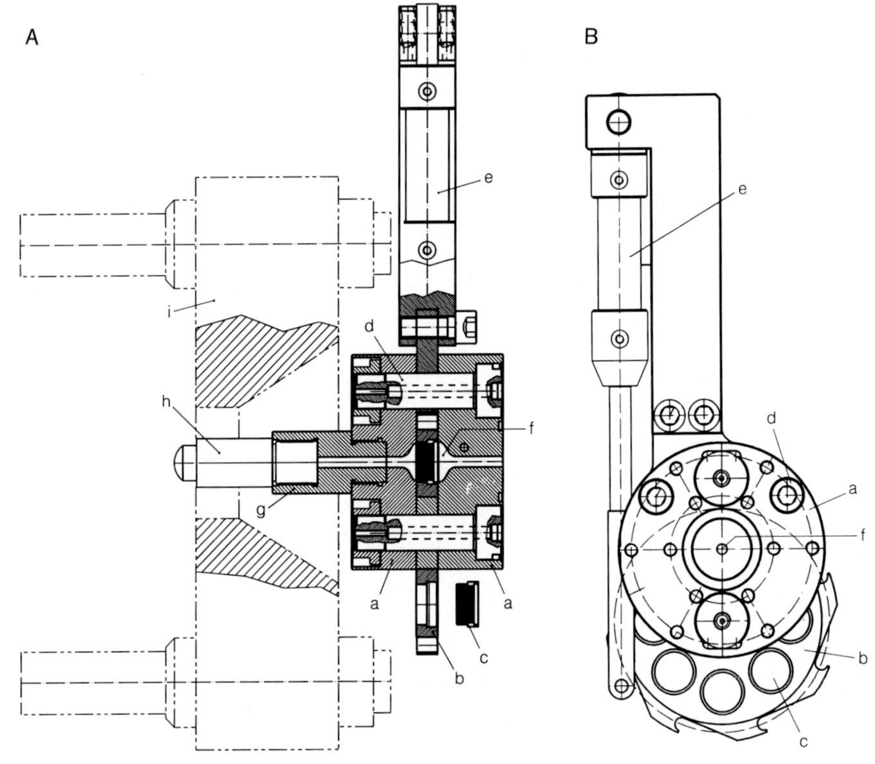

Fig. 64 Nozzle with revolving changeable material filters (System Gneuss)
A: Sectional view
a: Filter block, b: Filter disk, c: Replaceable filter elements with screen, d: Threaded bolt, e: Hydraulic actuator, f: Conduit with expansion towards the active element, g: Nozzle extension, h: Nozzle, i: Stationary machine platen
B: View of material filter in flow direction Symbols same as in A

Such filters are equipped with a separate band heater and a thermocouple for temperature control.

The nozzle presented in Fig. 64 is provided for a continuous operation with an injection molding machine. It is designed to withstand injection pressures up to 200 MPa according to its manufacturer [151]. A metal-to-metal sealing has been selected. The sectional view explains how the filter disks are turned and how they can be replaced at two free positions to permit a continuous filtering. These filters can assist if heavily contaminated material is reprocessed and save an additional compounding step.

Besides those nozzles already mentioned, there are a number of nozzles for special applications such as immersion nozzles for direct gating, cold-runner nozzles for thermosets and others [147].

3.6.4 Internally Heated Nozzles

Recently nozzles are coming out which are heated with heat pipes, also called thermal pins. They are characterized by their relatively small diameter, uniform heat distribution over their whole length and can be made up to a length of 800 mm according to a manufacturer [152, 154]. Such a length can be very useful with thick machines platens and hot-runner manifolds may be made shorter because the nozzle can immerse more deeply into it.

Heat pipes were granted a patent already in 1942 and employed in space technology since the 1960's. They are hermetically closed cylindrical pipes of stainless steel (Fig. 65) [149] filled with a liquid heat conductor, the composition of which depends on the temperature of use. If the pipe is heated at one point, the fluid evaporates. The heat is adiabatically transported to the cooler condensation zone, where the vapor releases its latent heat and is condensed again. The fluid travels by capillary action along a wick back to the heated zone. The wick consists of wire cloth or sintered metal.

Heat pipes require the consideration of certain design guide lines as not to reduce their effectiveness [149, 153]. They should be installed horizontally or with the cold end upwards. Never must the cold end be placed below the horizontal, otherwise the thermal conductivity is decreased. Gravity is involved herein.

Nozzles with heat pipes can be employed without separate heating up to a length of 250 mm. They transport the heat from the barrel to their tips. Longer nozzles for larger machines with about 530 mm in length are equipped with a band heater next to the barrel (Fig. 66).

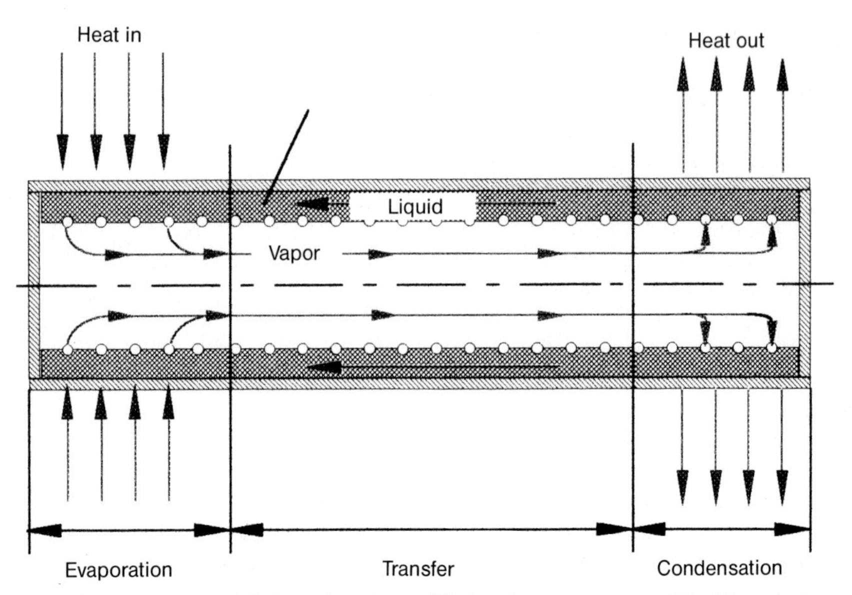

Fig. 65 Schematic design of a heat pipe: the equilibrium between vapor and liquid results in an effective temperature balance

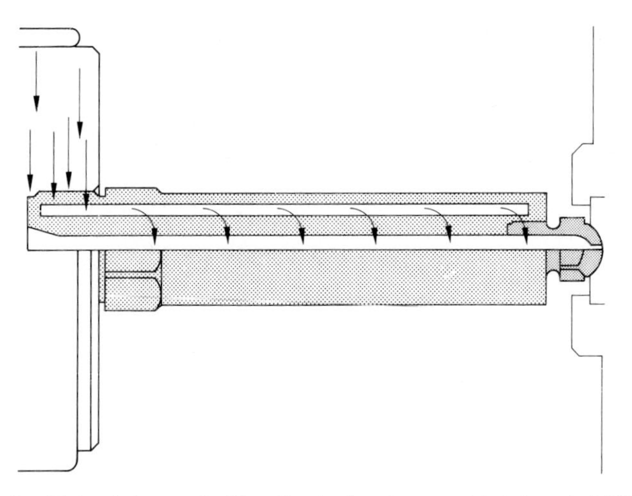

Fig. 66 Nozzle with heat pipes and without heater band, screwed into barrel. Additional band heater for greater length is needed (System Kona)

3.7 Accessories for the Injection Unit

Standard equipment with standard functions (without controls) includes:

- stainless steel feed hopper with shutoff against barrel, sight glass, cover, vent, provision for draining,
- rotatory screw drive,
- hydraulically (pneumatically for mini equipment) actuated injection cylinder,
- controllable carriage movement (nozzle retraction eventually time delayed),
- complete plasticating unit (barrel, barrel head, band heaters, open nozzle, screw, nonreturn valve), exchangeable (Fig. 67A),
- screw cooling installation (Fig. 68) with lateral (version A) or end inlet and exit (version B) of coolant,
- safeguard of screw against rupture,
- non-reversing lock on screw,
- screw return device before and/or after feeding,
- cooling jacket in feeding section with sight control,
- mechanism for alignment of nozzle,
- optical control of screw position,
- back pressure relief,
- intrusion hookup (optional),
- safe guards in accordance with regulations.

Special equipment (without controls) includes:

- plasticating unit for PVC (Fig. 67B), thermosets (Fig. 67C), and elastomers (Fig. 50),
- selection of rotatory screw drive (electromechanical or hydraulic),
- option of accumulator connection for injection,
- shut-off nozzles,
- variable pull-in force for careful treatment of nozzle and sprue bushing,
- variable speed control.

It may be completely acceptable for standard equipment as well as special accessories for injection molding machines to deviate from this more generalized list. The extend to which accessories are employed will depend on the scope of the task and economic feasibility. The purchase of a complete package is rarely necessary and often uneconomical.

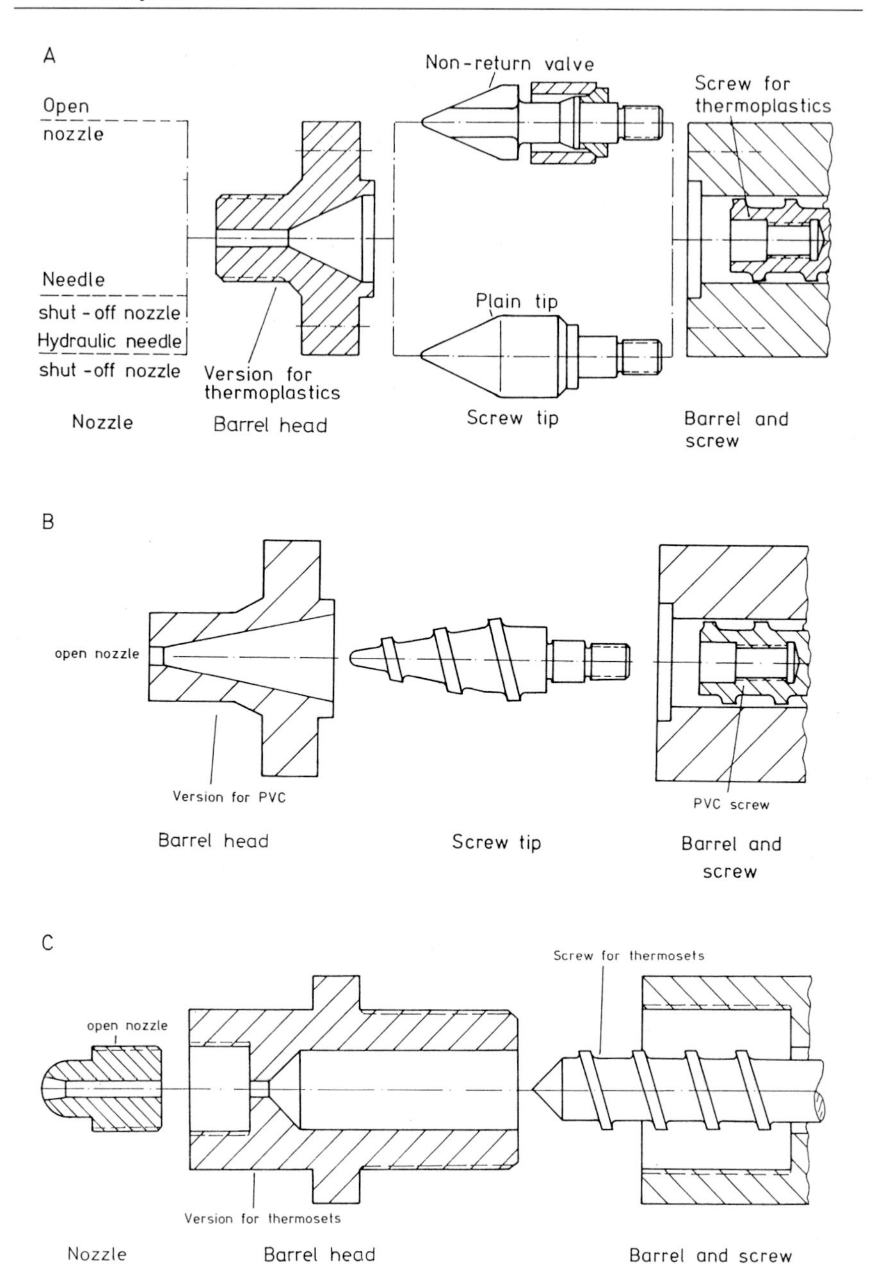

Fig. 67 Plasticating unit for thermoplastics with cylinder head, non-return valve, and nozzle A: For processing thermoplastics, B: For processing rigid PVC, C: For processing thermosets

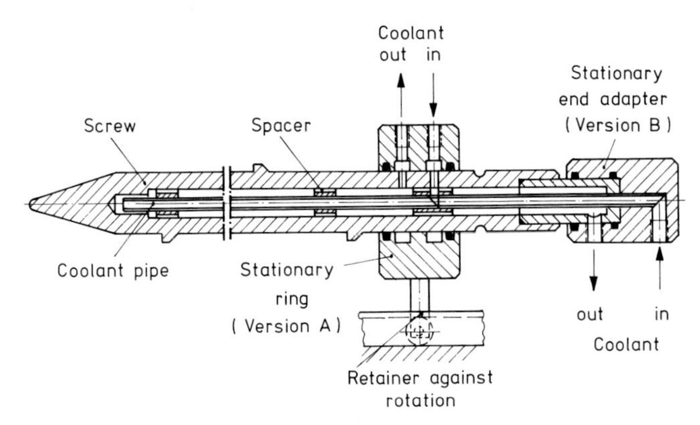

Fig. 68 Hollow-cored screw with coolant pipe and lateral coolant entry (A) or end entry (B)

4 The Clamping Unit

The clamping unit of an injection molding machine accommodates the injection mold. It provides the motion needed for closing, clamping, and opening, and produces the forces which are necessary to clamp and open the mold. Its principal components are the tie bars, the stationary and the movable platen, and the mechanism for opening, closing, and clamping.

The clamping unit together with the mold form a closed system of forces. There are three basically different design concepts:

- mechanical clamping systems (positive locking with toggles actuated by a lead screw or hydraulic cylinder),
- hydraulic clamping systems (force locking with hydraulic cylinder),
- mechanical-hydraulic clamping systems (force locking with hydraulic cylinder, mechanical travel).

The clamping system is the mechanism that keeps the mold effectively closed during the injection and holding-pressure stages. Modes and design concepts of the function are summarized in Table 15.

Older statistics on the use of injection molding machines in the Federal Republic of Germany show that 85% of all machines have a clamping force of a magnitude up to 3000 kN [155]. Today this situation should have changed in favor of heavier machines.

To some degree, there are considerable differences among the systems in the way the effective forces act during closing and clamping. The toggle system is generally superior in the speed of movements, compared to other systems with equal power input. The difference is about 10 to 20% or, to formulate it differently, a fully hydraulic machines needs 10 to 20% more power input than a toggle machine of the same size and with the same dry-run speed. The ejection forces of the toggle system, on the other hand, generally are rather modest. Within the scope of comparative productions, no differences in product quality have been found, though. Thus, there is presently no qualified objective basis for comparison, to assess advantages and disadvantages of mechanical clamping systems versus fully hydraulic ones. Both the toggle clamp and the hydraulic clamping system have their merits. Neither has a demonstrated superiority over the other. The tendency to hydraulic clamping units, of course, can easily be justified with lower manufacturing costs [157].

A slight advantage of hydraulic clamping may be found in an easily controlled stroke for coining, although the small elastic deformation of a loosely clamped toggle can also be used as a coining stroke. As a technique, the so-called coining method is very interesting [158, 159, 160] and can be accomplished basically with every machine if it is equipped with relevant controls.

Table 15 Frequently Employed Clamping Systems for Injection Molding Machines

Clamping unit		Closing and opening	Clamping force	Maximum force buildup	Mold height adjustment
Mode of clamping	Design				
Mechanical	Vise	Mechanical with spindle*	Mechanical	Mechanical	Mechnical
	Single-toggle	Mechanical or hydro-mechanical	Mechanical or hydro-mechanical	Mechanical	Mechanical
	Double-toggle	Hydro-mechanical	Hydro-mechanical	Mechanical	Mechanical
Hydraulic	Direct hydraulic	Hydraulic	Hydraulic	Hydraulic	Hydraulic and automatic
	Hydraulic with rigid mechanical look	Hydro-mechanical	Hydraulic	Hydraulic	Hydraulic
		Hydro-mechanical	Hydraulic	Hydraulic	Mechanical
	Hydraulic cylinder on swivel arm	Hydro-mechanical	Hydraulic	Hydraulic	Hydraulic
	Toggle with hydraulic cylinder	Hydro-mechanical	Hydraulic	Hydraulic	Hydraulic

* Mini-machines only

4.1 Mechanical Clamping Systems

Today, actuating the toggle joints for closing and opening is mostly done hydraulically. In earlier years and after several starts, mechanical systems did not appear to have much of a chance, but have become employed again in modern machines. By and large, the hydraulically actuated toggle is used in a variety of design concepts. The single-toggle clamp with double acting hydraulic actuator is typical for small machines up to 500 kN clamp force (Fig. 69). The actuating cylinder can be connected directly to the toggle with crosslinks or is pivoted at the tail stock platen or the machine support. Both, this design and the corresponding double-toggle design require the smallest possible space although the latter is hardly used anymore. The actuator is small and therefore uses little oil; it can operate fast and is a provision for a rapid movement of the clamping unit [163] (Fig. 70). Such a toggle clamp permits only relatively short opening strokes, and the length of the links and the available head space for mounting the actuator are decisive. Maximum speed and minimum force are at the end of the stroke.

The double-toggle clamp is preferred for machines between about 1000 and 50 000 kN clamping force. The most often employed toggle-clamp unit with five pivot points is shown in Fig. 71. It is favored above the four-point toggle (Fig. 72) because of its long opening stroke and little over-all size (floor space). As a rule, this toggle clamp is operated by a central double-acting hydraulic actuator, coaxially positioned in the machine axis. Most of the time, a mechanically adjustable spindle provides for continuously variable stroke limitation of the movable platen.

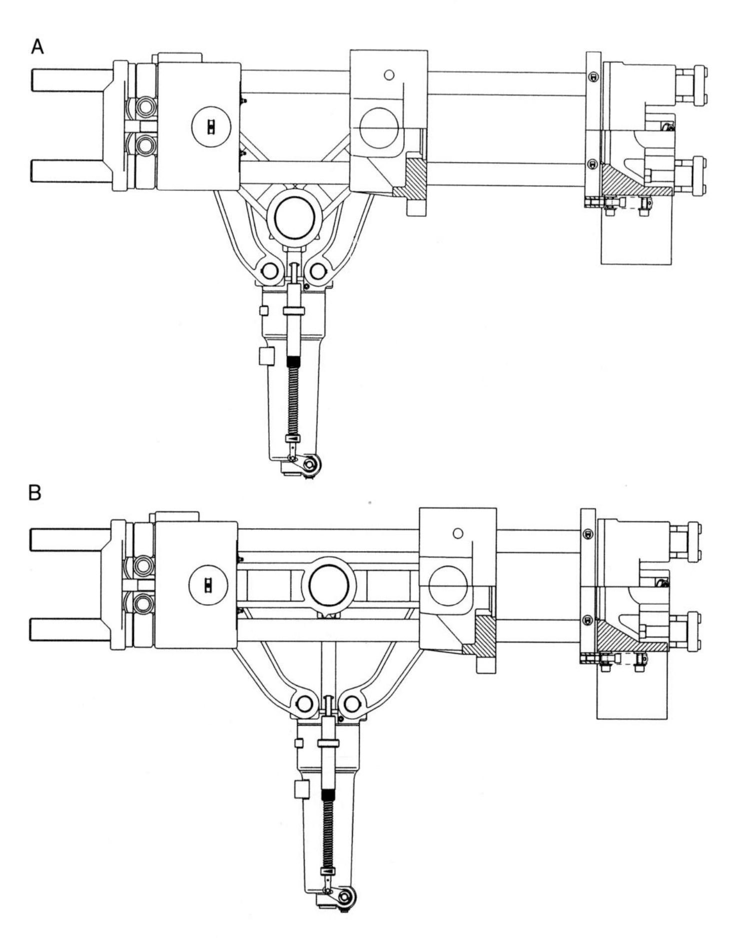

Fig. 69 Single-toggle clamp (System Arburg)
A: Open position, B: Clamped position

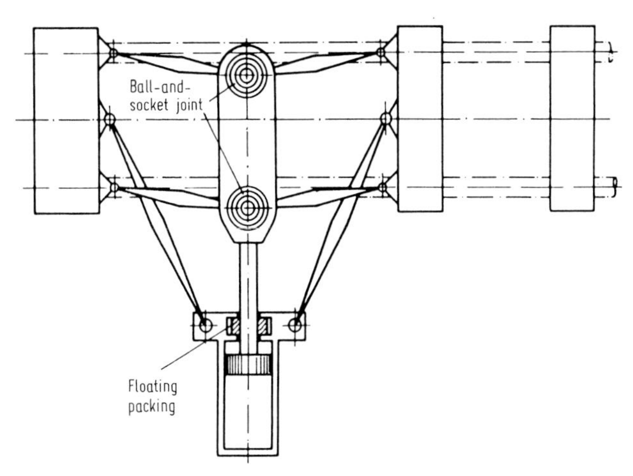

Fig. 70 Double-toggle clamp system with lateral actuating cylinder

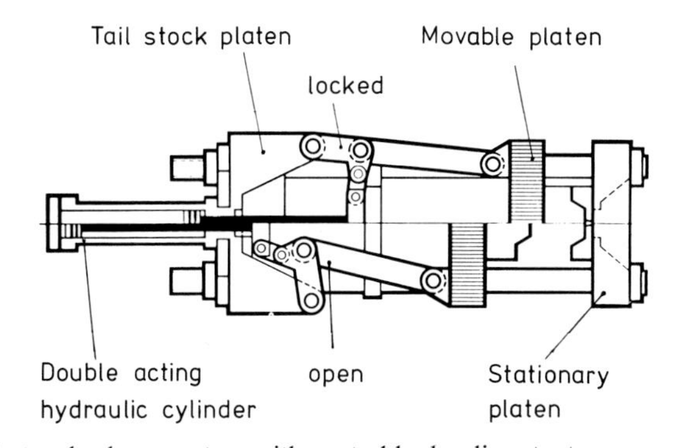

Fig. 71 Double-toggle clamp system with central hydraulic actuator

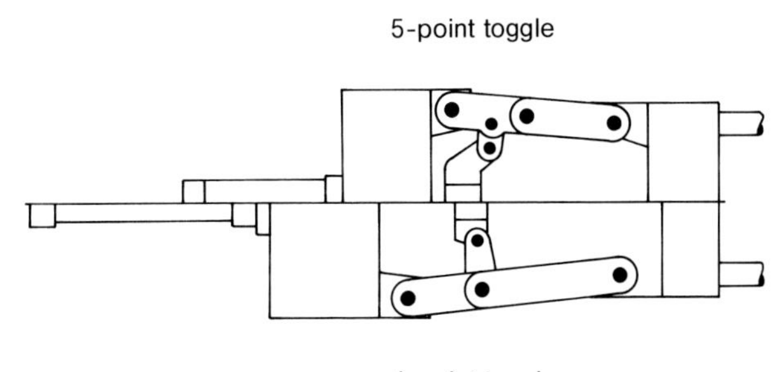

Fig. 72 Toggle clamp system corresponding to Fig. 65. Four-point toggle in comparison with five-point toggle with equal opening stroke

A special design of a toggle clamp is presented with Fig. 73. This concept combines the good characteristic of a short toggle with the long stroke of a hydraulic cylinder. It is of no importance for the function whether the toggle links are arranged side by side or one above the other. The latter arrangement permits easy access to the ejector. To guarantee a long service life and to reduce wear, hardened bolts and bushings are used for the pivots.

The adjustment to different mold heights is done with a mechanical adapting device that moves the tail stock platen along the tie bars. The four lock nuts are given equal turns. This operation is frequently achieved by a central mechanism with a gear rim (Fig. 74) or with a chain drive that simultaneously moves the nuts. This mechanism also makes it possible to adjust the clamping force for each particular case. Equipment that measures the elongation of the tie bars is of advantage for the protection and control of the system and the reproducibility of the setting (monitoring of closing and clamping force).

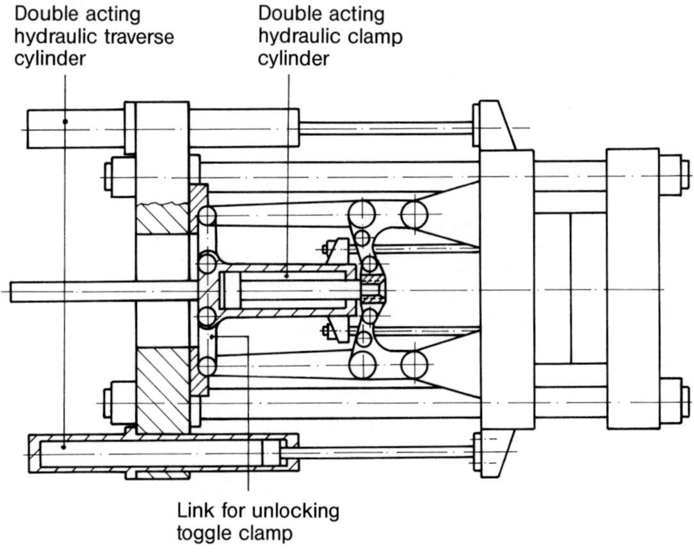

Fig. 73 Hydraulic-mechanical clamp system (Courtesy Triulzi)

A

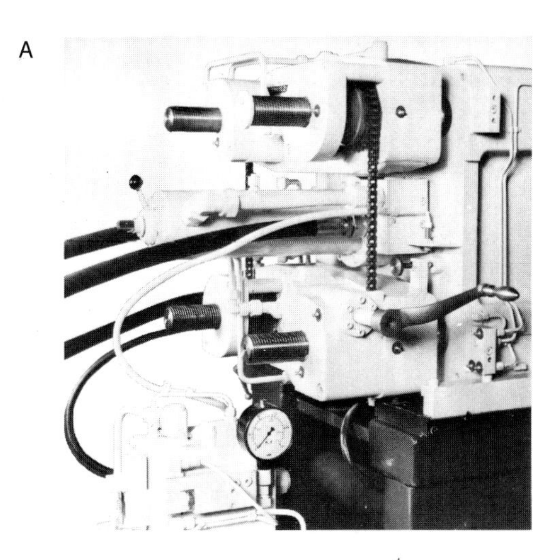

B

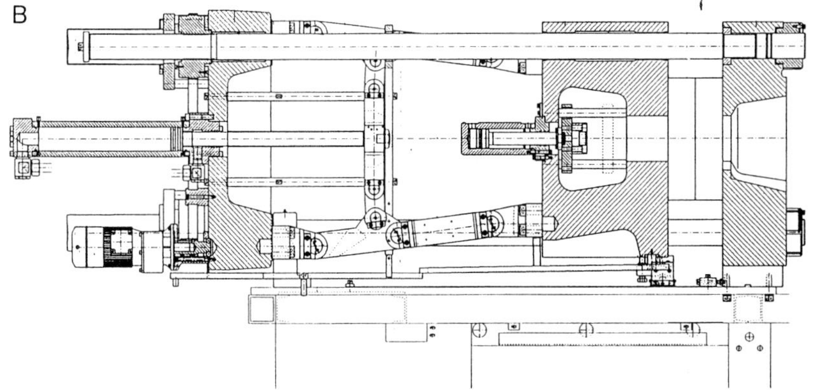

Fig. 74 Central daylight adjustment
A: Locknuts are driven by chain (Courtesy Mannesmann-Demag Kunststofftechnik),
B: Central gear drive (System Engel)

4.1.1 Clamping Sequence

The toggle clamp system is characterized by an almost ideal kinematic velocity feature. During the closing stage, the movable platen is accelerated from a slow start to maximum speed and then decelerated to slow motion with increasing link extension. The two mold halves meet smoothly. Thus, this system is fast but protects the mold and saves energy. One generally strives for a velocity sequence in accordance with Fig. 75, which shows a relatively wide velocity plateau in the middle range of the

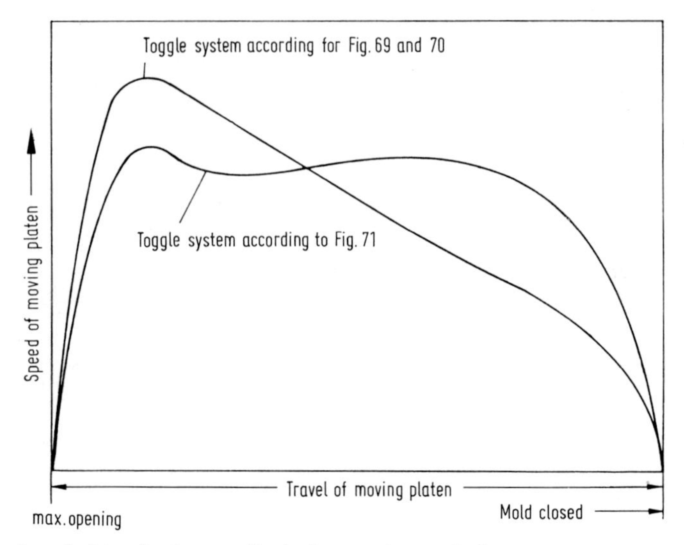

Fig. 75 Speed of toggle clamp units during maximum stroke

Table 16 Average Speed of Movable Platen. Maximum speed of toggle clamps higher by a factor of 1.4 – 1.8

Clamping force (kN)	Maximum speed of closing (mm/s)	Minimum speed of closing (mm/s)	Maximum speed of opening (mm/s)	Minimum speed of opening (mm/s)
≤ 500	600 – 1300	80 – 100	600 – 1100	70 – 100
510 – 1000	500 – 1200	80 – 100	500 – 900	60 – 100
1001 – 5000	500 – 800	60 – 80	500 – 700	50 – 80
5001 – 10000	500 – 700	60 – 80	500 – 700	40 – 80
> 10000	500 – 600	40 – 60	400 – 600	40 – 60

stroke. The average maximum and minimum speed for the closing and clamping stroke are presented in Table 16. The average speed is determined as

$$v_{av\,c} = \frac{max.\ stroke}{t_{c\ min.}} \tag{10}$$

where $v_{av\,c}$ is the average speed during closing (mm/s), t_c min. the minimum closing time (s), and the stroke is in mm, or as

$$v_{av\,o} = \frac{max.\ stroke}{t_{o\ min.}}$$

with $v_{av\,o}$ the average speed during opening (mm/s), t_o min. the minimum opening time (s), and the stroke in mm.

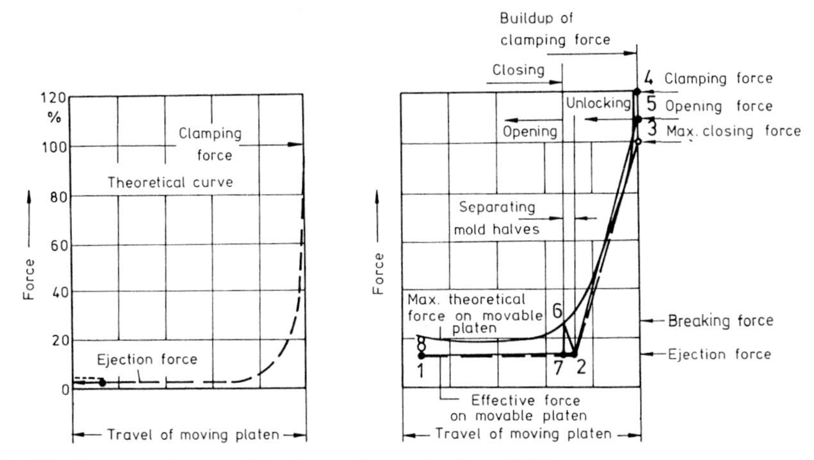

Fig. 76 Forces at the movable platen of a toggle clamp [164]

Several rather different forces take effect sequentially in the clamping system during mold closing and filling, breaking, and opening. Among those of major importance for the machine operator are the clamping force, the always greater total force resulting from clamping and reactive forces due to cavity pressure, and the breaking and ejection forces. Making use of the diagrams of Fig. 76, the generation of the various forces is explained as follows [164]:

In the left-hand graph, the relationship between clamping and ejection forces is presented accurately, while the curve in the right-hand graph is distorted at the right to show more clearly the individual forces.

Closing stage: Line 1–2 indicates the actual force on the movable platen during closing; it is smaller than the theoretical force due to frictional losses. Build-up of the clamping force: Mold halves touch at point 2; line 2–3 shows build-up of clamping force. The point of contact with the curve representing the maximum theoretical force indicates that the maximum clamping force is gained. The gradient 2-3 correspond to the spring characteristic of the clamping unit and represents it graphically (Sect. 4.1.2).

Build-up of total force: During injection, the load on the clamping unit decreases due to the reactive force from cavity pressure, curve 3-4. Thus, the total force is the sum of clamping and reactive forces (Sect. 4.1.2).

Clamping stage: The clamping unit is relieved by the decrease in cavity pressure during cooling, curve 4–5. In extreme cases point 3 and point 5 are identical.

Unlocking stage: Due to the kinematics of the toggle system, a fraction of the available hydraulic power is sufficient to initiate unlocking. As soon as the links leave their locked end position, they snap back almost like a spring, curve 5-2. Means of shock absorption are needed.

Breaking the mold: Curve 2–6 shows the increase in force on the movable platen that occurs as the mold is broken. At point 6 the maximum force, which is kinematically possible, may be reached. Opening stage: Curves 6–7 and 7–8 illustrate the theoretically possible force, reduced by frictional losses, which acts during opening.

Ejection stage: Ejection takes place between points 7 and 8. With mechanical ejection, the ejection force can rise to the maximum theoretical opening force.

4.1.2 Clamping Force of Toggle Machines

The maximum clamping force is reached when the toggle links are fully extended and straightened in their end position. This force F_C is built up in accordance with the spring characteristics of the tie bars. It is attained when closing and locking are completed and is equal to the sum of all tensile forces in the tie bars. The maximum clamping force $F_{C\,max}$ is the maximum force a machine can produce. This is still the most important characteristic for the specification and rating of an injection molding machine, because it is common to adjust other performance data to it and compare machines on this basis. Even the price of a machine is related to its clamping force. This approach deserves an urgent improvement, because other criteria are, at least, of equal priority.

The European standard (Euromap 7) proposes a method of measuring the clamping force. The maximum clamping force $F_{C\,max}$ produces an elongation of the tie bars of the magnitude $a_{tb\,max}$. The force-deformation diagram generated by a machine with a mechanical clamp system, such as a toggle clamp, is shown in Fig. 77 [168]. To simplify matters, the presentation is confined to the deformation of tie bars and mold. The clamping force $F_{C\,max}$ affects the mold and compresses it by the distance $a_{mo\,max}$. During

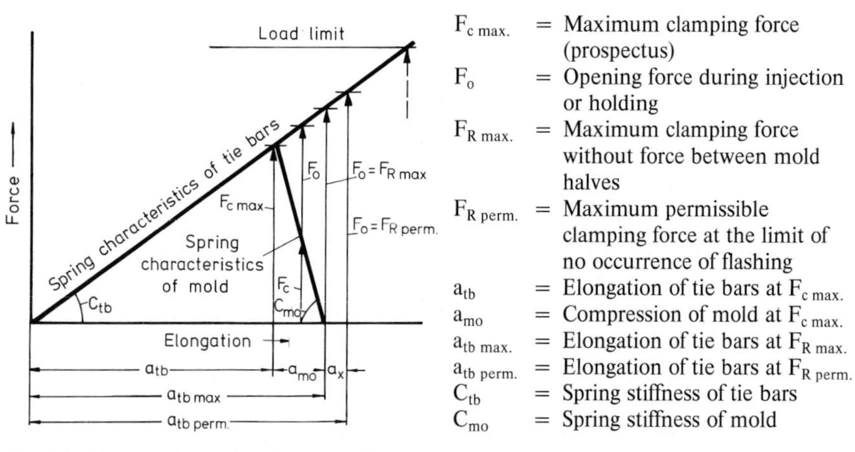

Fig. 77 Force-deformation diagram of locked clamping unit and mold

injection into the mold cavity, a reactive force F_R is produced, which is the product of the average cavity pressure multiplied by the cavity area projected on a plane through the parting line [169]. This force decreases the compression of the mold a_{mo} and increases the elongation of the tie bars a_{tb}. As soon as flashing occurs, this action has reached its practical limit somewhere between $F_0 = F_{R\,max}$ and $F_0 = F_{R\,perm}$. During this action, the elongation of the tie bars increases first to $a_{tb\,max}$. This causes the mating mold surfaces to separate by a distance a_x. The permissible value of a_x depends on the plastic material and on melt and mold temperatures. It is of the order of 0.05 and 0.1 mm. The usable force reserve of a toggle clamp system is of the magnitude of 10% of the clamping force. With this, the load limit of the tie bars should not yet have been reached. It is the duty of the machine designer to ensure this requirement by proper dimensioning length and diameter of the bars.

The character of the force-deformation diagram is defined by the spring characteristics of clamping system and mold [157, 393, 394]. The magnitude of these spring characteristic has a substantial effect on the resulting stresses. This will be demonstrated with the example of Fig. 78. A rigid clamp unit is compared with a less rigid one. In the first case, short but thick tie bars are employed, while in the second case, long and thinner bars are assumed [168]. The corresponding force-deformation diagrams demonstrate that the rigid unit permits a higher total clamping force (reactive force) before the mold halves begin to separate. At this moment, the stresses in the tie bars are higher than in the less rigid unit ($F_{R1\,max} > F_{R2\,max}$). Thus, one can conclude that rigid bars call for a greater safety factor than yielding bars. This is confirmed by the fact that rigid bars generally break more easily than yielding (thin) ones, if they are frequently overloaded by incorrect operation (equal flash in both cases). Reproducibility and precise

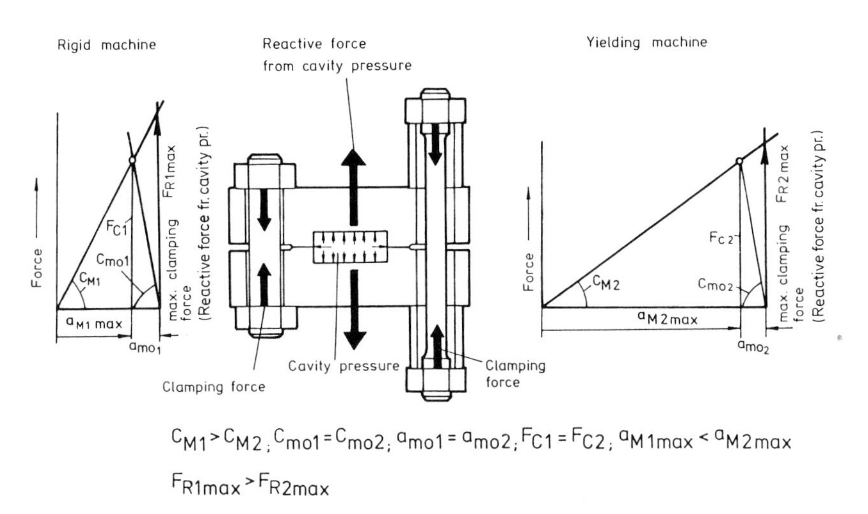

Fig. 78 Force–deformation diagrams for (left) a rigid and (right) yielding tie bars [168]

adjustment of the clamp force are certainly not improved with an extremely rigid design, because the force is produced within a relatively small distance and subject to great variations (Fig. 77). It is left to the reader's imagination to visualize the effect of a so-called yielding mold with the help of Fig. 78. The greater deformation of such a mold results in higher stresses in the bars, just as small deformations of rigid bars do, and therefore promotes bar rupture.

To set the desired clamping force, a suitable distance between mold platen and tail stock platen is selected and adjusted with mostly a central mechanical adjusting device (Sect. 4.1). Such an adjustment is particularly wise if the bars are subjected to heat and ensuing elongations or the machine is operated under critical conditions with only a negligible reserve of clamping force.

Another critical point is the deflection of machine platens, primarily the stationary platen. For reasons of strength, little thickness is sufficient most of the time. If the plate would be dimensioned with respect to strength only, this would result in too much deflection, leading to mold wear and flashing. With respect to quality, a dimensioning is desirable based on maximum permissible deflection. An acceptable value would be a deflection of < 0.2 mm related to 1 m distance between tie bars. Some manufacturers offer respective equipment [167, 168]. Others allow a deformation which is larger by a factor of 2 to 3 [172, 173]. The problem of deflecting machine platens, of course, also exists in machines with hydraulic clamping units.

4.1.3 Opening Force

The opening force is the force produced by the clamping unit during the opening stage. The forces available over the full stroke are smaller during opening than during closing (Fig. 78). The nominal opening force is that force which remains attainable for opening the mold after clamping and subsequent unlocking. In practice this force serves to break apart the two mold halves, which may stick together tightly due to the friction between molding and cavity surface or movable mold components. One can recognize that this force begins to act when the forces of the toggle have declined to a fraction (usually 15 to 30%) of the clamping force. Large machines generally perform somewhat better than small ones (Table 17).

The toggle forces are fully sufficient to break open the mold. Ejection is often provided during mold opening by an ejector bar, against which the movable platen moves and which activates the ejection system of the mold. The magnitude of the ejection force which is exerted upon the ejection system in the mold is important. This operation uses the opening force of the toggle which is available after part of the opening stroke has been completed. It cannot be more than the minimum opening force, which is only 1 to 2% of the maximum clamping force in the most adverse case (Table 17).

This force may be too small in some cases, but it can be increased, utilizing kinetic energy, by moving the movable platen with the corresponding mold half with high

Table 17 Opening and Ejection Forces of Injection Molding Machines

Clamping force	Rated opening force (Breaking force) (kN)		Minimum opening force (Ejection force) (kN)		Ejection force (kN)
	Toggle clamp	Hydraulic clamp	Toggle clamp, direct	Hydraulic clamp	Hydraulic ejector
500	75 - 150	50 - 60	7 - 15	50 - 60	15 - 25
1000	150 - 250	60 - 100	15 - 20	60 - 100	30 - 50
5000	800 - 1200	100 - 200	75 - 100	100 - 200	100 - 150
10000	-	350 - 400	-	200 - 400	150 - 220

speed against the ejector bar so that ejection is made possible. The location of the mechanical ejector in toggle machines corresponds with that of the hydraulic one shown in Fig. 91a, position B. Ordinarily, hydraulic ejectors are considered standard today. They make the forces listed in Table 17 or larger ones available for ejection. Pneumatic ejectors are only met in exceptional cases.

4.2 Hydraulic Clamping Systems

A common characteristic of all hydraulic clamping systems is the hydraulic cylinder, which is generally positioned in the center. Its ram can frequently travel the entire length of the closing or opening stroke for the movable platen. In such a case the ram is solidly attached to the moving platen (Fig. 79). To keep the energy for moving the oil within limits, in machine of more than 1000 kN clamping force the oil is sucked into the hydraulic cylinder from a tank. Thereby, intake of air has to be avoided.

The positive motion is provided by a central or by two or more laterally mounted small high-speed traverse cylinders, which are connected directly with the hydraulic pump. The possibility of tilting an eccentrically loaded movable platen is reduced by the supporting effect of the ram. In theory, the entire stroke of the ram can be used for mold height adjustment. It is decreased only by the opening stroke needed for the mold. This is the reason why these machines list only the maximum open daylight. The mold height adjustment is especially simple. Fig. 80 depicts a hydraulic clamping system with a mechanical locking function. Two or more lateral high-speed traverse cylinders produce closing and opening. In the closed position, a central spacer rod, which is mounted to the movable platen, is locked by cross bars, and the short-stroke clamp cylinder is pressurized. Mold height can be adjusted with the appropriate travel of the clamp cylinder ram, by changing the length of the spacer rod mechanically or adjusting corresponding stops. All these adjustments can be carried out easily. This design concept is retained even if more than one spacer rod and a corresponding number of hydraulic cylinders are used. A rotating pressure platen with cutouts can

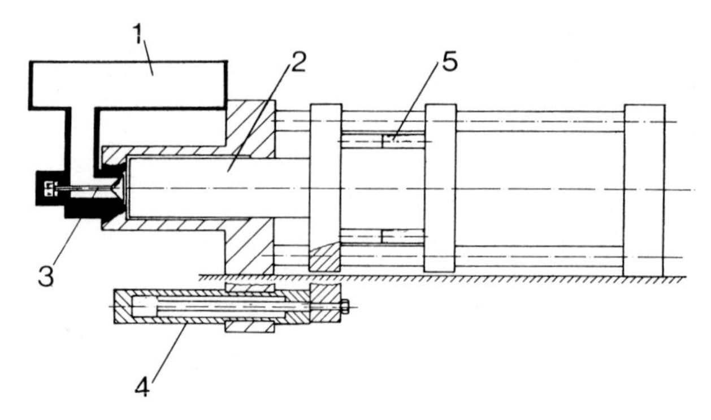

Fig. 79 Clamping unit with central hydraulic clamp cylinder and separate traverse cylinders. 1: Oil tank. 2: Central clamp ram. 3: Intake valve. 4: Double acting traverse cylinder. 5: Spacer for two-platen system

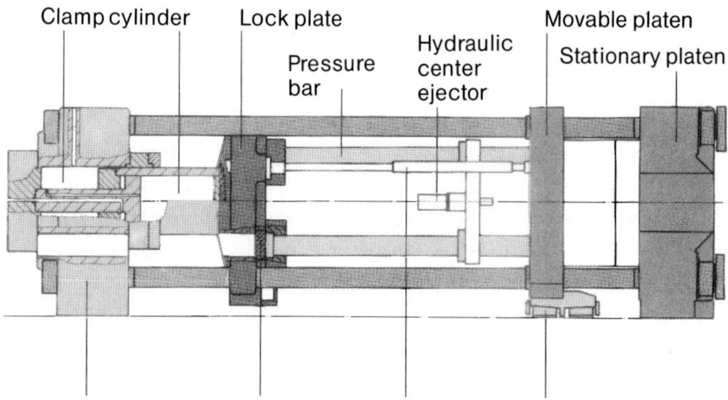

Fig. 80 Clamping unit with central hydraulic cylinder and mechanical lock (System Kraus Maffei)

also be employed for locking. It is turned into a position such that it provides stops for the spacer rods. Before opening, the platen is turned again so that the rods can move through the cutouts (Fig. 84).

There are machines that have the central clamping cylinder mounted to the movable platen. Because of its short guide length, it can easily tilt, and the movable platen looses its parallelism with the stationary one [105]. They are not manufactured anymore.

A clamping unit with central clamping cylinder and likewise central high-speed traverse cylinder is depicted with Fig. 83.

For smaller machines up to 500 kN clamping force a manufacturer offers a clamping unit with a flow-rate multiplier [14]. This multiplier, connected to the central pump, delivers for a short time a large quantity of oil to rapidly move the machine platen. A separate small pump provides the pressure during the clamping stage (Fig. 81). Another design concept uses a pivoted, short-stroke hydraulic cylinder with integrated mechanical mold-height adjustment, which can be moved into and out of position. The whole system revolves around thc rear upper tie bar (Fig. 82A).

A successfully employed design is shown with Fig. 85. It uses a clamping unit with four clamp cylinders acting in the same direction. The tie bares are guided in the stationary platen and carry pistons. Closing and clamping are accomplished by applying pressure to the annular piston surface. For opening, the pressure acts on the reduced area of the opposite surface. Oil can flow economically from one side of the piston to the other without the action of a pump. Short overall length and easy access to the space underneath the clamp unit are of advantage. The accessibility of the nozzle is slightly restricted.

At the K'89 in Duesseldorf an injection molding machine without tie bars was presented. The clamping unit is carried in an appropriately rigid, U-shaped framework (Fig. 86). The load-bearing cross section was taken ten times that of the otherwise common tie bars. The special rigidity is needed because the machine platens, guided only on one side, would tend to loose their parallelism sooner than with a guide by tie bars. On the other hand, this rigidity of the framework proportional to its cross section (spring characteristic) results in low resilience against reactive forces. Thus, flashing is easily prevented. To ensure this plane parallelism under the effect of reactive cavity forces the movable platen is pivoted horizontally in its center. Parallelism can be

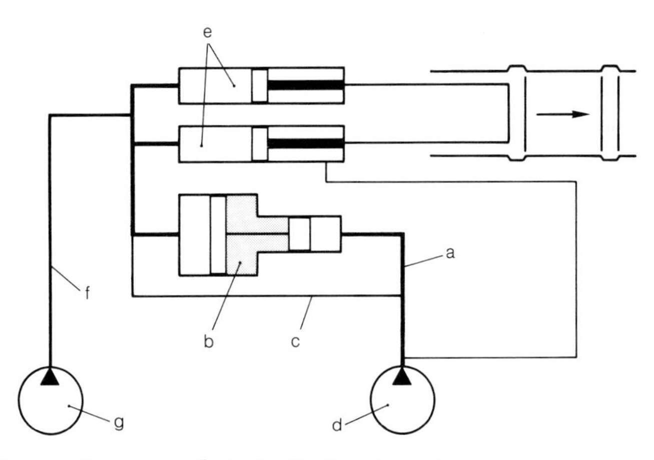

Fig. 81 Diagram of operation of a hydraulic clamping unit
a: Rapid clamping with volume multiplier b, c: Direct connection between pump d and clamping cylinder e to raise clamping force, f: Pressure is kept by separate pump g

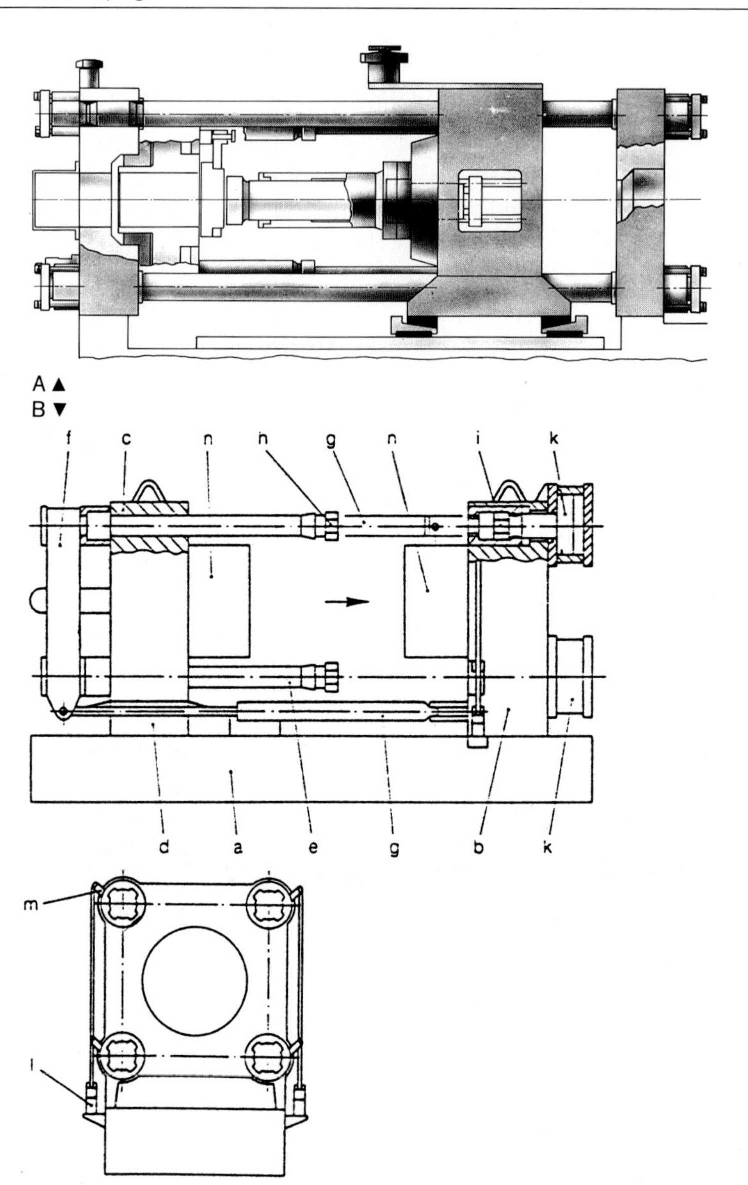

Fig. 82 A: Clamping unit with central, laterally pivoting clamp cylinder (System Battenfeld) B: Clamping unit in short design with chuck locking (System Hemscheidt)
a: Undercarriage, b: Stationary platen, c: movable platen d: Slide, e: Bar, f: Tail stock, g: Rapid-motion cylinder, h: Chuck, i: chuck sleeve, k: Clamping cylinder, l: Lock cylinder, m: Locking lever, n: Mold

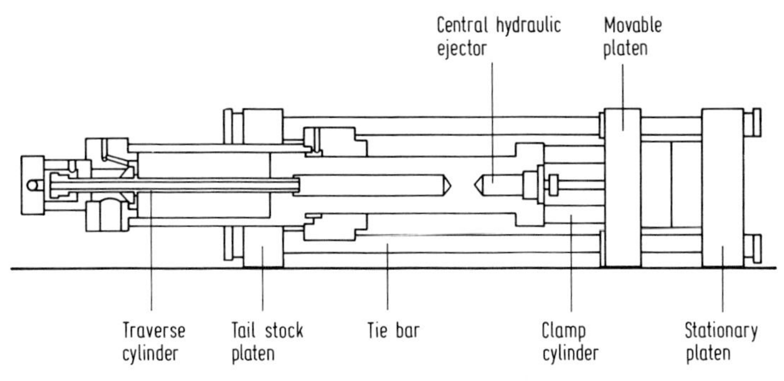

Fig. 83 Clamping unit with central clamping and central rapid-traverse cylinder (System Krauss Maffei)

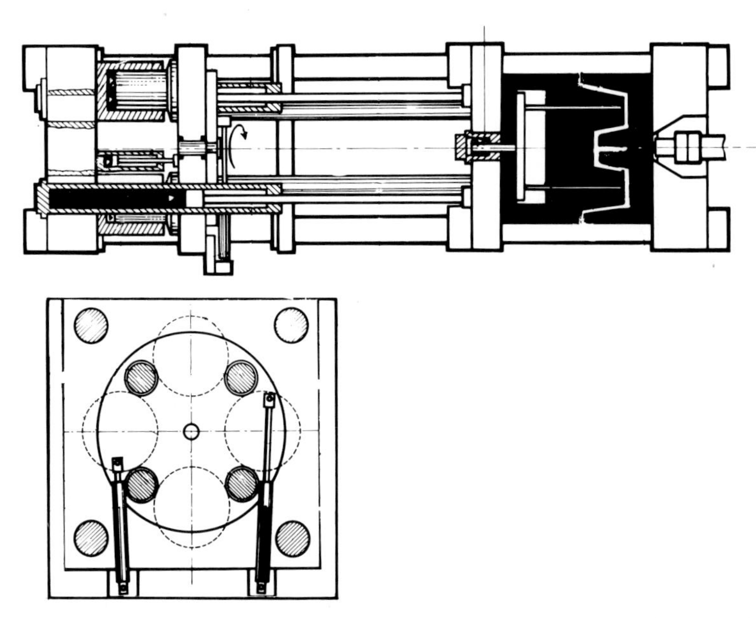

Fig. 84 Clamping unit with several clamping cylinders and revolving lock plate through which support bars can pass (System Billion)

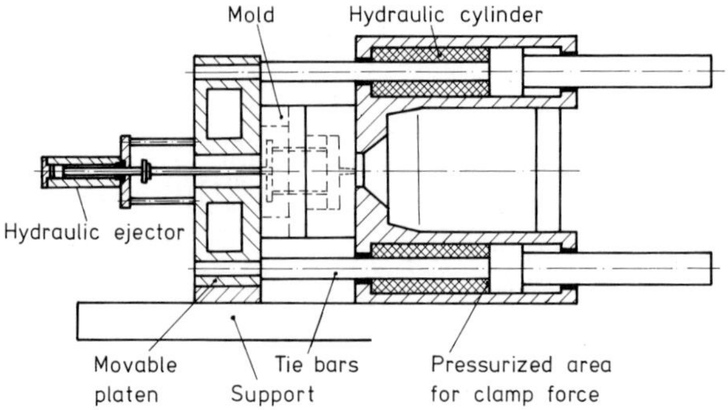

Mold
Hydraulic cylinder
Hydraulic ejector
Movable platen
Support
Tie bars
Pressurized area for clamp force

Fig. 85 Detached clamping unit witch movable tie bars hydraulically actuated [170]

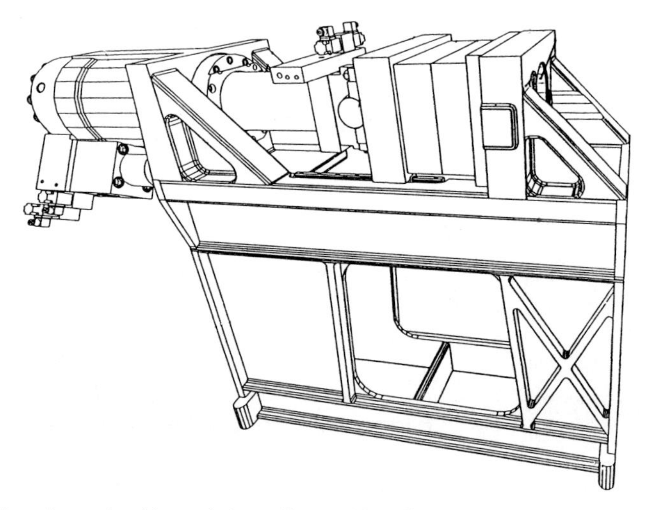

Fig. 86 Clamping unit without tie bars (System Engel)

corrected with a set screw if needed. These simple means should secure a careful handling of the mold in practical operations.

A deflection of machine platens is possible almost exclusively in the horizontal and is in the range of permissible values. Especially interesting is the option to exchange machine platens to eventually accommodate larger molds.

This design also provides free accessibility to the clamping unit, and set-up of molds, mounting devices for core removal, demolding with special ejectors, part removal with robots and installing of connectors can be done in a simple manner [172,173]. The success of these machines on the market is excellent and gives evidence of a unique approval.

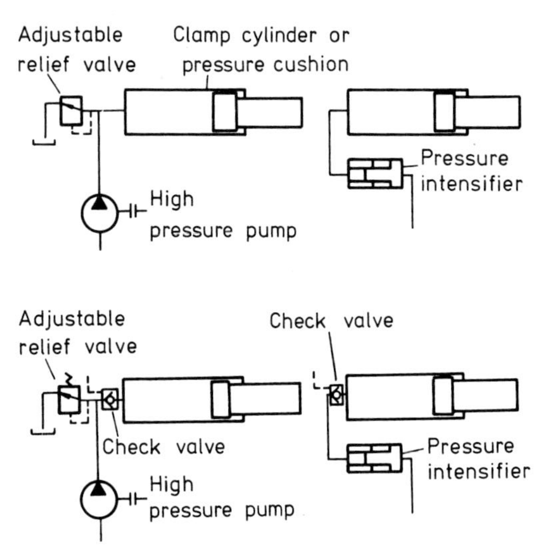

Fig. 87 Methods of producing clamping pressure for injection molding machines with fully hydraulic clamp. Top: Systems with source of constant pressure for pressure cushion. Bottom: Systems with shut-off valve in return line

Another interesting concept, which is just being tested, uses detachable tie bars. In the open position, it generates great freedom for mounting and set-up of molds and removal of large parts. With regard to the clamped position, refer to Sect. 4.1.2 and the effect of short tie bars. Tie bars of any length, however, are still more rigid then even a small volume of oil, which is needed to provide the final clamping force. Of course, this small volume of oil exhibits little compressibility and, therefore, good performance during the clamping stage. All machines with a hydraulic clamping unit and fully hydraulic opening and clamping actions operate with almost constant speed along the entire travel of the movable platen. This differentiates them from toggle clamp machines. It takes a considerable amount of control to slow down the motion into a programmed position. Even if travel and clamp systems are separated, more time is consumed for closing and opening than with a toggle design, provided that the pump output is the same. For this reason, the toggle system is better suited for high-speed machines. On the other hand, it is easier to reproduce and maintain the clamping force with fully hydraulic machines. Generally, one chooses one of the four options of generating the clamping force shown in Fig. 87. They are distinguished as follows:

- build-up of clamp pressure and maintaining maximum pressure with a high-pressure pump (top left),
- build-up of clamp pressure with a high-pressure pump and maintaining pressure with a check valve in the return line (bottom left),

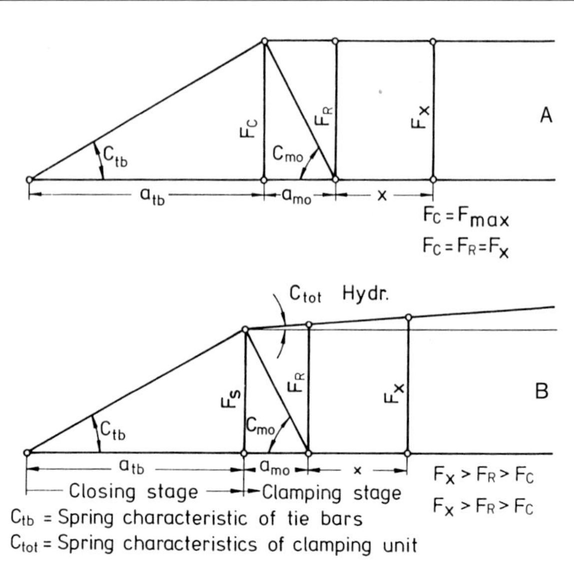

Fig. 88 Force – deformation diagramms for injection molding machines with fully hydraulic clamp. a: System without check valve in return line. b: System with controlled check valve in return line (Refer to Fig. 87)

- build-up and maintaining clamp pressure with a pressure intensifier (top right),
- build-up of clamp pressure with a pressure intensifier and a check valve in the return line (bottom right).

With all four systems, the clamp force can be adjusted simply and precisely to any desired value. It only requires an adjustment of the hydraulic pressure. The selected force remains constant during the entire clamping stage (Fig. 88A), but there is no reserve capacity available. Closing and clamping force are of the same magnitude. Elongation a_{tb} of the tie bars and the forces F_{tb} on them remain unchanged during injection. The resilience of the mold is compensated by the oil pressure. An overloading of the system is impossible. For reasons of energy efficiency, it is better to shut off the return line after the clamp force has been built up. Then the effects of forces change, and the total clamping force can become higher than the closing force. Fig. 88B shows the increase of thc total force in accordance with the spring characteristic of the whole system. The oil is compressed in response to the reactive force. Thus, the elongation of the tie bars increases. The rise of the total force over the closing force depends essentially on the compressibility of the hydraulic fluid and its volume behind the ram in the closed position. Table 18 presents the reserve capacity of the clamping force if the mold halves open 0.1 mm through flashing. The varying oil pressure is expressed in mm of oil column. The figure for a toggle clamp unit is added for comparison. For an exact comparison, the spring characteristics of the mold and the total system should be included [155].

Table 18 Increase in Total Clamping Force with Varying Oil Pressure Expressed as Height of Oil Column and for a Toggle Clamp

	Height of oil column*				Toggle clamp
	2000 mm	1000 mm	500 mm	100 mm	
Increase of force over clamp force for 0.1 mm elongation of tie bars	0.15 - 0.2%	0.3 - 0.4%	0.6 - 0.8%	3 - 4%	~10%

* Height of oil column in hydraulic clamping units up to 2000 kN clamping force and in clamping units with mechnical locking function.

4.3 Combined Clamping Systems

Such systems are rather expensive to build and only remainders are still on the market. They are characterized by separating closing and opening operations from producing the clamping force. The toggle system is generally used for long strokes because it is very fast, and the kinematics of its motion are of advantage (Fig. 89). The build-up of clamping pressure is achieved with a hydraulic cylinder. Some toggle machines were provided with up to four hydraulic cylinders (Fig. 90). As far as known, such machines are not manufactured anymore.

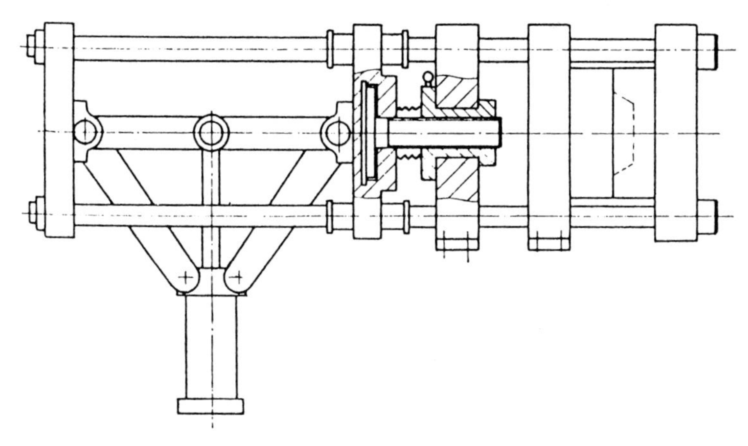

Fig. 89 Clamping unit with a combined mechanical (toggle)-hydraulic (flat oil cushion) clamping system and central mechanical mold height adjustment

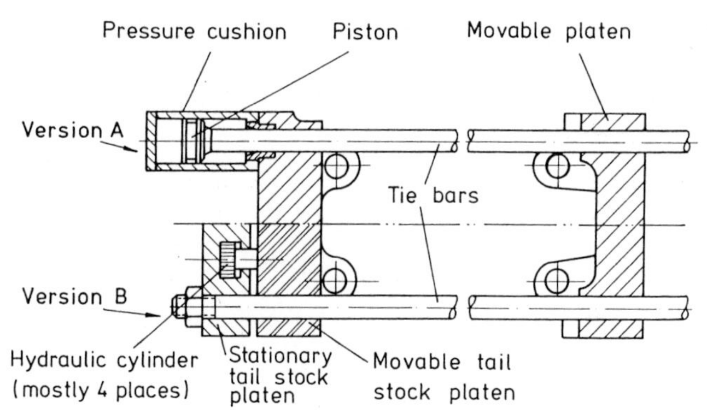

Fig. 90 Clamping units with a combined hydraulic-mechanical (toggle) clamping system. Version A, Piston at the end of tie bar. Version B, Short-stroke hydraulic cylinder acting upon a separate tail stock platen

4.4 Accessories for the Clamping Unit

The range of equipment on the clamping side of machines, which meet today's needs, comprises the following accessories, which can be divided into two segments. One can rarely dispense with standard accessories, while the benefits of investing in one or more of the items listed as special accessories have to be evaluated in each case.

Standard equipment:

- machine setup with slow motion,
- variable speed for independent closing and opening,
- speed control for careful clamping (mold protection),
- positioning the movable platen along the entire stroke, actuated separately for more than 5000 kN clamping force (day light adjustment),
- safety device (hydraulic or electrical) to protect against uncontrolled movements,
- adjustment of clamping force from zero to maximum, read-out in real physical units,
- mold protection with low pressure,
- central ejector mechanism, mechanical, or better hydraulic,
- impulse control of ejector,
- central lubrication with control,
- safety guards on all sides in accordance with safety regulations,
- safety gates with window, front and rear,
- gate safety devices, electric and hydraulic,

- actuated gates for relief of operator in machines of more than 5000 kN clamping force,
- easy removal of at least one tie bar in clamping units with four tie bars, for mounting oversized molds.

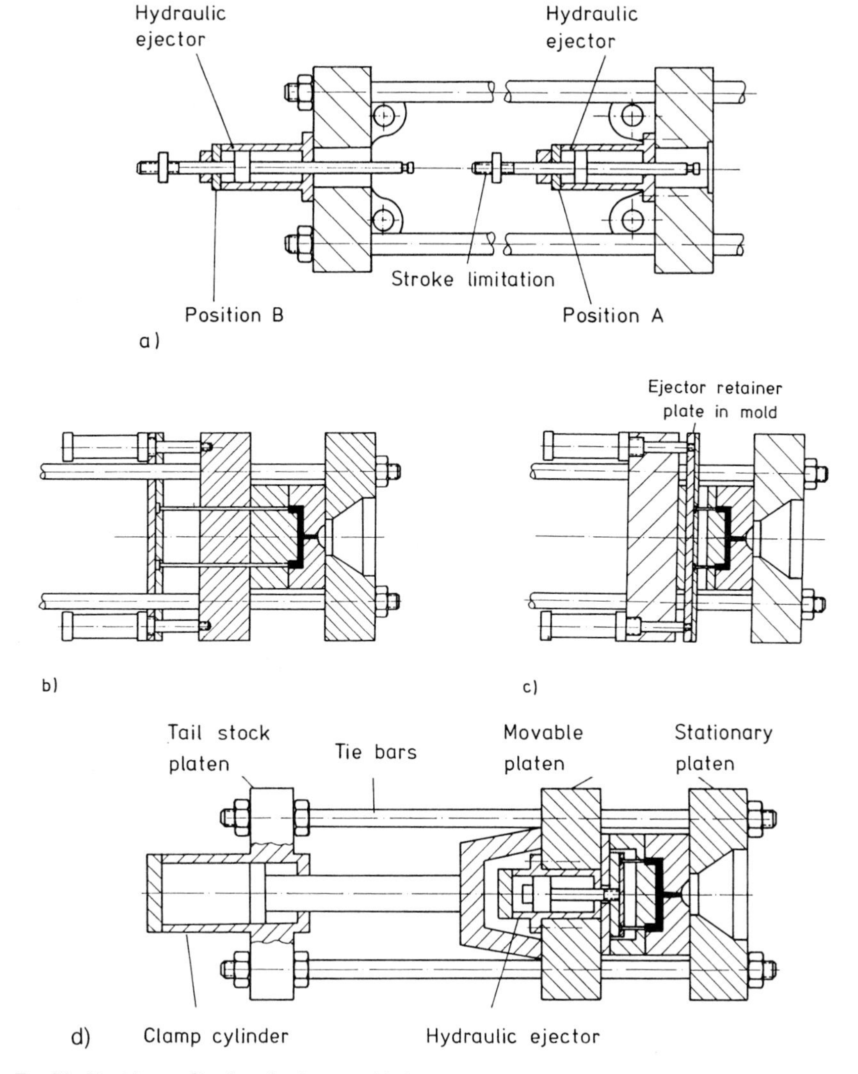

Fig. 91 Positions of hydraulic ejectors. (a) Central position in movable platen or tail stock platen. (b) Side position, actuating ejector retainer plate. (c) Side position, actuating ejector retainer plate in mold. (d) Central position in movable platen or ejector box

Optional accessories

- mechanical protection against unintentional closing (supplement),
- limitation of opening stroke with shock absorber,
- hydraulic or pneumatic ejector system (Fig. 91) with adjustable stroke and speed, in center or on both sides of clamping surface,
- equipment for hydraulic cam action (multiple), parallel or in sequence,
- device for retraction of mechanically actuated ejector,
- cooling lines for machine platens,
- automatic central lubrication with control,
- adjustable setup for coining process,
- electric or hydraulic operation of mold height adjustment; for machines with more than 5000 kN clamp force, this is standard equipment,
- swivel arm with hoist (only up to about 2000 kN clamp force).

5 The Drive Unit

The standard drive for an injection molding machine is the single electrohydraulic drive, which consists of an electric motor and a hydraulic pump. Central energy supply with water hydraulics or mechanical drives, although wide-spread some time ago, are only of historical interest. A pneumatic drive is used exclusively for mini plunger-type machines.

A "drive" can consist of electric motor, transfer elements (e.g. couplings), energy transformer (e.g. hydraulic pump), and gear transmission. The kind of use (rotation or translation) as well as the number of uses in one machine decide on the individual arrangement in each case.

5.1 The Electric Drive

5.1.1 Methods of Operation and Torque

The drives of injection molding machines are, in their majority, subjected to changes in the moment of inertia, either time-dependent during continuous or torque-dependent during intermittent operations. There is a characteristic dependence on time in the function of an injection molding machine (Fig. 92). In a selected and then fixed periodical sequence the build-up of a moment of inertia is repeated. Torque-depend-

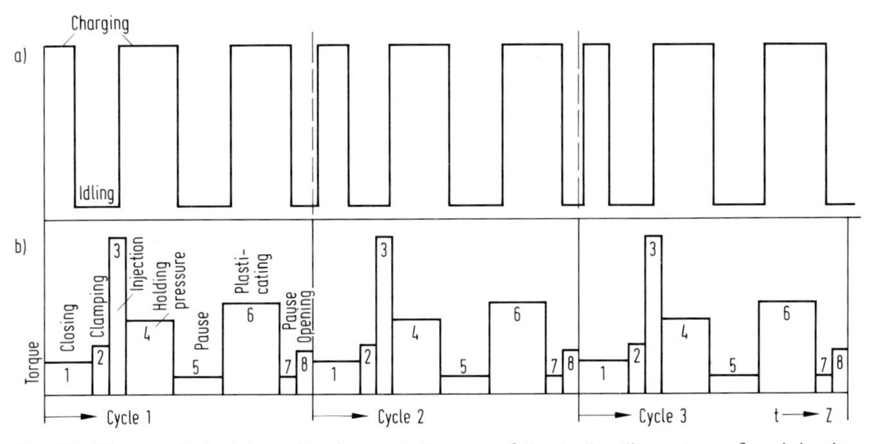

Fig. 92 Torque of electric motor for central pump of the hydraulic system of an injection molding machine
a: Principle pattern of charging an accumulator supplying all pressure-consuming devices
b: principle pattern of direct supply by pump

ent changes occur during feeding. No basic difference exists between the direct drive with an electric motor or with a hydraulic pump as energy transformer. The energy consumption of a direct electric drive decreases to zero, though.

5.1.2 Environmental Conditions

Besides a number of general requirements concerning the supply of electric energy and its use, other considerations and standards have to be observed. The National Electric Code (NEC), which is also adapted by the Occupational Safety and Health Act (OSHA), aims at the assurance of an electrically safe human environment, which will be fostered by following NEC rules. They are the governing authority in case of local differences and represent the national standard.

Standards set by the Institute of Electrical and Electronic Engineers (IEEE) deal particularly with the rating of electrical apparatus.

In shops, where combustible dust presents a special fire or explosion hazard, sources of helpful information and recommendations are provided by the National Fire Protection Association (NFPA).

5.1.3 Electric Motors

5.1.3.1 Polyphase Induction Motors

Most of the time, polyphase squirrel-cage motors are used as power generator because of their constant speed during continuous operation and their rugged construction. This is the classical case of an electrohydraulic drive where electric energy is converted into hydraulic energy and a hydraulic pump or a combination of pumps are coupled with the motor.

These motors have a starting torque of about 1.5 times the full load torque at rated voltage. The locked rotor currents vary between four and seven times full-load current. With motors of larger size, more than 370 kW (200 hp), reduced voltage may be necessary to meet starting-current restrictions. A common method is to use a compensator or autotransformer. In the starting position, a connection is made in Y across the line supplying the motor with reduced voltage. According to the starting current, the starting torque is reduced to about 1/3 of that from direct connection. Since hydraulic pumps are not started under full load, this method is the preferred one.

Motors can operate with a higher locked rotor torque for a short time. This is called working with an overload margin. The name plates of such polyphase induction motors carry a code and a design letter. The latter (A to D) provides information about torque characteristics. Thus, the upper limits of the locked rotor torque of a 100 hp (75 kW) motor are (A) 125%, (B) 125%, (C) 200%, and (D) 275% of the rated torque. This is particularly advantageous if load surges occur during intermittent operations. Consequently, an overload is permissible during the short- time injection stage and an electric motor of 30 kW can drive a hydraulic pump up to 48 kW.

On the other hand, the efficiency curve for the range of rated loading is fairly flat and the economics of asynchronous motors are not considerably reduced if it is over-

Table 19 Converter Circuits for Three-Phase AC Motors

Converter type	Converter with direct-voltage intermediate circuit	Converter with direct-current intermediate circuit
Control method	Frequency change and voltage control, pulse-width control to provide sine-shaped current	Frequency change and current control
Frequency range	2.5 . . . 120 Hz	5 Hz . . . 50 Hz 5 Hz . . . 87 Hz
Load range	1.5 kVA . . . 171 kVA	16.5 kW . . . 50 kW 50 kW . . . 1050 kW
Typical rpm range	1 : 200	1 : 20
Motor type	Three-phase squirrel-cage motor	Three-phase squirrel-cage motor
Motors per converter	1 or several	1
Applications	Feeding devices, extruders, conveyers, fans, mixers, machine tools	Blowers, fans, pumps, mixers, extruders, conveyer belts, presses, screws

sized for reasons of safety. By switching poles, two speeds in a ratio of 2:1 are readily obtained with induction motors. Thus, 3000 rpm can be reduced to 1500 rpm and 1500 to 750 rpm. Motors with two windings even allow four speeds. One should keep in mind that this a compromise and may sacrifice desirable characteristics. Anyway, it may occasionally be sufficient for the screw drive of simple injection molding machines.

Synchronous polyphase motors are used as the main drive primarily in the power range of more than 100 kW. Below this, the rugged asynchronous squirrel-cage motor dominates because of its price advantage. The supply for synchronous motors is taken from a synchronous converter according to Table 19. It is integrated in the motor and delivers direct current for the armature winding.

5.1.3.2 Speed Control of Polyphase Motors

The induction motor inherently is a constant speed motor. The rotor speed is:

$$N = 120\,f\,/P \quad (rpm)$$

where f = frequency (Hz) and P = number of poles

To change the speed, a frequency converter is needed. With the exception of slow-speed motors, one always employs intermediate-circuit converters.

One can distinguish:

1. Converters with direct-voltage intermediate circuit rectify the line current with diodes to a constant direct voltage, which is smoothed by a condenser parallel to the output. Then power transistors form positive and negative square waves of the desired frequency.
2. Converters with direct-current intermediate circuit operate with a thyristor rectifier (silicon controlled rectifier (SCR)), which produces, with the help of an inductive reactance, a direct current proportional to the requested torque. Then the direct current is inverted by means of GTO Thyristor (gate turn off thyristor.) to alternating current of the desired frequency and supplied to the motor.

So far, the most important converter technique has been the concept of the direct-voltage intermediate circuit with sine-shaped pulses of direct current. The generated square waves are divided into many individual impulses in a 20 kHz timing sequence. Control of each pulse width is done in such a manner that a largely sine-shaped current results.

Today sine-modulating converters are available with solid-state control, which is easy to operate. Noise and losses are minimized so that common squirrel-cage motors can be used (Fig. 93).

Most recently, so-called insulated-gate bipolar transistor chips (IGBT) are used for speed control. They are very energy efficient and need only little space. The latest development in this direction is the MOS-controlled thyristor (MCT), which supposedly has higher current-handling capacity and even smaller electrical losses. Conclu-

Fig. 93 Three-phase-current drive system: Squirrel-cage motor – Motor with transistor – Inverter (Courtesy BBC, Mannheim)

sive experience has not yet been made. Standard polyphase motors with converters have a problem in the speed range $N < N_{rate}$. During extended length of operation, they must not run with the rated torque. This condition requires special equipment with, among others, separate cooling. Such machines are ideal to drive screws since they maintain the selected speed very precisely independent of the torque. In fully-electric injection molding machines, where the holding pressure has to be maintained without further rotation of the motor such motors are not only employed as rotatory screw drive but also to actuate clamping unit and injection.

5.2 The Electromechanical Drive

Electromechanical drives disappeared from the market at about 1970 because of their obvious functional problems. In 1985 information from Japan reports of smaller machines with 50 and 100 kN clamping force [176]. Theses machines do not offer any new concept with the exception of an electric servo drive. All motions are executed with gear drives, which are put into operation by couplings as required. Linear motions are achieved with lead screws. Such transmissions are known since about 1965 but this technique was abandoned because there were problems on the clamping side with breaking forces if self-locking could not be overcome due to the effect of the reactive force.

Most recently, considerably improved electromechanical machines are put on the market. They were already successfully tested. About six years ago, the first commercial modern machines, commonly called all-electric injection molding machines, were produced by Japan's Fanuc in conjunction with Cincinnati Milacron [15]. These machines transform the rotational movement of an electric servo motor into a linear motion with a ballscrew. The development of three-phase brushless servo motors and ballscrews went hand in hand with the progress in production, and reliable drives could be employed for commercial machines of 500 to 3000 kN clamping force [15, 17]. The toggle of the clamp unit is actuated by a ballscrew. So is the ejector with its own motor. On the injection side, extremely wide and stiff cogged V-belts transmit the motor force to the ballscrews, which perform the injection or immediately to the screw for feeding (Fig. 94). Such all-electric machines appear to have a clear advantage in precision of programmed operation.

At the K'92 two German manufacturers showed developments of their own in a modified design. Battenfeld presented a 500 kN version of its CDK series, but this machine has not yet been commercialized. Kloeckner Ferromatik displayed its Elektra 100, which is comprehensively portrayed in [16]. It features a four-point, twin-toggle clamping system, which is actuated by a rack-and- pinion gear drive. A double-rack gear box with an upstream two- stage spur-gear set is used (Fig. 95). The advantage of this system, as it is said, is a theoretically limitless speed for clamping and opening. A separate motor is used for the crank-slide driven ejector (Fig. 96). Ejection force and speed are programmable. The former is the highest at the onset of ejection

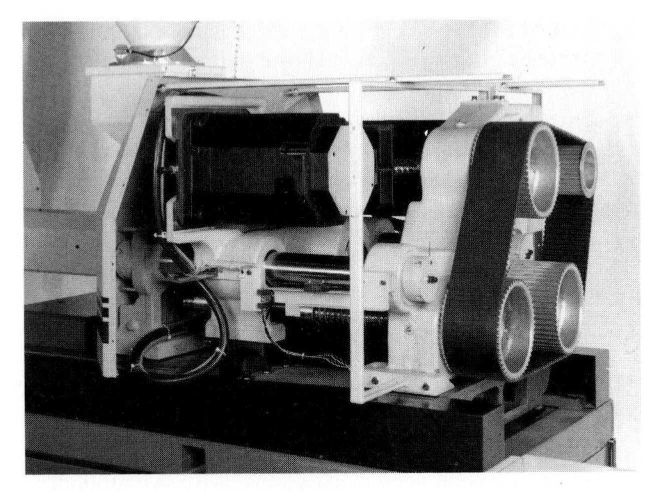

Fig. 94 Drive unit of an electromechanic injection molding machine (System Cincinnati Milacron)

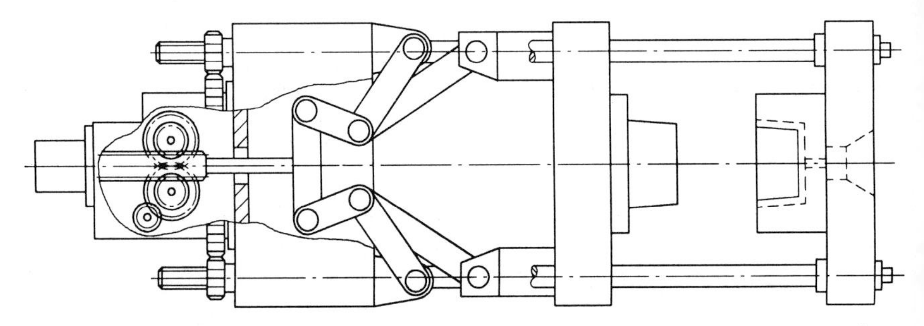

Fig. 95 A double rack drives the linkage in a fully electric injection molding machine with four-point toggle

while the latter increases in the progress but slows down to zero in the end position. This machine accomplishes injection with a patented crank drive, too. It is said that it permits a high injection speed of maximum 200 mm/s. This is sufficient for molding thin-walled technical parts. Whether or not the reduced pressure of 120 MPa in the middle range of the screw travel is still adequate, has to be shown in practice (Fig. 97). The movement of the injection unit is accomplished with a ball screw.

All machines are presently undergoing practical testing. So far there is common agreement, though, about the energy-saving potential of all-electric machines. Losses from energy conversion are eliminated. Besides this, an increase in operating time can be expected [15, 16, 17, 18].

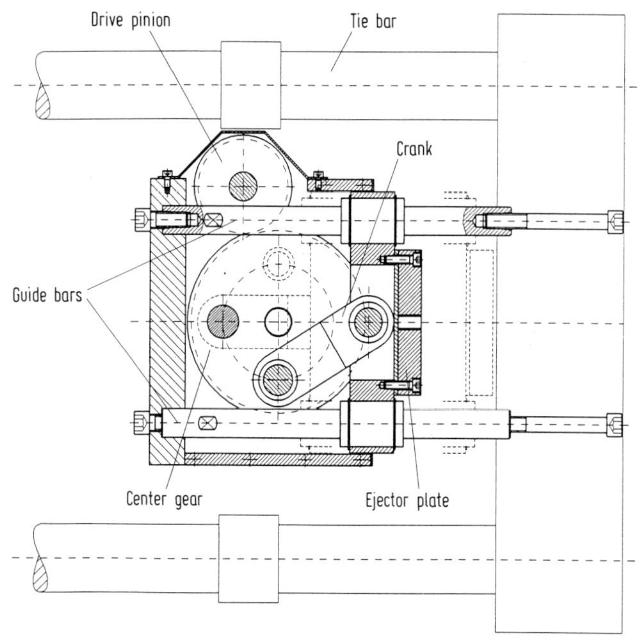

Fig. 96 A crank disk drives the ejector, initially with high ejection force, then with high velocity [16]

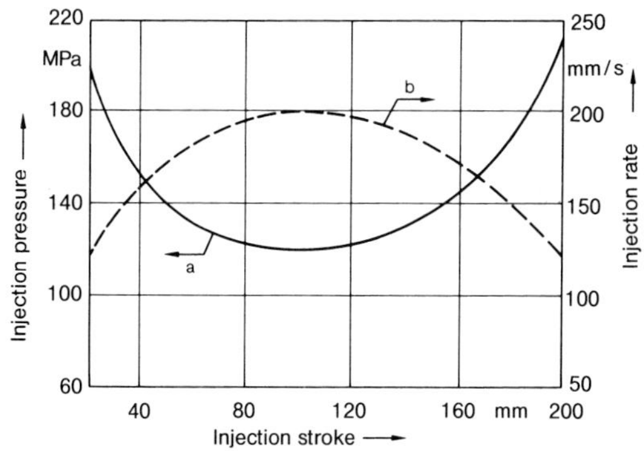

Fig. 97 Variation of injection pressure (a) and injection rate (b) along injection stroke (non-linearized)

5.3 The Electrohydraulic Drive

To convert electrical into mechanical energy one employs hydraulic pumps. Vane pumps, radial-piston pumps, axial-piston pumps, also as variable displacement pumps, gear pumps and internal gear pumps are available. Basically similar conversion from the energy of pressure to mostly rotatory motion are made with hydraulic motors.

5.3.1 Hydraulic Motors

Hydraulic motors, as far as they are employed in processing plastics, are always part of an electrohydraulic drive. One can distinguish between linearly operating and rotatory hydraulic motors. Linearly working motors are generally called hydraulic cylinders. They convert hydraulic energy (energy of pressure) into work in a linear manner (strokes) or transform it kinematically, e.g. with toggles, into linear motion.

In the following, only rotatory hydraulic motors will be discussed. They are used in plastics processing almost exclusively for rotating the screw, primarily in injection molding machines.

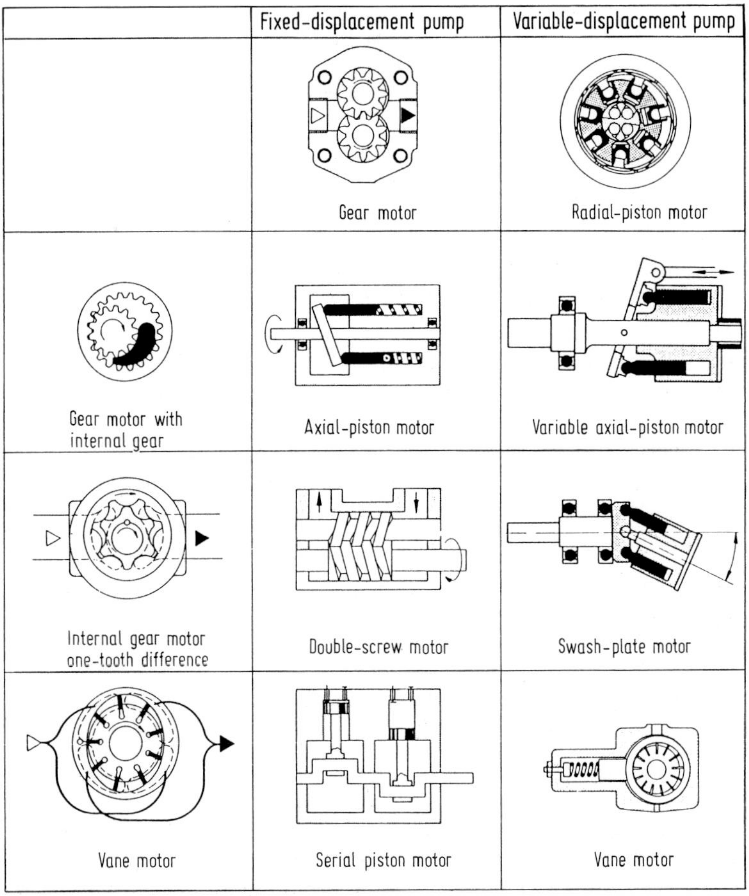

	Fixed-displacement pump	Variable-displacement pump
	Gear motor	Radial-piston motor
Gear motor with internal gear	Axial-piston motor	Variable axial-piston motor
Internal gear motor one-tooth difference	Double-screw motor	Swash-plate motor
Vane motor	Serial piston motor	Vane motor

Fig. 98 Classes of hydraulic motors [178]

Table 20 Performance Data of Hydraulic Motors

Pump type	Capacity cm^3/min	Torque Nm	Pressure range MPa	Speed rpm
Gear motors	5 - 300	14 - 900	15 - 22	4000 - 800
Vane motors	50 - 2300	130 - 6600	14 - 20	1500 - 200
Swinging-vane motors	60 - 750	140 - 1600	14.5 - 21	1000 - 600
Internal-gear-motors	12 - 800	18 - 2000	14 - 18	1600 - 200
Axial-piston motors	10 - 2000	40 - 8000	25 - 35	6000 - 2000
Radial-piston motors	25 - 23 000	90 - 85 000	25 - 31.5	900 - 80

Hydraulic motors are characterized by the following features:

- Continuous speed control over a wide range,
- Almost constant torque over the whole speed range,
- Simple torque limitation with check valves (protection against overload and fracture due to twisting off),
- Starting torque only slightly smaller than operating torque,
- Low inertia promotes fast starting and braking (beneficial for exact feeding with intermittently operating screws),
- Low weight-to-efficiency ratio and compact design (good with direct drive) improves acceleration of moving masses of injection unit (may promote quality),
- Efficient utilization of the rated capacity since the motor is only supplied with energy during feeding.

As far as design is concerned, one can distinguish among the following basic concepts (Fig. 98 and Table 20):

- Gear motors,
- Vane motors,
- Swinging-vane motors,
- Internal-gear motors, and
- Piston motors (axial- and radial-piston motors).

Gear motors are of secondary significance because of their excessive slip and, consequently, low efficiency. Vane and swinging vane motors are losing importance more and more since their efficiency decreases progressively with the increase in working pressure. Internal-gear motors are primarily used in small and medium-sized machines because of the limited capacity. Axial-piston motors as a high-speed hydraulic drive for small and medium-sized machines have been in use successfully. For larger machines of more then 50 mm screw diameter, they have to be combined with mechanical transmissions in order to achieve the required torque. This, however, has a negative effect on efficiency, noise level, costs, and size. Among the motors listed in Table 20, the radial-piston motor was especially reliable as direct drive for injection molding machines. These motors with hydrostatically relieved mechanism excels by their high mechanical and volumetric efficiency and their extremely low noise during operation (Fig. 99). The pistons transmit the torque immediately to the drive shaft (mostly the screw) with piston rod and cam and without any turn. They can be employed without step-down gearing as drives for the smallest up to the largest screws known today. Radial-piston motors are available with constant and with step-wise variable displacement, and most recently with continuously variable displacement, too.

The step-wise variable-displacement radial-piston motor permits a good adjustment to the speed and torque requirements with an efficient use of the power input.

The power input is converted into torque and speed in reasonable steps (possible ratio of variation 3:1). The hydraulic power of the drive remains constant but it is feasible to combine high torque with low speed and high speed with low torque by switching accordingly. Such a drive covers a wide range of demands. There remains a range of uses, though, that cannot be covered with constant power (Fig. 100B). The employment of step-wise variable- displacement motors offers considerable advantages over the fixed-displacement motor, the same technical parameters assumed, because of their small size, lower weight and costs, and smaller operational expenses [194].

The motors just discussed here can practically meet all the demands of an injection molding machine. Since the feeding stage, however, needs a relatively great amount of energy, a careful dimensioning of the pressure supply (hydraulic pump) and the screw drive (hydraulic motor) with respect to energetics is required. The smallest losses occur on the supply side of the system if flow rate and pressure are adjusted to the consuming item or better, the consuming component to the supply pressure.

In the first case, the pressure is best supplied by a controllable pump. This does permit additional users in parallel only with more added technical devices to the hydraulic system.

In the second case, a variable-displacement radial-piston motor can be used as a direct drive (Fig. 101).

This variation renders it possible to automatically accommodate instantaneous loading conditions almost free of losses or to preselect displacement or torque for a certain operating situation (Fig. 100C). Such drives meet all torque requirements without problems (e.g. PE: T = 1, PC: T = 2.4-2.8) and one can dispense with exchanging drives or plasticating units.

Setting the displacement can be done manually with a handwheel, a hydraulic cylinder, or with a proportional valve (Fig. 102).

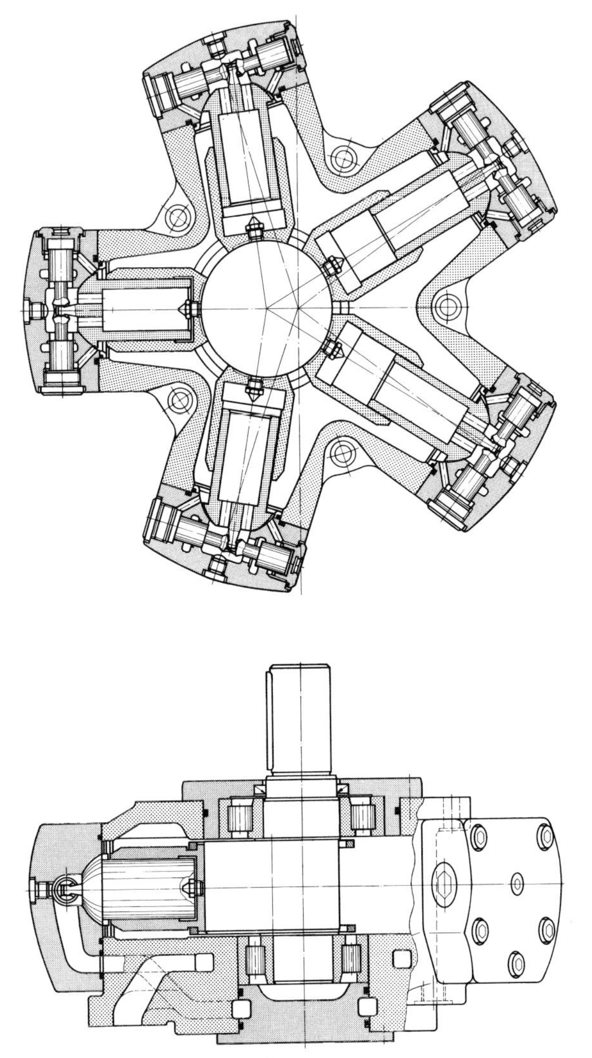

Fig. 99 Radial-piston motor (System Pleiger)

Of course, a modern screw drive needs a solid-state control and display, preferably in a standard format, since e.g. speed is an important processing parameter, which has to be displayed. A torque read-out would be desirable but is available only on special demand.

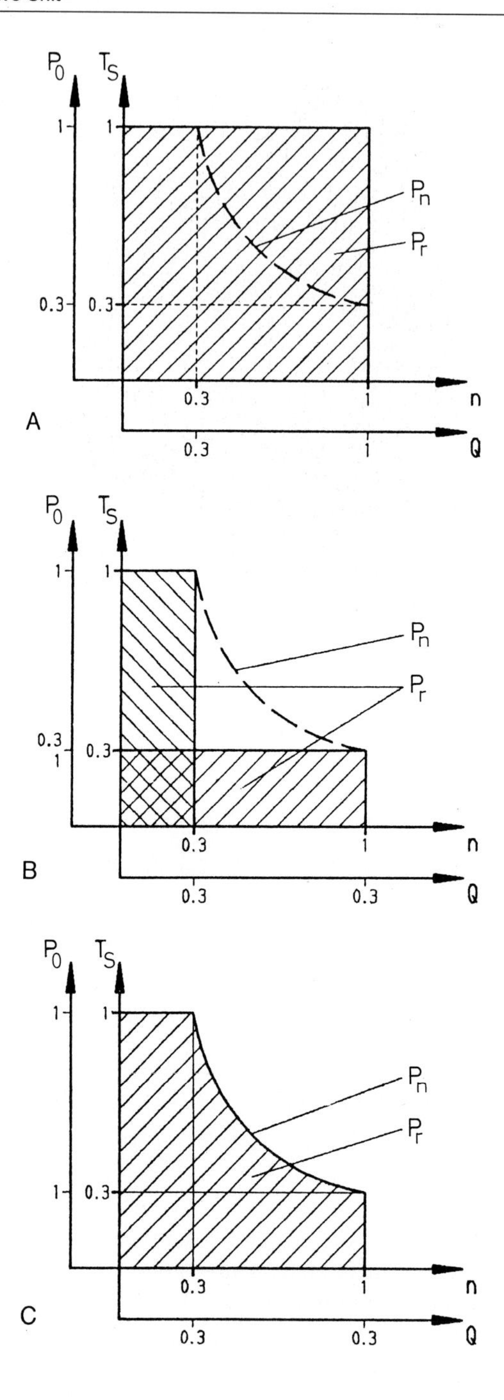

P_0
T_S
1
0.3
P_n
P_r
0.3
1
n
Q
A
B
C

Fig. 101 Continuously variable radial-piston motor (System Pleiger)

◀ *Fig. 100* Variation of torque and rotational speed with different hydraulic motors
A: Fixed-displacement motor
B: Switching motor
C: Variable-displacement motor
P_O = Operating pressure,
T_S = Screw torque [Nm],
n = Revolutions per minute of screw [min^{-1}]
Q = Flow rate [l/min]
P_r = Rated power [kW]
P_n = Power needed [kW]

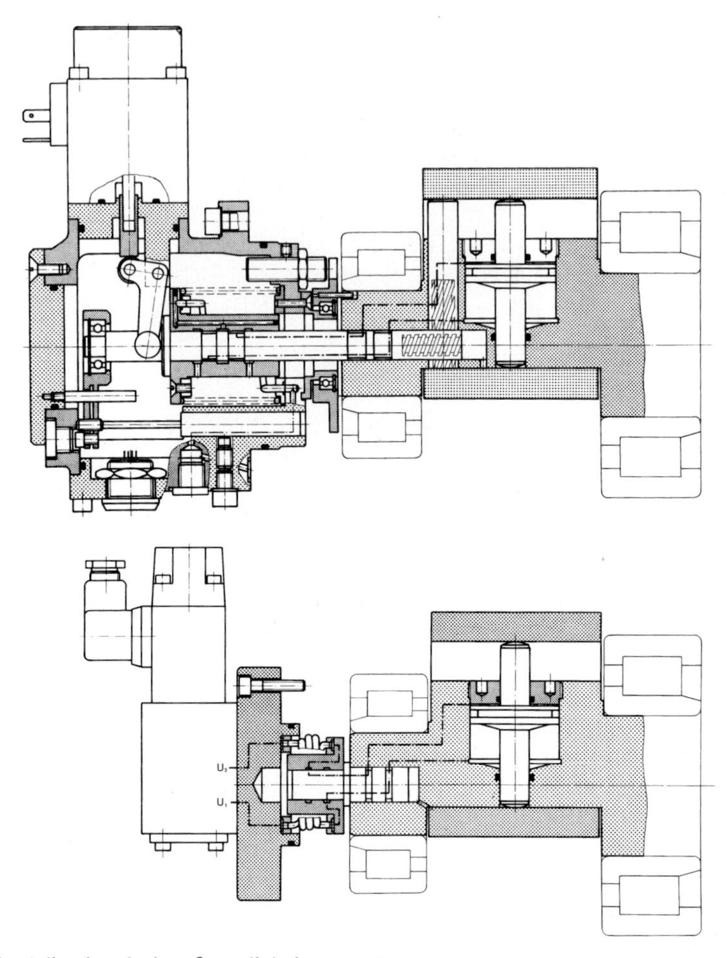

Fig. 102 Adjusting devices for radial piston motors

5.3.2 Hydraulic Pumps

Hydraulic pumps are an integrated component of the drive unit in injection molding machines. The direct single drive is standard design. Central water-hydraulic systems as energy supply are rare and can be met in only a few shops. Hydraulic pumps convert the mechanical energy of the almost constant-speed rotations of an electric motor and the corresponding torque into hydraulic energy:

$$k_1 T_1 \times N = k_2 \times V \times p_H$$

with k_1 and k_2 as constants, T_1 as torque of the motor shaft, N as speed, V as displacement of the pump, and p_H as hydraulic pressure. One can recognize that, by varying V or p_H one has two options to accommodate the demands of energy-consuming devices. This simple adaptation has made the hydraulic pump the preferred drive unit, wherever there is a periodic and local variable energy demand as is the case with clamping units, injection pistons, ejector systems, and others. Most common pump design in plastics processing is demonstrated with Fig. 103 to 110. Diagrammatic symbols for the various types are summarized in Fig. 111 (25).

The noise level of pumps is of particular concern (Table 21). The maximum permissible level is presently 85 dB(A) in Europe. Many molding facilities have reduced this limit to 75 dB(A) as an in- house guide line in anticipation of stricter future regulations. In the United States, OSHA (Occupational Safety and Health Act of 1970) requirements call for a noise level limit of maximum 90 dB(A) at a distance of 3 ft. from the noise source. A reduction to 85 dB(A) is being considered and is, therefore, already requested by large sectors of the industry. Generally speaking, an injection molding machine with a noise level of more than 80 dB(A) can be called noisy and one with less then 75 dB(A), quiet [14]. The data are valid for full load. Almost all pumps on the market realize the lower level.

Great importance is also attached to efficiency at high pressure. The total efficiency should be about 90% or higher at maximum operating pressure. High efficiency of the pump and low losses from throttle effects contribute to little heating of the oil. This not only is favorable with respect to energy utilization, but also enhances the service life of various elements in the hydraulic system and the reproducibility of the machine operation. The pump efficiency n_{tot} (Table 21) should not be confused with the energy efficiency of the hydraulic system as a whole. The system efficiency of injection molding machines is between 6 and 25% [194, 206]. The closer to full load the machines operate, the better is the efficiency.

Hydraulic pumps in injection molding machines should always be positioned below the fluid level. This can be done in the tank or, as well or even better, outside the tank. Such an arrangement provides a positive potential pressure differential. This differential, plus the dimensioning of pipelines for a maximum flow velocity of 1.5 m/s and the use of a minimum of well-rounded elbows reduce the negative pressure to a minimum and contribute to better energy utilization. A filter in the intake line should be used only if the higher pressure differential from contamination can be recorded clearly and reliably.

The drive unit should always be mounted in such a manner that no unacceptable noise is conducted through pipelines and machine support. This can be achieved rather easily by rubber cushioning and hose connections near the source of energy, that is the pump. Modern hydraulic drives should have preheating equipment, which provides the temperature level intended for continuous operation already before the start up of production. They should also be furnished with efficient oil heating and temperature control during operation. Thus, problems resulting from viscosity variations are eliminated from the start and uniform production capability enhanced.

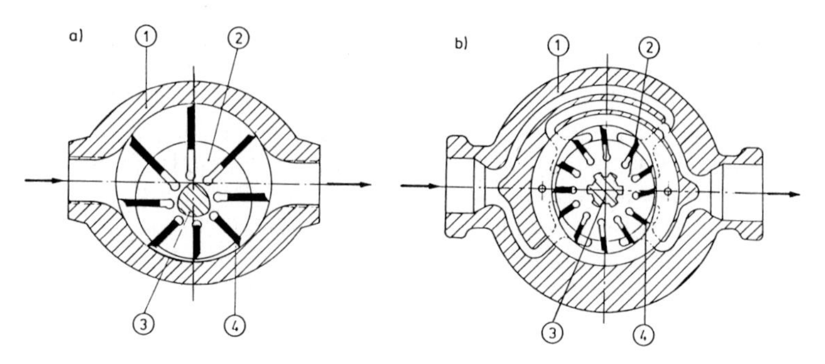

Fig. 103 Rotary vane pump (System Bosch)
a: Guided-vane pump with eccentric rotor, b: Guided-vane pump with cylindric rotor and oval body, 1. Body, 2. Impeller, 3. Drive shaft, 4. Vanes

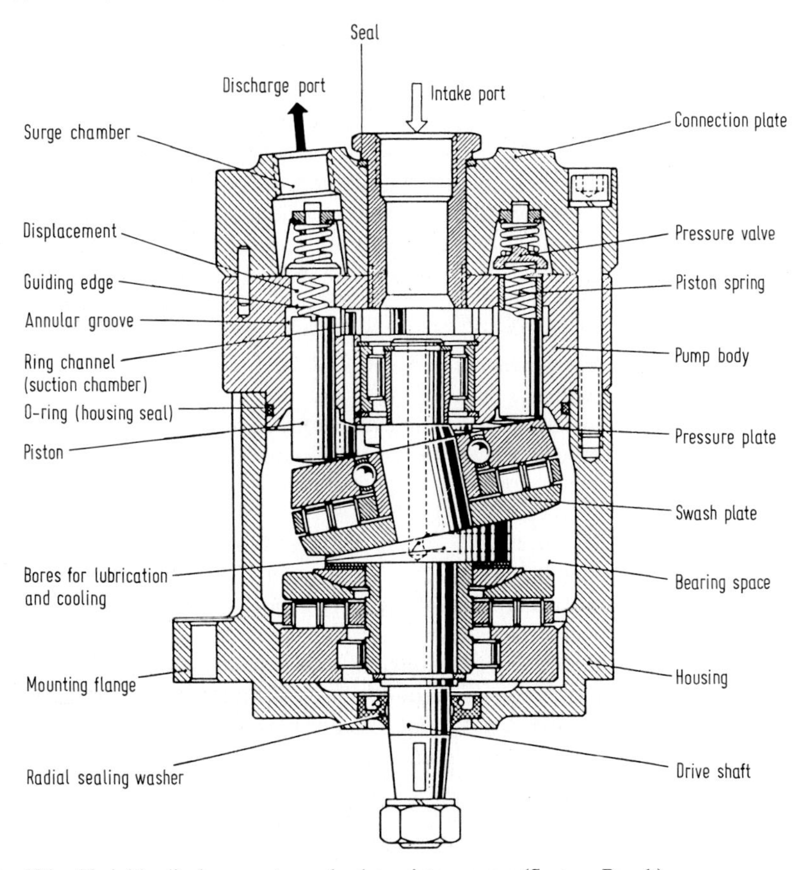

Fig. 104 Variable-displacement swash-plate piston pump (System Bosch)

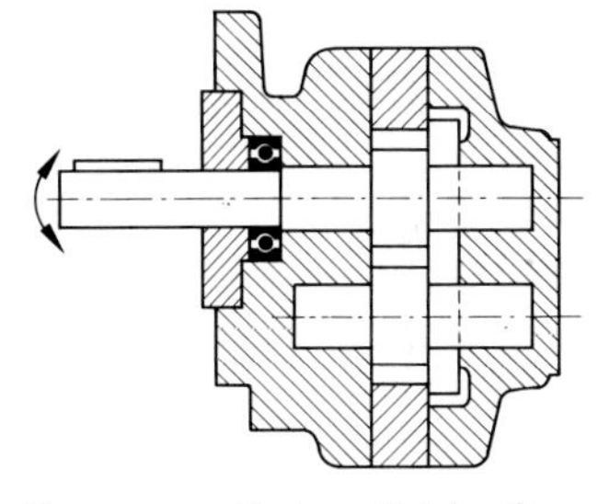

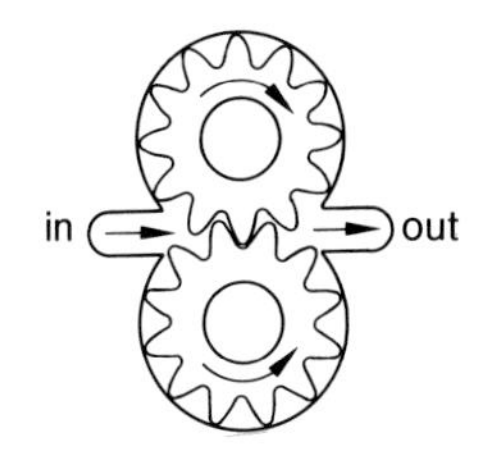

Fig. 105 Gear pump (System Reichert)

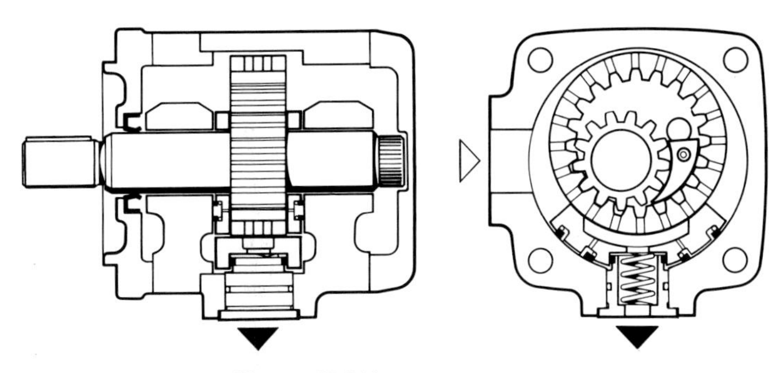

Fig. 106 Internal gear pump (System Voith)

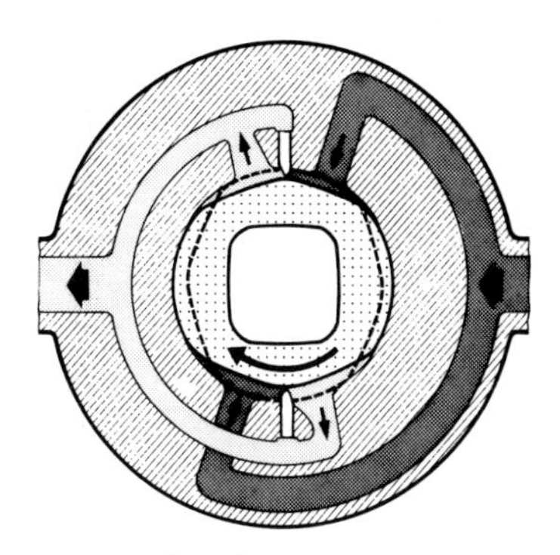

Fig. 107 Rotary cam pump (System Sauer)

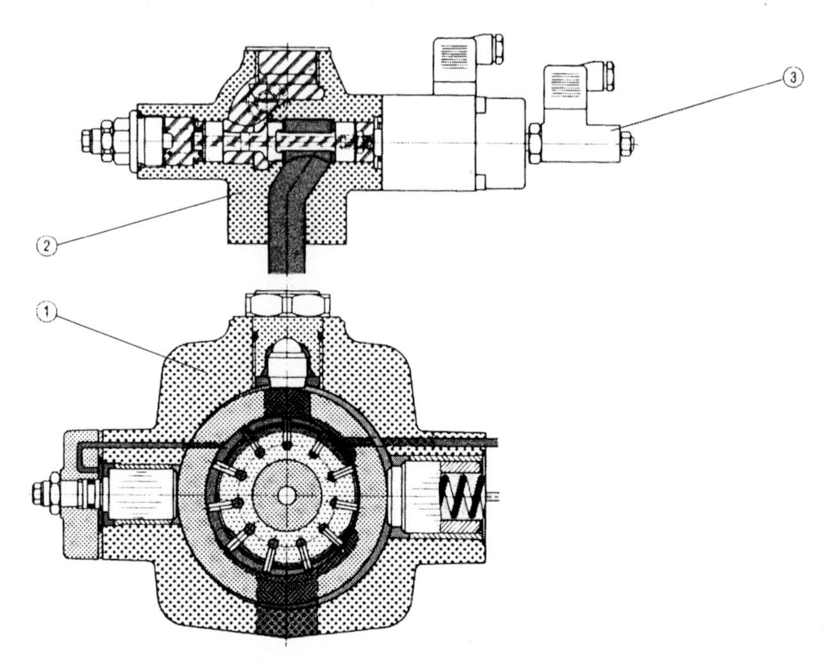

Fig. 108 Variable-displacement vane pump (System Mannesmann-Rexroth Hydromatic)
1: Pump body, 2: Port control for pressure adjustment, 3: Inductive displacement control for feed back

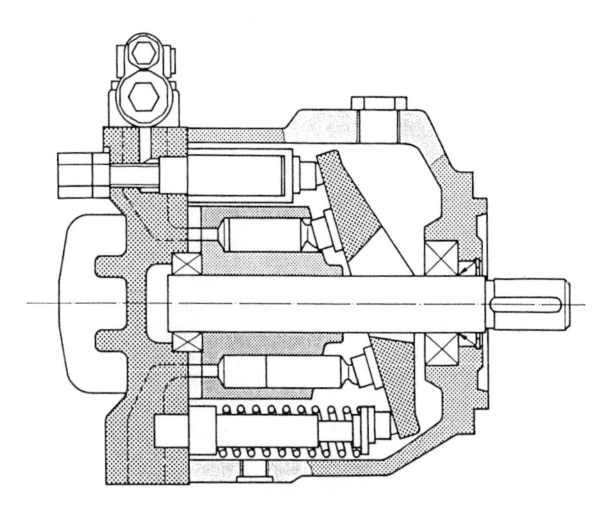

Fig. 109 In-line variable-displacement piston pump with slanted disk (System Mannesmann-Rexroth Hydromatik [25])

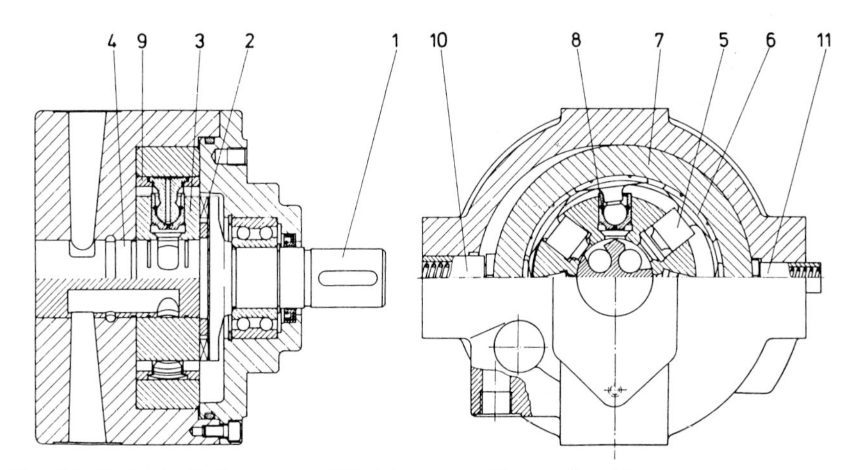

Fig. 110 Variable-displacement radial-piston pump (System Bosch)
1: Drive shaft, 2: Cross-type disk coupling, 3: Body carrying pistons, 4: Control pin, 5: Piston, 6: Pressure guides, 7: Drive ring, 8, 9: Thrust rings, 10, 11: Adjustment pistons

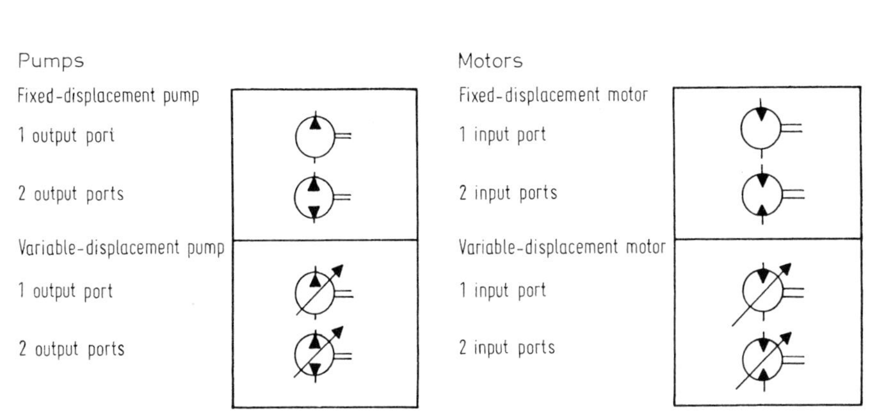

Fig. 111 Diagrammatic symbols for fixed-displacement pumps and motors and variable-displacement pumps and motors

Table 21 Efficiency n_{tot} and Sound Pressure dB(A) of Various Pumps and with Different Pressure at 1500 rpm

Features		Fixed-displacement pumps				Variable-displacement pumps			
		Vane pump	Axial-piston pump	Radial-piston pump	Circumferential gate-valve pump	Gear pump	Internal-gear pump	Radial-piston pump	Vane pump
n_{tot}	7 MPa	76–85	85–87	85–87	91–92	77–80	90–92	82–86	76–82
	14 MPa	75–84	90–92	87–90	88–90	76–78	88–90	88–90	78–86
	21 MPa	74–82	91–93	87–90	87–88	74–76	87–89	86–88	-
Noise level dB(A)	7 MPa	60–70	78–80	76–80	~ 62	74–80	70–73	65–67	65–68
	14 MPa	72–75	80–83	77–81	~ 65	75–82	~ 76	69–71	65–70
	21 MPa	76–80	80–86	79–83	~ 67	76–85	~ 77	72–74	-

5.4 The Hydraulic System

All power-consuming units (cylinders, motors) of an injection molding machine are mostly regulated by proportional pressure- and flow-control valves. High demands are made on an exact metering of the produced forces and velocities. Since the execution of some functions requires great energy and the system inherently has high losses, a careful design of the electrohydraulic drive is mandatory. One differentiates between functions of motion and of work. Motion, which calls for a large oil-flow rate but only low pressure, includes all the displacements of the clamping and injection units and plastication. Holding pressure and the build- up of clamping pressure are work functions, occasionally with high pressure demands, but requiring only a small oil volume. The injection stage generally calls for both, a large oil volume and, at times, high pressure (Fig. 113).

A block diagram of the hydraulic system (Fig. 112) shows the power-consuming units in an injection molding machine. Fig. 114 gives an account of supply and control of an injection unit in principle. The rotatory screw drive is a hydraulic motor. Screw speed is determined by a proportional flow-control valve (1) with the four-way directional control valve (6) in position a. The backpressure caused by the rotating screw moves the injection piston to the right. Oil is released through a controlled pressure-relief valve, a combination of (4) and (2). The magnitude of backpressure is set by valve (2). The four-way directional control valve (5) is in position b. During injection a pressure-reducing valve, a combination of (3) and (2) determines the injection pressure. The four-way directional control valve (5) is in position a. Injection speed is controlled by the proportional flow-control valve (1). The four-way directional control valve (6) is in position b. The holding pressure is set with valve (2). This drive system can be modified by appropriate selection of electric motors, hydraulic pumps, and kind of flow and pressure controls. With regard to function and utilization of energy, it is mandatory to

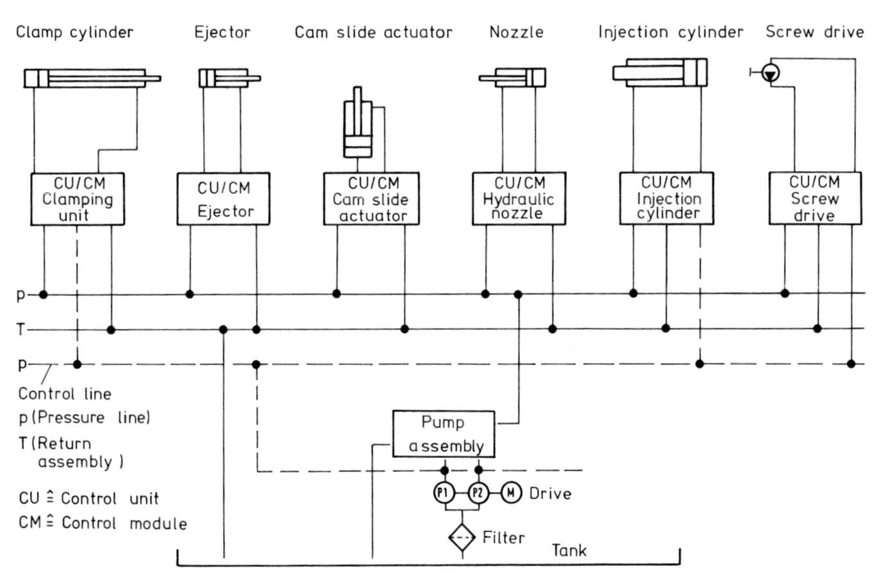

Fig. 112 Block diagram of the hydraulic system of an injection molding machine

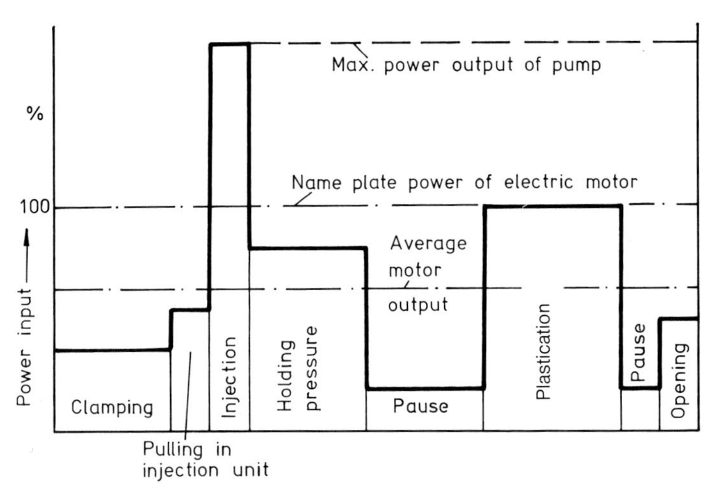

Fig. 113 Power input of an injection molding machine during a typical cycle in comparison with the name plate power of the electric motor and with the maximum pump output during motor running above capacity

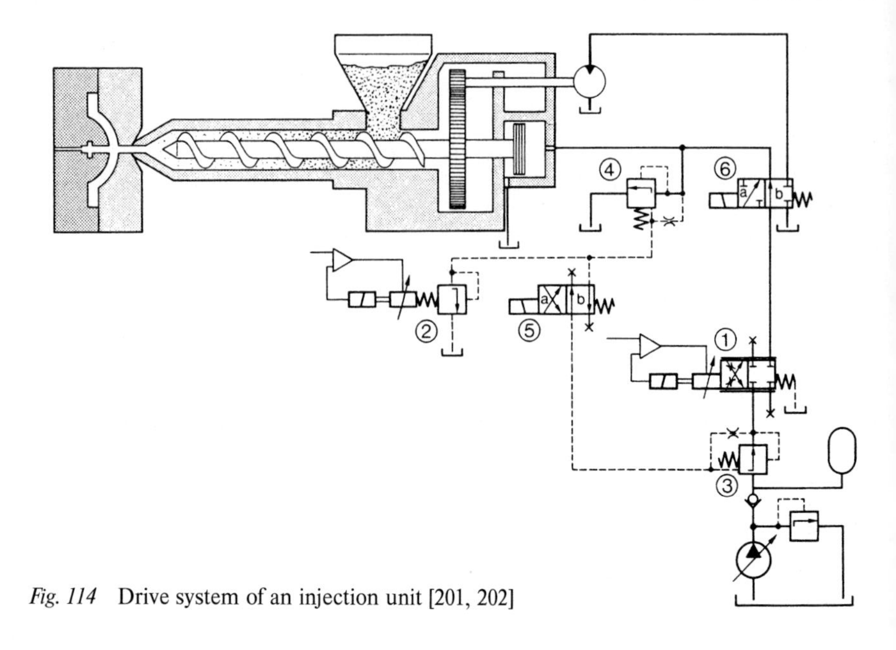

Fig. 114 Drive system of an injection unit [201, 202]

strive for an adjustment of the power input to the respective demand by appropriate control. Power losses should be kept as low as possible.

Fig. 115 provides a chart of partly realized, partly conceivable drive systems of the injection unit (feeding and injection). Assuming that the pump operates during the whole cycle against the maximum pressure, then losses occur in pressure- and flow-control valves. These losses are equal to the whole difference between the demand of each step of the process and the maximum power input (Fig. 116).

The amount of power loss, not considering efficiency, is:

$$E_L = p_P \cdot Q_P - p_O \cdot Q_O$$

with p_P = pump pressure, Q_P = oil-flow rate of pump, p_O = pressure demand of operation, and Q_O = oil-flow rate of operation.

If this correlation is applied to each stage of a cycle, according to Fig. 115, then it becomes evident that:

- Power input to the pump of the same magnitude as the power demand for injection is uneconomical. Solution: Use of accumulator for injection or operating electric motor above capacity, maximum 160%.
- Power input has to meet the demands of holding-pressure and plasticating stages, because both are relatively lengthy.

- Reduction of power input to the pump is indispensable during stages of low power demand. Injection molding machines are operated with only 35% efficiency on an average. The power loss may be between 40 and 95%.

Machine manufacturers and processors are thinking more and more about energy savings in injection molding. A number of studies provide informations about actions directed to this goal [202 to 221]. Some proven solutions will be discussed here without going into the details of all existing or conceivable possibilities (Fig. 117).

The most favorable solution for a drive control with respect to energy conservation would be to supply as much power as needed for each stage during a cycle. There are two routes to this. The first one takes too much effort; it would require a large number of control elements to constantly adjust pressure and flow rate to the ever-changing demand during one cycle or various cycles. This does not even consider that the product of all single efficiencies would reduce the resulting positive effect by a considerable degree. The second route requires a continuous adjustment of the hydraulic pump to the demand, and control of the oil pressure.

The following drive systems are commercially important:

a) Single fixed-displacement pump with two-way flow-control valve;
b) Single fixed-displacement pump with three-way flow-control valve (eventual injection with accumulator);
c) Two-stage fixed-displacement pump with three-way flow-control valve (eventual injection with accumulator);
d) Variable-displacement pump;
e) Speed control (oil flow control) with speed-controlled electric motor and fixed-displacement pump [175];
f) Accumulator for use with all or the majority of actuators.

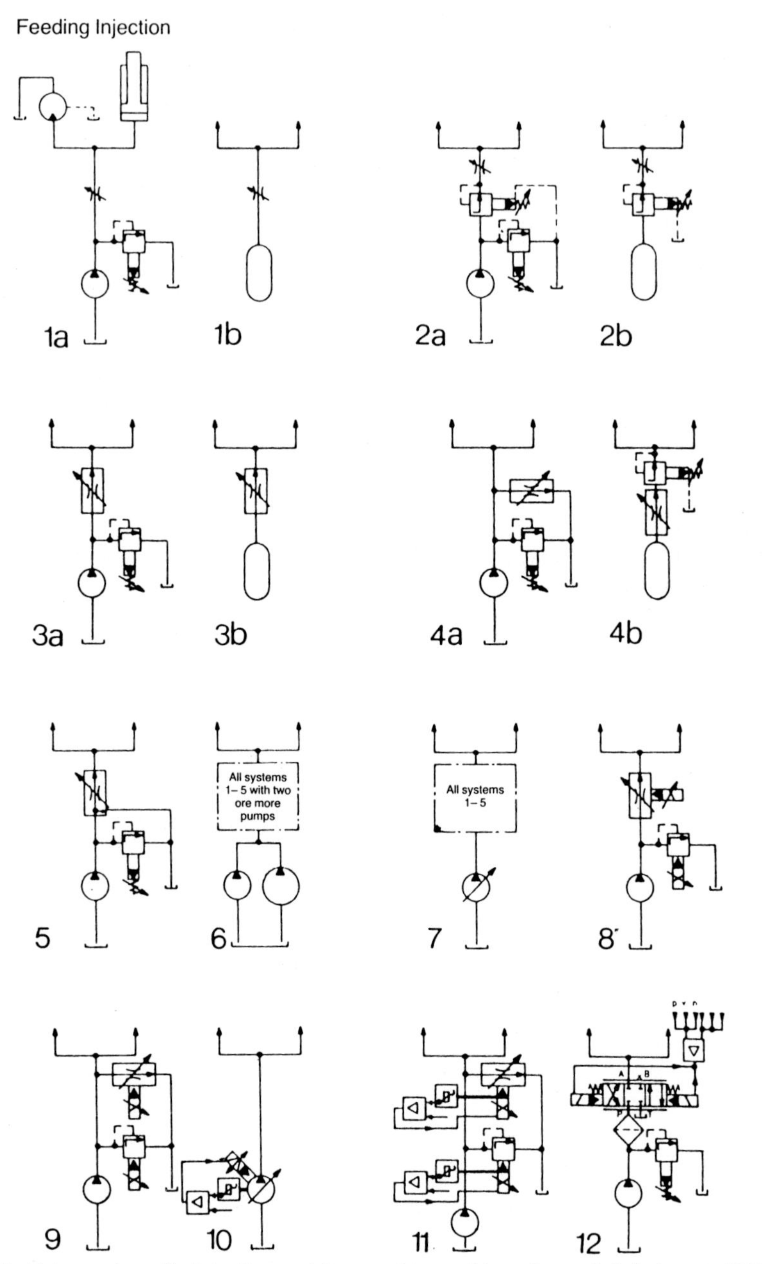

Fig. 115 Drive systems for injection molding machines with analog and digital control [213]. Description see page 131

Common analog systems:

(1a) Fixed displacement pump with throttle and pressure relief valve.
(1b) Accumulator with throttle.
(2a) Fixed displacement pump with pressure relief valve, pressure-reducing valve, and throttle.
(2b) Accumulator with pressure-reducing valve and throttle.
(3a) Fixed displacement pump with pressure relief valve and flow control.
(3b) Accumulator with flow control.
(4a) Fixed displacement pump with pressure relief valve and flow control in by-pass.
(4b) Accumulator with flow control and pressure-reducing valve.
(5) Fixed displacement pump with pressure relief valve and three-way flow control.
(6) All systems 1-5 combined with two or more pumps.
(7) All systems 1-5 combined with a variable displacement pump.
(8) Fixed displacement pump with electric remote control of pressure relief valve and of flow control.
(9) Fixed displacement pump with electric remote control of pressure relief valve and of flow control in by-pass.
(10) Variable displacement pump with electric remote control.
(11) Fixed displacement pump with electric remote control of and feedback from pressure relief valve and flow control in by-pass.
(12) Fixed displacement pump with pressure relief valve, pressure and flow control servo valve and microfilter.

Common digital systems:

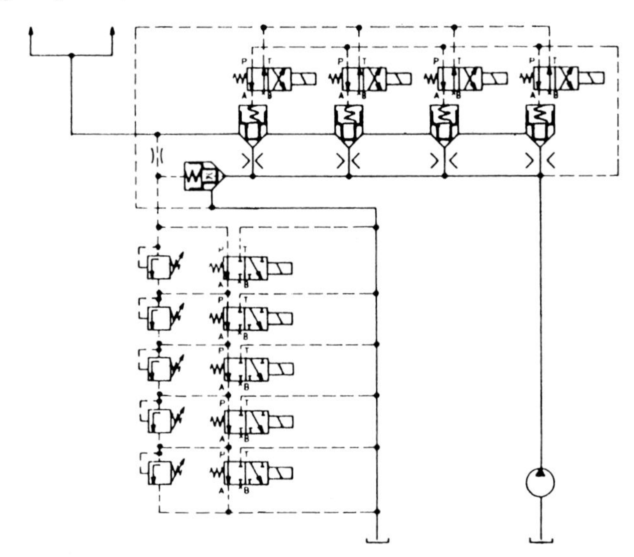

(13) Fixed-displacement pump with digital thrcc-way flow control, with cartridge valves and digital pressure relief valve.

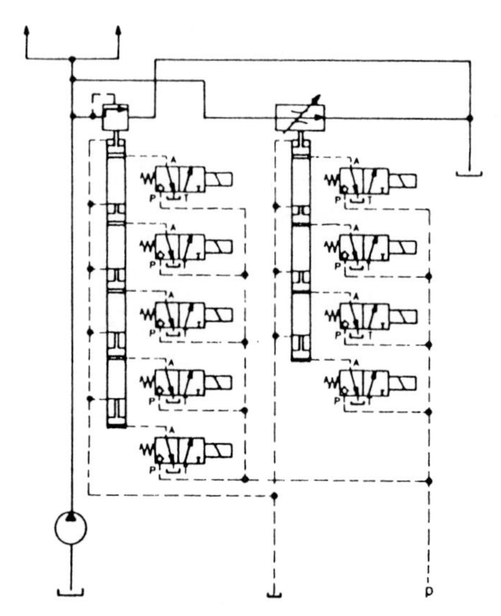

(14) Fixed displacement pump with pressure relief valve and flow control in by-pass, operated by pilot valves in binary code.

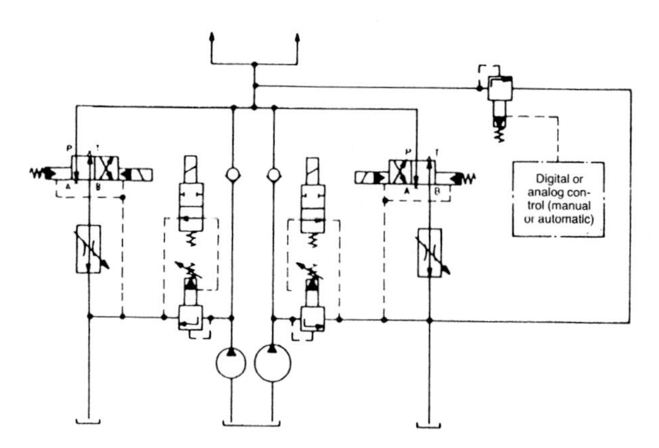

(15) Two fixed displacement pumps with a defined output ratio and with two variable flow controls in by-pass which unload defined volumes to tank.

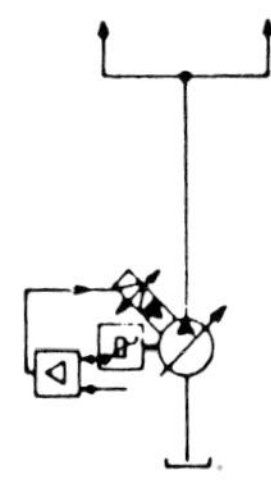

(16) Variable displacement pump with digital electric control.

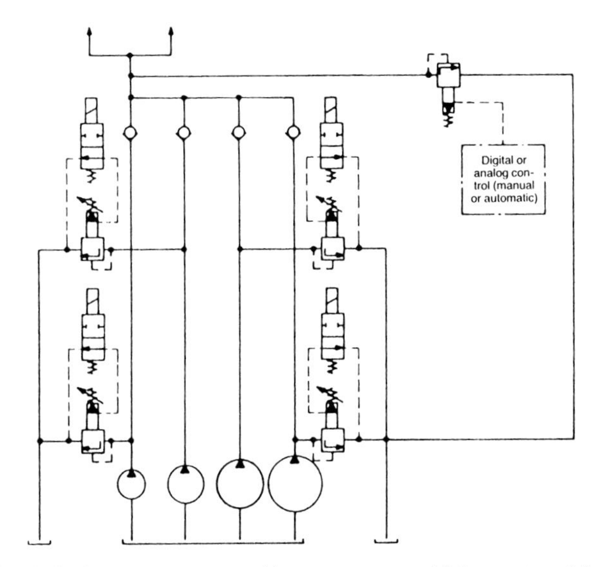

(17) Four fixed displacement pumps with output rates which are staged in a geometrical progression and can be combined as needed.

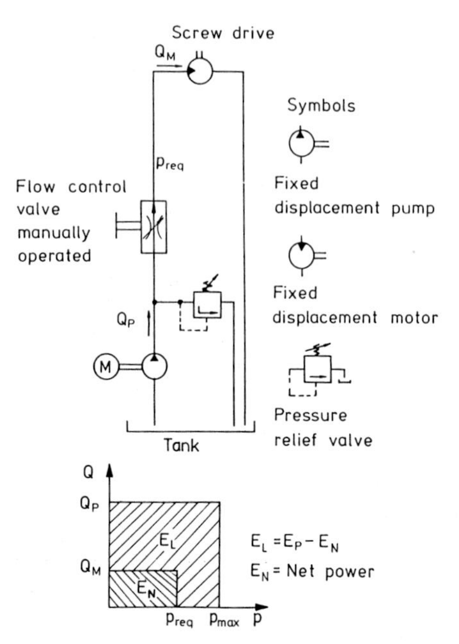

Fig. 116 Hydraulic screw drive system with manually variable throttle in main line and maximum pressure relief valve in secondary line [212]

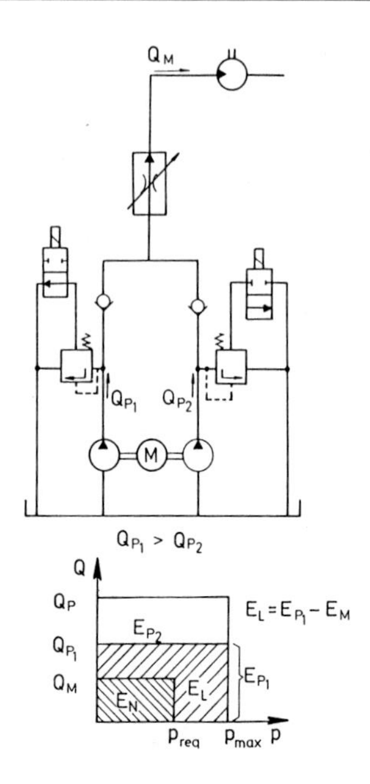

Fig. 117 Hydraulic drive system of an injection machine with two fixed displacement pumps and manually operated flow control valve in main line

5.4.1 Fixed-Displacement Pumps

The fixed-displacement pump supplies a constant power output:

$$E_P = Q_P \times p_{max} \times n_e \tag{12}$$

If, at times, only part of this power output is required as in Fig. 113, then valves must be provided for power control. This is depicted in Fig. 116 for a hydraulic motor as screw drive. Flow control and pressure-reducing valves ensure a power supply to the motor of the magnitude

$$E_M = Q_M \times p_{req}. \tag{13}$$

This results in a considerable power loss during normal operation:

$$E_L = E_P - E_M \tag{14}$$

Such a poor utilization of energy with a two-way flow-control valve and a fixed-displacement pump can only be justified for mini machines today. A better adjustment to power demand can be achieved by arranging the flow-rate control in a by-pass or with a three-way valve. In the latter case, the pump generates only the required working pressure. Nothing else can be operated with this pump if it needs a higher pressure than the one determined by the pressure-control valve. This is of disadvantage to the overall operation and is not used anymore with modern machines.

An improvement can be made with two-stage or multi-stage pumps. Such pumps are employed as modules, sometimes with incorporated controls. This solution is chosen for machines with more than 500 kN clamping force as standard. Two pumps are sufficient to operate the clamping unit and supply the rotatory screw drive at the same time. Sometimes one can do without a conventional directional control of the oil flow during the cycle if each of several pumps is positively assigned to its own hydraulic circuit. The supply of each circuit with only the pressure needed, results in distinct energy savings. During intermittent periods of no demand, the unemployed pump returns the oil to the tank without pressure. During injection, however, all pumps can act jointly on the injection cylinder.

The performance diagram of a drive with two fixed-displacement pumps is presented in Fig. 117. If the power loss from idling pump 2 is not considered, the power loss, compared with a single-pump drive, is reduced by

$$E_{P2} = Q_{P2} \times p_{max} \tag{15}$$

because

$$E_L = E_{P1} + E_{P2} - E_{P2} - E_M \tag{16}$$

where E_{P1} is the power output of pump 1 and E_{P2} output of pump 2.

Another improvement in energy efficiency is achieved by arranging a variable flow control in a by-pass or by a three-way flow-control valve. With the latter method, another unit in the system can be supplied. Since every pump circuit has its own pressure relief valve, a supply with the individually required oil pressure is feasible:

$$p_{P1} \neq p_{P2} \text{ or } p_{P1} = p_{P2} \tag{17}$$

With two pumps, the first step has already been taken towards a digital oil supply. Heavy machines often have sets of two three- or four-stage pump modules. Four pumps permit a digital flow-rate control of $2^4 - 1 = 15$ steps according to the binary code. The individual flow rates have to be suitably correlated (Fig. 115, design 16). Thus, a favorable and efficient energy adaptation to the demand can be achieved.

5.4.2 Variable-Displacement Pumps

A very good adaptation to energy demand can be obtained with a variable-displacement pump with fast response. The power loss is

$$E_L = E_P - E_e \tag{18}$$

The power loss can approach zero if, during the cycle, the power output of the pump E_P is precisely adjusted to the continuously varying effective power E_e, and the response time is less than 10 ms. This assumes a very exact flow control. An efficiency of 1 holds true only for the idealized case, which does not consider pump efficiency and losses in pipes and from branching. The total efficiency of variable-displacement pumps is between about 50% (with 20% displacement) and almost 90% (with 100% displacement). The supply of a hydraulic motor for a screw drive by a variable displacement pump is schematically pictured with Fig. 118. The problematic nature of these pumps, although most favorable from the viewpoint of energy efficiency, is due to their slow response and their high noise level. Digital control by electromechanical means can be achieved reliably and reproducibly. This provides a high degree of reproducibility for the whole process, which is a prerequisite for quality injection molding.

High noise level and increasing losses from no-load operation places limitations to this drive system in mid-sized and heavy machines [203, 214, 215]. Here, multiple fixed-displacement pumps have a definite advantage since a multi-stage arrangement of variable-displacement pumps is not used because of costs and, above all, dynamic problems.

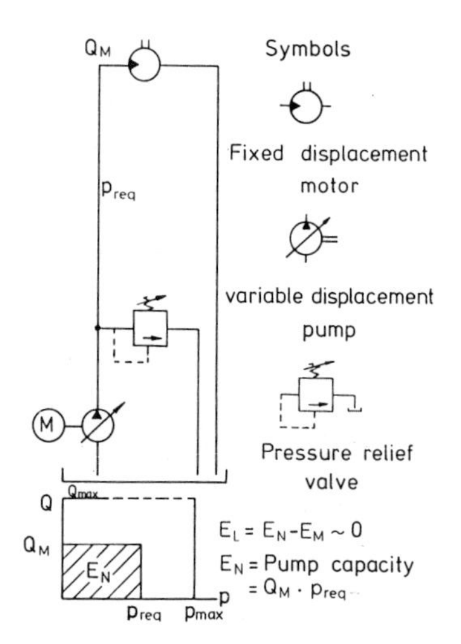

Fig. 118 Drive system of an injection molding machine: Power supply of a screw drive with variable-displacement pump [180]

5.4.3 Accumulators

Accumulators have found a firm place in modern injection molding machines. They have proven useful as pressure reservoirs with high output capacity for rapid injection. Most machine manufacturers offer accumulators as optional equipment for this task. They are generally of the bladder design, which, when combined with a manifold, can be accommodated as a module (Fig. 119). Such hydraulic energy storage is always useful if the ratio between maximum and average demand for power or oil volume is very large [216]. Many injection molding machines have been successfully equipped with a combination of accumulator and pump drive. The accumulator can supply power for several functions, such as movement of the injection carriage, action of the hydraulic ejector, and, above all, injection piston. In some cases, accumulators are directly associated with constant-pressure functions. In contrast to a variable- or fixed-displacement pump, the accumulator suffers a pressure drop during injection. Thus, the capacity of the accumulator has to be chosen in accordance with its function and in such a way that the pressure remaining after discharge (after injection) corresponds with the maximum pressure demand of the remainder of the cycle (compression and holding pressure). This pressure should be sufficiently high to meet the requirements of Table 3 and provide adequate holding pressure.

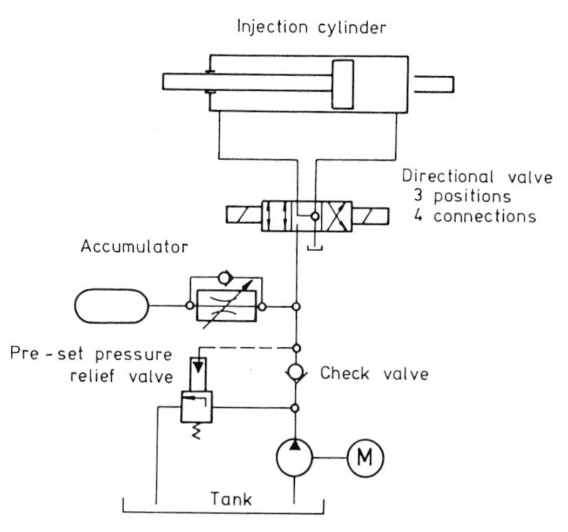

Fig. 119 Hydraulic drive system of an injection molding machine with accumulator for injection [180]

There are three possible ways of achieving this:

a) the accumulator pressure is appropriately high;
b) a smaller screw diameter is selected, which provides the same pressure reserve;
c) the cross section of the hydraulic cylinder is increased, as compared to the pump drive. This results in a high theoretical injection pressure like b) but with larger screw diameter.

Fig. 120 illustrates that maximum pressure and volume of the accumulator have a definite effect on the available pressure level. If the maximum injection pressure is to be effective at the end of the injection stage, the maximum accumulator pressure should be about 25% above the remaining pressure after about 12% volume drain. The capacity limits can be established approximately as follows:

Accumulator drain ~ 10% -> remaining pressure ~ 180 MPa
Accumulator drain ~ 20% -> remaining pressure ~ 150 MPa

Although some attention must be paid to the available injection pressure from an accumulator, the associated problem can be resolved easily by correct dimensioning.

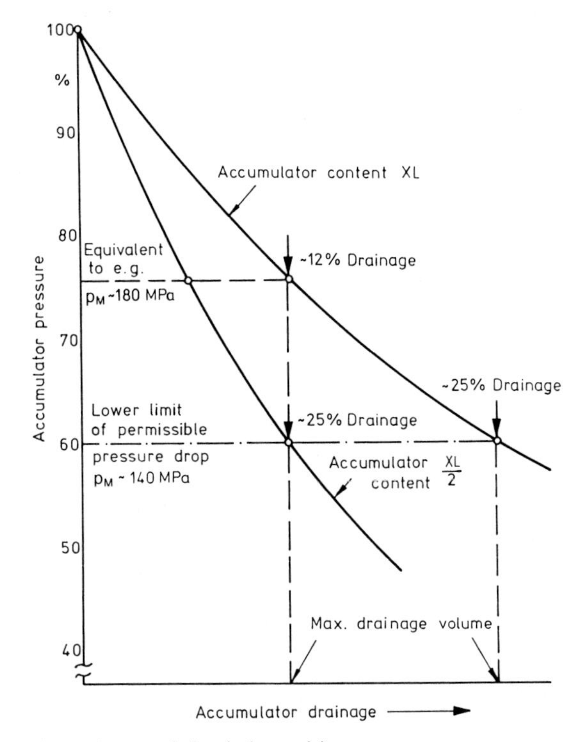

Fig. 120 Accumulator characteristics (schematic)

The oil intake from the accumulator is controlled either by a proportional or by a servo valve independent of pressure.

5.5 Hydraulic Controls

Whenever the power output $E = Q \times p$ cannot be changed by varying the pump output, and this is normally the case, valves must be provided. Besides their function as plain capacity control, they have to meet the demands on switching characteristics, reproducibility, and ease of control.

The generated flow of the hydraulic fluid has to be switched into another direction, divided, or blocked; its flow rate has to be changed or brought to a lower or higher pressure level. These tasks are performed by a number of valves with very different functions. Depending on their effect on the hydraulic system, they can be divided into directional-, pressure-, and flow-control valves. Flow-control valves can be subdivided into steady and unsteady ones.

It would go too far, at this point, to outline individual features and modes of function of all hydraulic control components. A summary is presented with Table 22. Only functions and responses of such flow- and pressure-control valves, which are instrumental in injection molding for a controlled effect on the process, shall be portrayed in more detail.

By now, almost all valves of an injection molding machine are actuated electromechanically by solenoids. Thus, a central setting is possible at the control panel of the machine and the reproducibility, as far as it can be affected by man, is considerably improved.

Table 22 Fluid Power Controls in Injection Molding Machines

Component	Symbol and function	Operation
Directional valve (electrically or electrohy-draulically actuated)	A B / P T — Four-way, three-position valve (center condition: pressure closed, A and B open to tank)	Control of machine operation, such as start, stop, reverse direction
	A B / P T — Four-way, three-position valve (center condition: all ports closed)	
Check valve (one-way flow or in either direction)	Check valve with spring	Blocking flow in one direction, eventually creating pilot pressure
	Pilot-operated check valve	
Flow control valve (flow rate control), pressure and temperature compensated	Throttle with fixed restriction	Speed control of actuators
	Throttle with variable restriction	
	P / A — Throttle with variable restriction, manually operated	
Flow control valve (flow rate control), pressure and temperature compensated	Flow control valve with fixed restricition and pressure compensation	
	Flow control valve with fixed restriction and pressure relief to tank	Speed control of actuators
	Flow control valve with variable restriction and pressure relief to tank	

Table 22

Component	Symbol and function	Operation
Pressure control valve (usually direct operated poppet valve)	Pressure relief valve Pressure-reducing valve, adjustable Pressure-reducing valve, adjustable, permitting reversed flow and unloading to tank	Pressure limitation (safety function); Pressure setting (injection pressure, holding pressure, etc.)
Servo valves (proportioning valve)	Example: four-way valve with infinite positioning, electrically operated and spring centered Example: five-way valve with infinite positioning, electrohoydraulic pilot operation and spring offset Example: four-way valve with infinite positioning, electrohydraulic pilot operation and spring centered	Closed loop control of injection and holding pressure through feedback of cavity or hydraulic pressure
Catridge valves	Various functions as above	Various operations but different design

5.5.1 Flow-Control Valves

In order to influence the oil flow, throttle and flow valves with two- or three-way functions are available. Although they can be operated manually, electromechanically controlled valves are employed most often. Such devices are:

- Directional-control valves for digital control.
- Proportional valves (good proportionality between direction and flow rate; mostly adequate dynamic features).
- Pilot-operated valves with a high dynamic range.

As actuators, either directly or through pilot operation, the following devices are employed:

- Solenoids (small hysteresis band, good linearity, small controlling force).
- Torque motors (good dynamic features, broad hysteresis band of 2.5 to 5%) [177]. (Torque motors have no significance in practice anymore.)

Most of the time, the force produced by electromechanical converters actuates a hydraulic intensifier.

Fig. 121 presents a summary of various flow-control valves which are in use today. The first three lines provide information about the kind of adjustment and oil flow-rate control as well as suitability for open- or closed-loop control, which are important features for the processor.

Throttle control of flow rate generally is no longer employed except where pressure dependency and high power losses are of little importance (small oil volumes). Pressure dependency is prevented if a pressure-differential valve is added. It balances the pressure and should be installed on the side where the larger pressure fluctuations can be expected.

A pressure-relief valve has to be installed parallel to such a pressure-compensated two-way valve to relieve the unused oil flow to the tank. A three-way valve has pressure compensation and throttle arranged in parallel and has less energy loss. The unused oil returns through the throttle to the tank.

Disturbances from leakage, viscosity variations, and pressure losses cause the actual flow through control valves to be different from the set flow-rate value. In very fast operations, e.g. injection, discontinuities occur because of a response time of about 200 to 300 ms [179]. Thus, for rapid functions, such as step-wise programmed injection or ejection speed, servo valves are recommended. They are spool valves which permit changes in flow rate as well as direction of flow and are small in order to achive a good dynamic behavior. They are expensive, because they have to meet high precision rquirements. Less costly is the pilot-operated proportional valve (Fig. 122) [181, 182]. It usually meets the demand of the hydraulic system of an injection molding machine with respect to eigenfrequency, hysteresis, and linearity (Table 24). Its quality level is disputed, though, and certainly not as good as to render servo valves superfluous for special requirements [182, 193, 198]. The so-called servo valves are fast

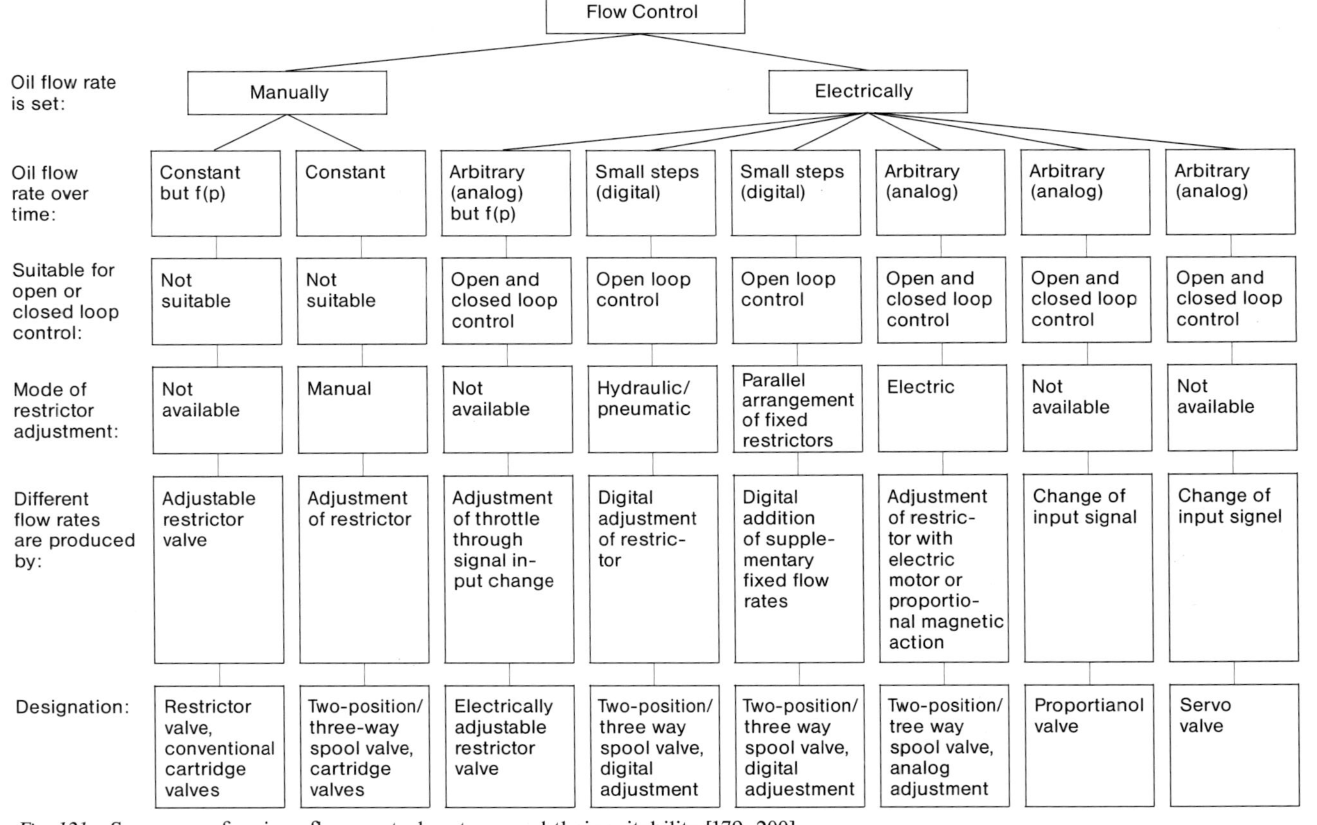

Fig. 121 Summary of various flow control systems and their suitability [179, 200]

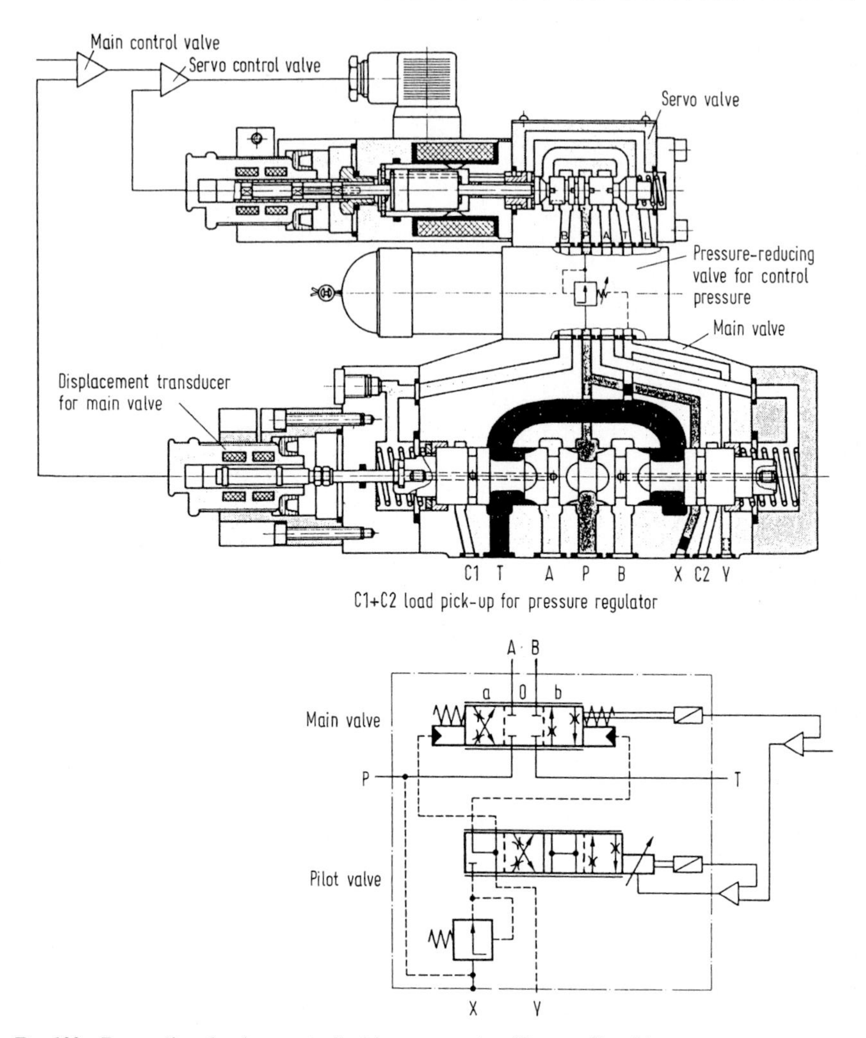

Fig. 122 Proportional valve controlled by servo valve (System Bosch)

operating proportional valves. The usage of this term is not always clear. As control in the main line, chosen for reasons of accuracy of the sequence control, they exhibit very high losses, which mostly depend on the size of the valve. Process control cannot do without these valves, though.

Table 23 Features of Servo, Proportional, and Digital Hydraulic Controls [182 to 185, 189]

Criteria	*Servo valve* Proportional valve, pilot-operated, in closed circuit	*Proportional valve* Position-controlled proportional valve in open circuit	*Digital control* Control by fixed value steps in binary code
Mode of function	A magnet, which actuates a valve, is supplied with a voltage proportional to the pressure. Actual pressure and flow rate are measured, compared with the set value, and adjusted.	A current is supplied to magnet. The magnet opens and closes a valve. Position of magnet is inductively monitored and adjusted.	Oil flow in digital mode with restrictors, which are opened or closed corresponding to the selected values. Addition of pressures or volumes in binary code.
Accuracy:			
Hysteresis (pressure)	< ± 0.5%	~ ± 1% (2)	~ ± 0.5%
Hysteresis (volume)	< ± 0.5%	~ ± 2%	< ± 0.5%
Accuracy of repetition	< ± 0.5%	= ± 1%	< ± 0.5%
Drift after startup	possible	occurs, re-adjustment recommended	slight
Switching time			
Pressure	30 ms	80 ms	60 ms
Volume	30 ms	100 ms	80
Effect of oil viscosity T_{oil} = const	slight	yes	yes

Instead of spool valves, cartridge valves are used on an increasingly large scale to control oil flow. They are positive-seating, two-port valves distinguished by an especially simple and lightweight design. They can be housed in control blocks as an essential component of modular design. Thus, the number of leakage points is reduced. If there is special interest in achieving short delay times and smooth switching, then cartridge valves meet these requirements very well because of their compact design and steady opening characteristics. They even present a number of additional advantages, such as leak-proof seats and protection against floating, low loss from heat, short switching periods, reliable switching performance, good hydraulic efficiency, and little noise. They are suitable for all modes of oil-flow control and are employed particularly in digital control.

5.5.2 Pressure-Control Valves

Fig. 123 presents a summary of options for designing and operating pressure-control valves. Only pilot-operated valves are considered. The first three lines provide important information about manual or electrical control. The significance of the electrically or electrohydraulically operated pressure valve is on the increase because the reproducibility of the pressure setting is very important and has generally taken priority over continuously variable control. The electrical components are the same as for flow-control valves. Such valves have a hysteresis band of about ± 3%. This can be reduced to less than ± 1% by superimposing a dither signal and using a position-control circuit with displacement transducer. Preset pressure-relief valves with narrow hysteresis bands are needed if pressure-control circuits are installed (variable holding-pressure control).

5.5.3 Digital Control Elements

Because of their nature, variable-displacement pumps, pressure-relief and control valves, and flow-control valves provide analog functions. If the demand for reproducibility outweighs the demand for high resolution, then digital setting and control are of advantage. This is usually the case. Nowadays, digital control components can be found for flow and pressure control in numerous variations. One has to distinguish between the digital operation of the individual pressure, flow-rate, or directional control and the direct sequential control of a set of valves in a digital mode. This control by steps uses the binary code:

$$N = 2^n - 1$$

where N is the number of steps and n the number of controlled elements.

The digital volume control is achieved by

- digital control of variable-displacement pumps (Fig. 124). This solution is economical because the pump output can be adjusted to the demand. The load dependency increases, however, and results in a drop of efficiency before the maximum load is reached. Advantage: Adjustment in small steps according to the electrical signal.
- combination of several pumps, the output of which can be added to the hydraulic circuit in a binary mode, depending on the demand. Advantage: Little load dependency. Disadvantage: Small number of steps N.
- digital setting of flow-control valves (pressure and temperature compensated).
- parallel arrangement of fixed restrictors or cartridge valves with predefined flow rates, pressure compensated (Fig. 125).
- digital control of pilot-operated valves (proportional valves).

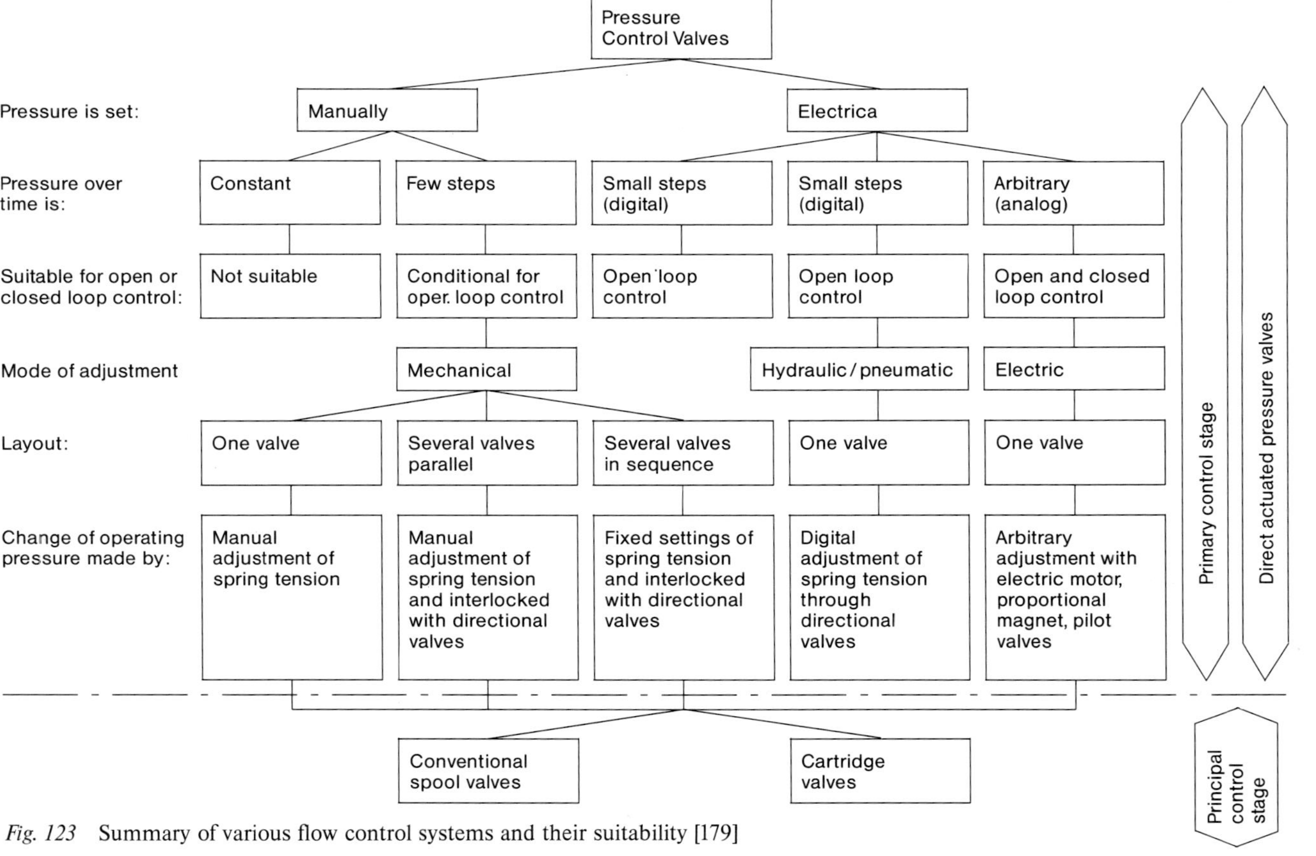

Fig. 123 Summary of various flow control systems and their suitability [179]

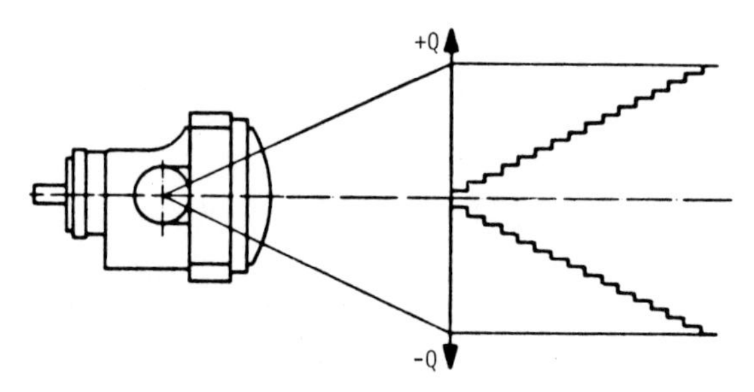

Fig. 124 Principle of variable displacement pump and digitally controlled flow rate steps. Digital adjustment of tilt angle [188]

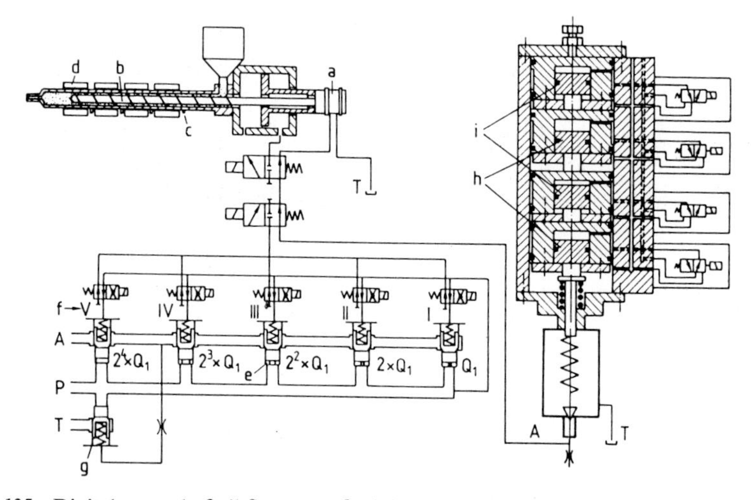

Fig. 125 Digital control of oil-flow rates for injection unti.
a: Screw drive, b: Screw, c: Temperature sensor, d: Heater band, e: Restrictors, f: Volume control, g: Pressure regulator, h: Activated valve, i: Valve not activated, A: Line to drive, P: Line from pump, T: Line to tank, Q: Flow rate

Digital flow control presents the capacity for an especially good reproducibility if changes in throttle losses from viscosity variations (temperature control) and wear (filtering) can be kept within narrow limits. Insufficient temperature control of the hydraulic fluid is a frequent deficiency of a hydraulic system which is not pressure compensated, and affects the reproducibility adversely. Digital flow control works without throttle noises and other unpleasant sound levels if cartridge valves are employed [191].

Digital pressure control is accomplished by:

- digital pilot valve setting which actuates the piston of a pressure-control valve in binary code;
- adding of pilot pressures in binary code to actuate a pressure-control valve;
- digital control of an electromechanically actuated pressure-control valve.

Fig. 126 pictures the injection side of an injection molding machine, which is equipped with a digitally controlled hydraulic system. Basically, the same control elements can be employed on the clamping side. Table 24 provides a comparison of hydraulic control elements in accordance with today's information and knowledge.

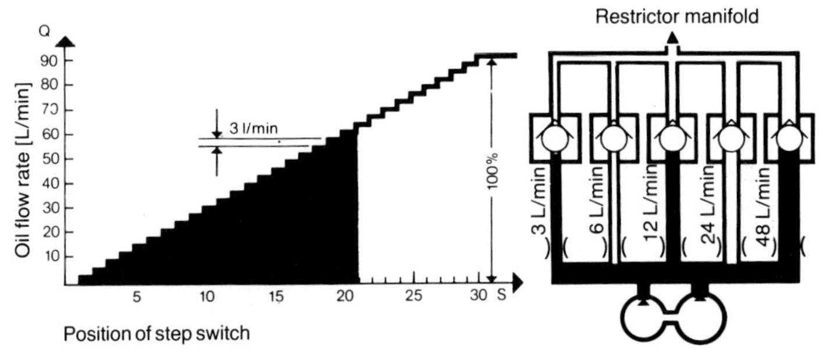

Fig. 126 Digital flow rate controlled by adding individual oil flow rates fixed with restrictors. Number of steps: $2^5 - 1 = 31$ (min. 3.23%, each step 3.23%). Courtesy Mannesmann Demag Kunststofftechnik)

Table 24 Energy Efficiency of some Hydraulic Screw-Drive Systems.
Q_{eff} = oil-volume demand of screw drive,
p_{eff} = pressure demand of screw drive, $N = Q \times p$

Drive system	(a) Single fixed-displacement pump with throttle control, or two-way flow control valve, or servo valve in main line	(b) Single fixed-displacement pump with three-way valve or two-way valve in by-pass
Efficiency diagramm	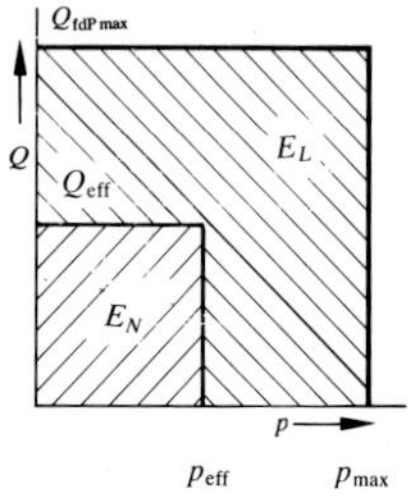 $Q_{fdP\,max}$, Q, Q_{eff}, E_L, E_N, p, p_{eff}, p_{max}	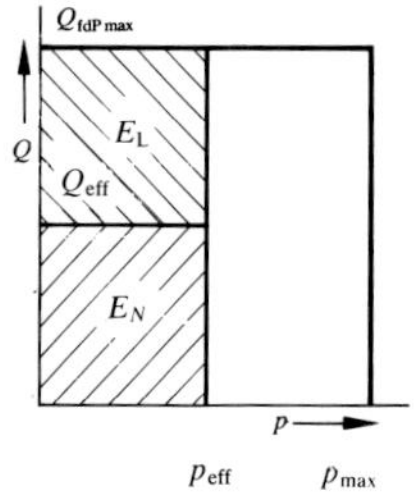 $Q_{fdP\,max}$, Q, Q_{eff}, E_L, E_N, p, p_{eff}, p_{max} Losses from throttling neglected
Theoretical efficiency n_e	$n_e = \dfrac{E_N}{E_{fdP\,max}}$	$n_e = \dfrac{E_N}{Q_{fdP}\,P_{eff}} = \dfrac{Q_{eff}}{Q_{fdP}}$
n_e with full load	1	1
n_e with 50% load (M_t = max, Q_{eff} = 50%) (P_{eff} = max, n = 50%)	0.5	0.5
n_e with 25% load (M_t = 50%, Q_{eff} = 50%) (P_{eff} = 50%, n = 50%)	0.25	0.25

Continued Table 24

(c) Double fixed displacement pump with three-way flow control valve $Q_{fdP1} = 4 \times Q_{fdP2}$	(d) Variable displacement pump with two-way flow control valve	(e) Variable displacement pump with three-way flow control
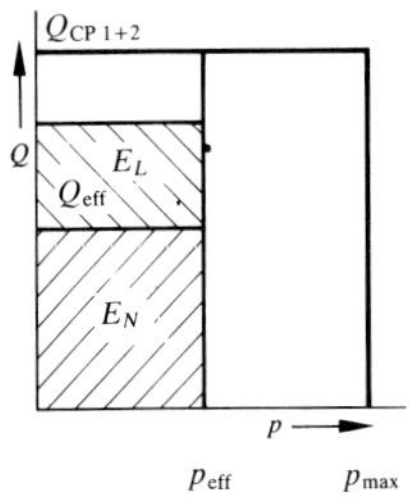 $Q_{CP\,1+2}$, Q, E_L, Q_{eff}, E_N, p, p_{eff}, p_{max} Losses from throttling neglected	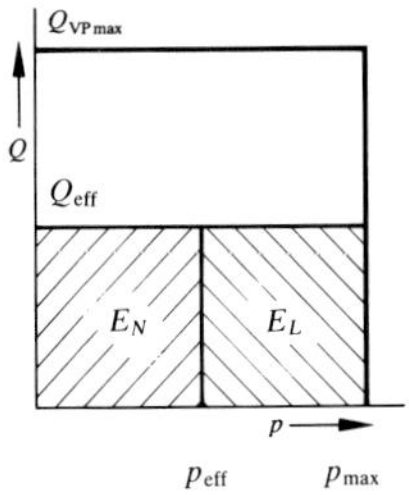 $Q_{VP\,max}$, Q, Q_{eff}, E_N, E_L, p, p_{eff}, p_{max}	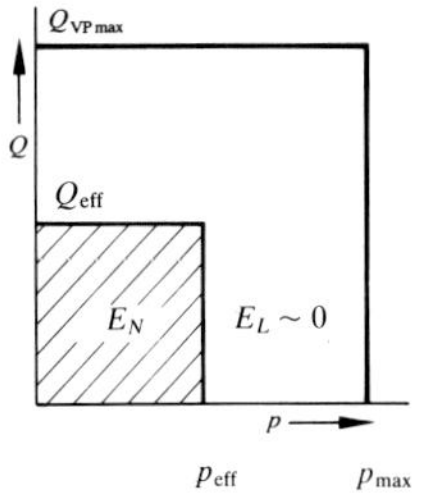 $Q_{VP\,max}$, Q, Q_{eff}, E_N, $E_L \sim 0$, p, p_{eff}, p_{max} Losses from throttling neglected
$n_e = \frac{E_N}{Q_{fdP1}\,P_{eff}} = \frac{Q_{eff}}{Q_{eP1}}$	$n_e = \frac{E_N}{Q_{eff}\,P_{max}} = \frac{P_{eff}}{P_{max}}$	$n_e = \frac{E_N}{Q_{vdP}\,P_{eff}} = 1$
1	1	1
0.625	1 (0.5 with $p = 50\%$, $Q_{eff} = \max.$)	1
0.8 (1 if $Q_{fdP1} = Q_{eff}$)	0.5	1

5.6 Monitoring Devices in the Hydraulic System

Wear, leakage and service life depend on pureness and temperature of the hydraulic fluid. One should maintain a constant quality of the fluid to ensure reliability and reproducibility. This influences the quality of the resulting product. A thorough filtering of the fluid prevents or reduces the wear of all movable components of the hydraulic system, such as pistons, seals, and valves. It increases their service lives. Even more important, the friction remains constant, and proliferation of internal leakage, which is difficult to observe, is avoided. Today, filters in the range of 5 to 10 μm are available and are suitable as full-flow filters. When operating a high-quality machine, great importance should be attached to the use of a fine filter.

Because of the usual clearance of less then 15 μm in their control elements, most hydraulic systems of injection molding machines belong to the class of sensitive hydraulic controls in the high-pressure range. The recommended size of fine-filters for such equipment is 2 to 5 μm. Thus, the following filter arrangement is necessary:

- filling of oil tank through a full-flow filter < 5 μm;
- intake filter in by-pass ~ 25 μm against coarse contaminants or effective sealing of the whole system to prevent contamination from outside;
- full-flow filter ~ 5 μm behind the pump or immediately ahead of sensitive valves, pressure control, eventual dual pressure filter for maintenance during operations.

As a matter of fact, today's effort for filtering does not meet the real needs with respect to the reproduction capability, service life, and leakage prevention of the systems.

Because temperature fluctuation of the hydraulic fluid results in an immediate variation in weight and dimensions of the final product, attention should be paid to careful temperature control. This is one requirement for achieving a constant quality in molding.

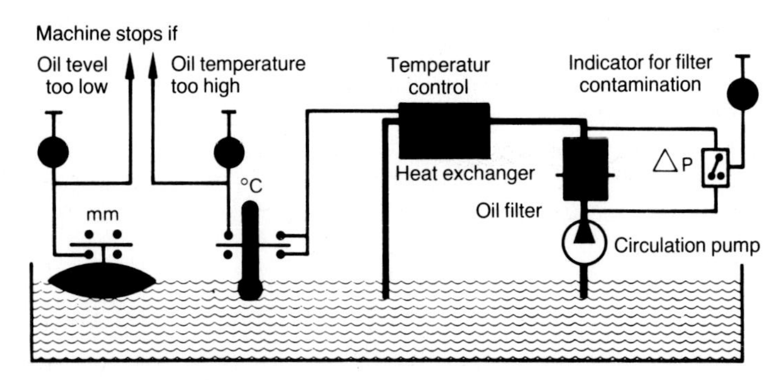

Fig. 127 Control system for oil level, oil temperature, oil filtering, and filter contamination (Courtesy Mannesmann Demag Kunststofftechnik)

Correct oil temperature has to be attained before production starts and should be kept constant within ± 20°C. The optimal temperature level is about 40 to 50°C. This calls for heating the oil either during startup of the machine or with a separate heater, and for an efficient temperature control circuit with heat exchanger. A complete monitoring system for the hydraulic fluid is presented with Fig. 127.

5.7 Noise Suppression in the Hydraulic System

In recent years, considerable progress has been made with regard to noise suppression. The principle cause of noise is the hydraulic system. Originators of loud noise are [223]:

- the fan of the electric motor,
- the hydraulic pump,
- the hydraulic motor,
- switching valves,
- the oil flow in narrow pipe lines and elbows,
- sound of impact from opening and closing the mold.

Pumps and motors are available today which operate exceptionally quietly over the entire load range (Table 21). Refer to section 5.3.2 (Hydraulic Pumps) about standards on noise levels. One has to distinguish, however, between the noise generation reported in a prospectus (pump tested in a quiet chamber) and the total sound emission of a machine, especially during feeding at high speed under full load. Since the processor has to anticipate more restrictive standards on noise pollution, he should only purchase machines with a sound emission not higher than 75 dB(A) according to a standard measurement.

Further reduction in noise generation can be accomplished with economically justifiable means by [191, 205, 206, 223]:

- reducing the speed of electric motors and simultaneously creasing the capacity of hydraulic motors for the screw drive;
- mounting drive units on vibration-damping elements;
- flexible supply and return lines for hydraulic drive units;
- employing valves without bouncing features by ramplike action;
- using adequately sized pipelines and bores in manifolds to reduce flow velocity;
- use of hydraulic motors and pumps with low noise level;
- employment of submergible pumps in the tank;
- encasing drive units in sound-absorbing housings.

The solution to a noise problem is of a complex nature. Thus, the mere use of an especially quiet pump is not a guarantee that a machine will create an adequately friendly environment. The whole machine has to be optimized with respect to noise emission.

5.8 Efficiency and Energy Consumption

5.8.1 Efficiency and Energy Consumption of the Hydraulic Drive

There is sometimes a considerable difference between power input and effective power. The following losses occur during dynamic power transmission:

a) Losses from friction between mechanical components. Heat of friction is generated (η_{mech}).
b) Losses from flow resistance (throttle effect) in pipelines and valves. Heat of friction is generated (η_{hydr}).
c) Losses from internal leakage through clearances (decompression) (η_{vol}).

The total efficiency is:

$$\eta_{tot} = \eta_{mech} \times \eta_{hydr} \times \eta_{vol} \tag{19}$$

To achieve a high degree of efficiency, it is necessary to use individual components with small losses. This is especially important for hydraulic pumps and motors. The efficiency is further reduced by valves, the length, cross section, and elasticity of pipelines, sharp elbows, compressibility of the hydraulic fluid, and external leaks. Thus, the efficiency can be expressed as

$$\eta_{tot} = \eta_{conv} \times \eta_{contr} \times \eta_{trans} \tag{20}$$

where η_{conv} is the efficiency of conversion, η_{contr} the efficiency of control, and h_{trans} the efficiency of transmission (Fig. 128). Each one of these efficiencies is composed of the individual efficiencies of the participating elements of conversion, transformation, and control. According to [204], one can assume the following individual efficiencies:

- cylinder with straight-line operation $\eta_{tot} = 61 - 87\%$
- slow-running hydraulic motor $\eta_{tot} = 51 - 83\%$
- fast-running hydraulic motor and gear drive $\eta_{tot} = 52 - 80\%$
- electric motor and gear drive $\eta_{tot} = 82 - 93\%$

However, the system of power distribution has the greatest influence on efficiency, since it determines which portion of the power output is used for each stage. The energy efficiency is

$$\eta_{tote} = E_e / E_E \tag{21}$$

where E_e is the utilized power and E_E the power input to the electric motor.

Reviewing Fig. 113 one recognizes that, most of the time, only a small portion of the drive power is utilized. On an average the rate of utilization is 20 to 50%. The system of power distribution determines whether 100% output is produced in all load cases or whether output is adjusted to demand [205]. In the first case, excessive power has to be canceled out by throttling about 80% of the output and converting it into heat, which must be removed from the oil tank by a heat exchanger. By and large,

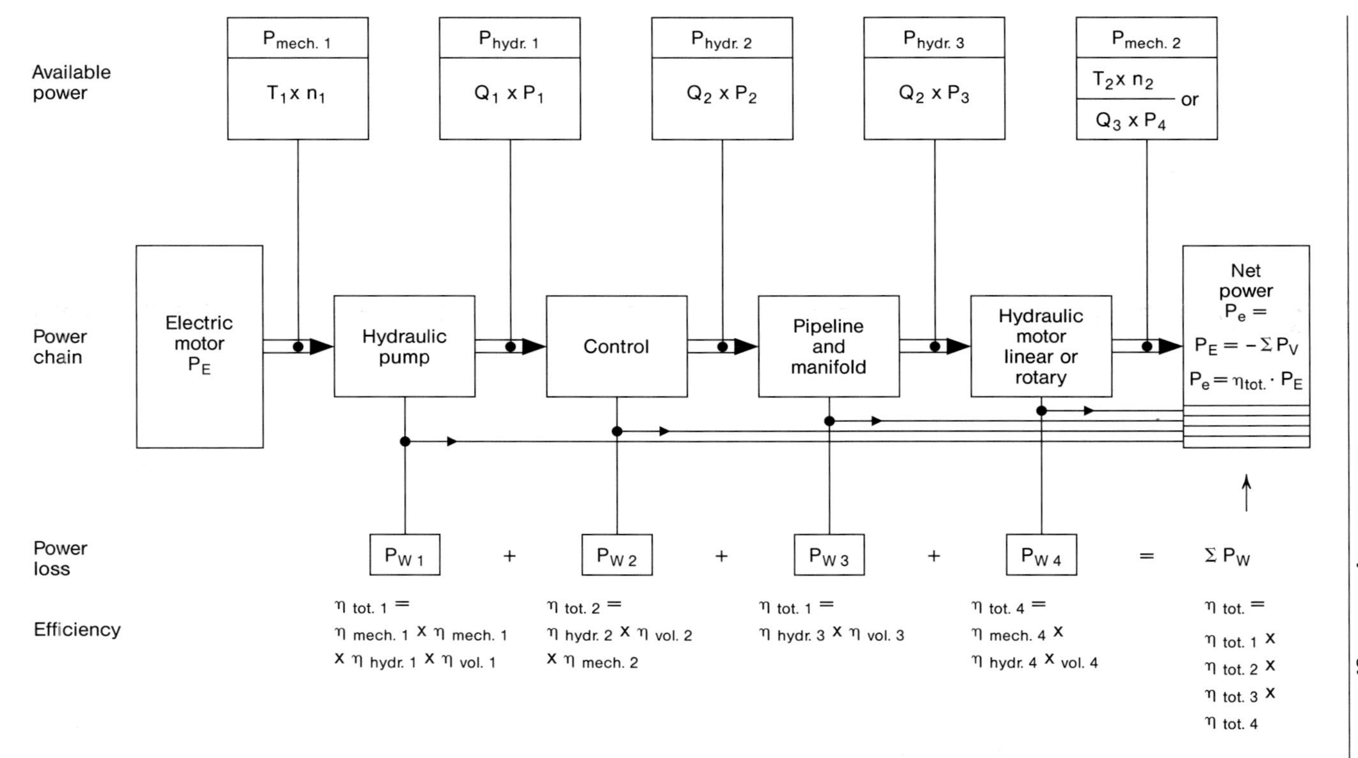

Fig. 128 Efficiency of a hydraulic drive for injection molding machines [204, 217, 222]

however, a variable power output is available, which follows the demand without much delay. Between these two extremes there are a number of practical intermediate solutions, some of which are presented in Table 24. The energy efficiency of all systems in Fig. 115 can be found accordingly. In these efficiency diagrams, the area

$$Q_{max} \times p_{max} = E_{max} \tag{22}$$

represents the power that is available after having passed the sequence shown in Fig. 128, that is

$$E_{max} = \eta_{tottot} \times E_E \tag{23}$$

The majority of older machines have poor energy efficiency, because they operate according to system a) in Table 24. Their energy efficiency is between 10 and 30% depending on utilization. Even so, new machines with lower clamping force (below 500 kN) and the same drive system have the same poor efficiencies. Drive systems with a more favorable power control are generally employed in medium-sized to large machines. Some of the more important systems are presented in Table 24 b and c. The efficiency of the controls is about 25 to 50%.

If the efficiency of the transforming system (valves, pipeline, elbows, filters) is assumed with about 75 to 90%, then the total efficiency (ntot) is

$$\eta_{tot} = \eta_{conv} \times \eta_{contr} \times \eta_{trans} =$$
$$(51-87\%) \times (75-90\%) \times (25-30\%) \times (90-95\%) = 8.6 \text{ to } 37.2\% \tag{24}$$

In view of the costs of energy, this result is not very gratifying. Steps to optimize the use of energy in the drive system of injection molding machines are urgently needed. To avoid pressure surge and to keep the injection characteristics optimal, considerable effort is called for. Fig. 129 demonstrates the characteristics of fixed-displacement pumps and accumulators. With a pump drive, the injection speed is almost independent of the load over a wide range, while the discharge of an accumulator is associated with pressure drop and decrease in speed. Generally, processors attach little importance to the pump characteristics. Consequently, there is little uniformity among the design concepts used by the machine manufacturers with respect to this feature, which is so important to the quality of a machine. Numerous publications have reported on the significance of the injection stage and particularly its constancy with respect to the control of the injection and holding-pressure stages. Unfortunately, the possibility of a direct correlation between quality and pump characteristics have not yet been investigated. Therefore, only generalized information can be provided here, concerning the given pump characteristic and steps to influence it.

Fig. 130 demonstrates the oil-flow rate of a pump as a function of the pressure in the hydraulic system as it may be required, for instance, for the injection cylinder. The nonlinearity of the Q-p curve is about 10% ($\Delta Q \sim 10\%$) at full load. If the flow control is not pressure compensated, as shown in Fig. 130A, the relative deviation increases as soon as the flow rate is reduced, e.g. for slower injection ($\Delta Q \sim 20\%$ with 50% of the maximum flow rate used as is done in this example). This increases the inconstancy of the injection speed even more. If, for instance, the speed deviation from a change in the viscosity of the hydraulic fluid or the melted material (temperature inhomogeneity) is already ± 2.5% at full load, then it can increase to ± 5% at half the injection speed. One can assume that this is reflected in a reduced quality of the molding. This addresses a problem for the hydraulic engineer: how to make a machine independent of internal and external interference during injection so that it operates with the same constancy shot after shot. Doing this by process control is only one possible answer. It subsequently corrects some of the deviations, but does not eliminate their cause.

There are also machines on the market that have the pump characteristic linearized by pressure compensation. The compensation pressure, however, is lost and not available for injection anymore. This pressure loss can amount to 1 to 7 MPa. Fig. 130B presents the Q-p characteristic of such pressure compensation. The pressure drop through compensation causes the unavoidable drop in performance to occur already between 50 and 80% of the maximum injection pressure (p_{max}) instead of 80 to 90% in uncompensated systems. Furthermore, the usable oil-flow rate is clearly diminished. The advantage of an improved constancy of the injection speed is consequently

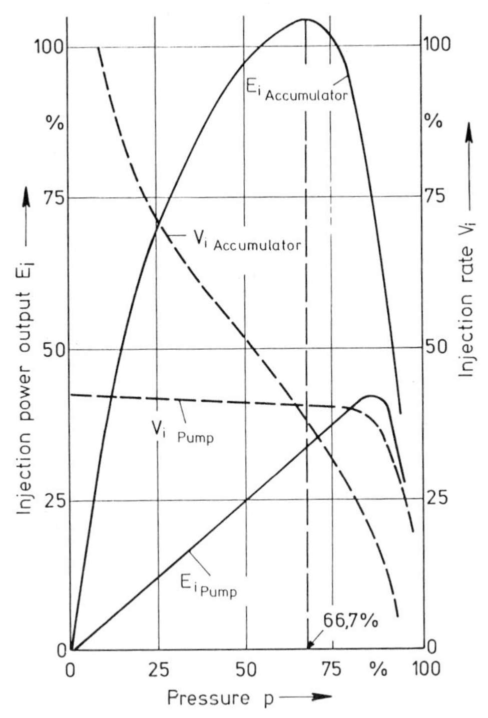

Fig. 129 Injection power output and rate versus hydraulic pressure from pump or accumulator

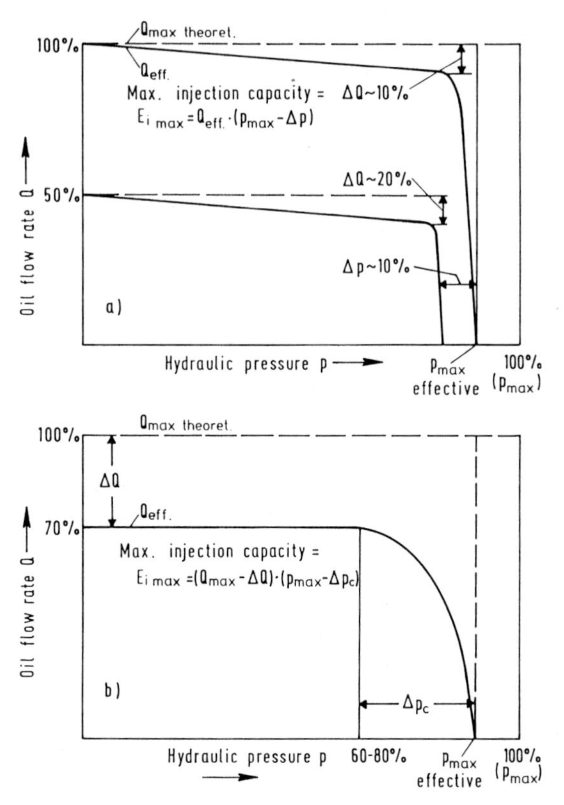

Fig. 130 The dependence of oil flow rate Q on pressure p.
(a) Q-p characteristic not liniearly corrected and constant Q. Injection rate may be between −10% (upper curve) and e.g. −20% at half the maximum rate.
(b) Q-p characteristic linearly corrected with oil flow rate loss (ΔQ) and pressure loss (Δp). The efficiency in the linear range is drastically reduced by 40–75%

paid for by a drastic reduction in energy efficiency. This generally holds also true for machines equipped with pilot-operated valves.

Because a considerable portion of the rated pump pressure is lost by pressure compensation, special consideration are called for during the design of the system (Table 3 and Fig. 12). Operating an injection unit near the maximum injection pressure, as it is listed by the manufacturer, should be avoided in all cases where medium or high quality is required. One proceeds according to the following example:

Required effective injection pressure within the linear range $p_{eff} = 150$ MPa. Hence

$$p_{max} = \frac{150\ MPa}{0.85} \quad \text{(without pressure compensation)} \tag{25}$$
$$= 176\ MPa$$

$$p_{max} = \frac{150\ MPa}{0.75} \quad \text{(if compensation pressure is 25\% of maximum pump pressure)} \tag{26}$$
$$= 200\ MPa$$

Without knowing the particular requirements, it does not appear to be wise to select a screw diameter which provides an injection pressure much below 200 MPa.

Other control systems for pressure and speed, which are not presented here, skew the speed characteristic even more than the pump characteristic would indicate. This reduces the efficiency and causes the injection stage to be greatly dependent on the viscosity. This results in adverse speed variations.

Digital systems have a fundamental advantage. Their relative error is always nearly constant, assuming that any interferences, which primarily influence the viscosity, are always of about the same magnitude. People cannot change this error by subjective action and reaction. The number of steps necessary for practical operation can be attained easily. On the other hand, these systems require additional effort in designing the dynamic behavior in such a way that no inadmissible discontinuity occurs during switching. While set and actual speed values coincide well with analog control, digital flow control can result in deviations (Fig. 131), which influence the injection process adversely. Valves with particular switching response take care that no pressure or speed surges occur.

If accumulators are employed, one generally considers their efficiency during charging and the possibility of using their large capacity for injection. Fig. 129 illustrates the difference between the discharge of an accumulator and the output of a pump. While the accumulator always has maximum output at 66.7% of its maximum pressure, the output drop of a pump, in the most favorable case, may not occur below 80% of the maximum pressure. Occasionally, this more favorable characteristic of a pump during injection is used to reject the accumulator drive. This is justified only if the accumulator is relatively small and the injection-power curves (E_i) intersect to the left of the maximum pump output, as shown in Fig. 129. This is usually not the case. The controlled sequence of operations should nevertheless be arranged in such a way that injection cannot start with an already partly discharged accumulator. If the accumulator supplies the injection stage, it must not be used for clamping and carriage movements.

The efficiency of an accumulator drive is good if the power input of the pump is small in comparison with the maximum capacity of the accumulator. This can be the case only with a machine that operates with a low sequence of cycles. Such machines can be fast within a cycle, during clamping, opening, or injection, but have to have a long dry cycle. The latter specifies not only the basic velocity of a machine, but also it

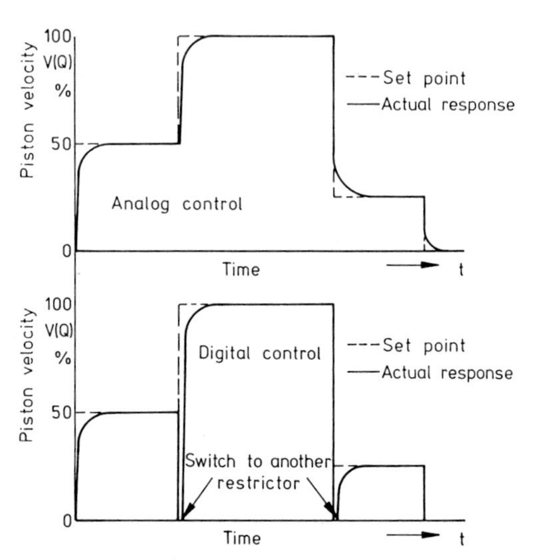

Fig. 131 Effect of analog and digital flow rate control

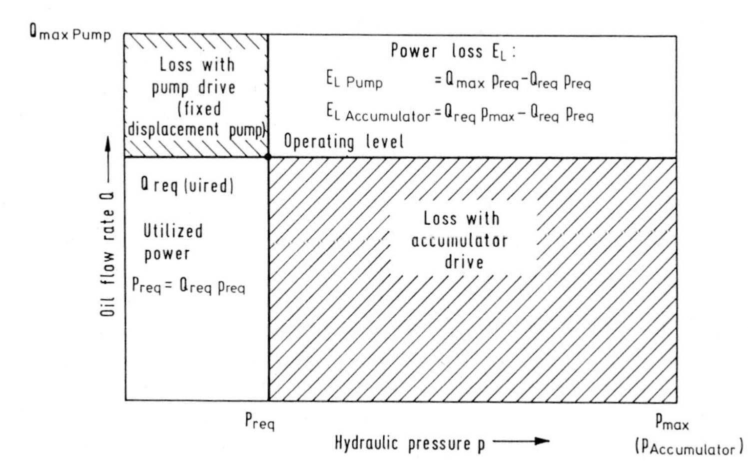

Fig. 132 *Q-p* diagramm and power loss with fixed displacement pump and accumulator drive at an arbitrary operating level [69]

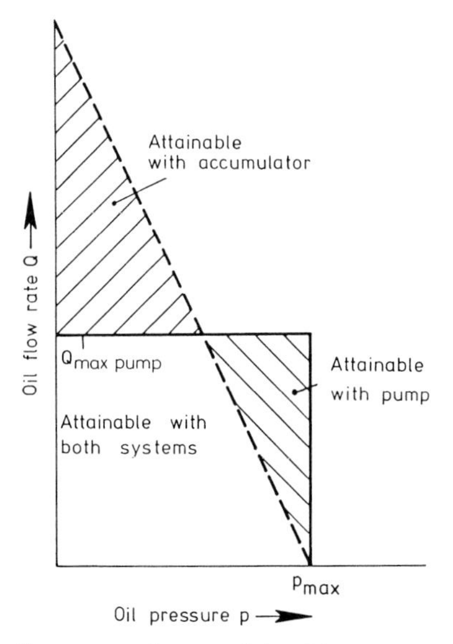

Fig. 133 Power ranges for pump and accumulator drive

continuous speed rating. A high continuous speed calls for a fast charging of the accumulator, which requires a powerful pump drive. With this, the advantage of a low total power rating is lost, and it is better to supply the machine directly with a pump.

The operating point of an injection molding machine influences the efficiency to a large extent. According to Fig. 132, there is an operating range in which the accumulator works with smaller losses than an optimized pump control. This is the case if high pressure and low oil-flow rates are needed. Under such conditions, however, one usually does not employ an accumulator.

The speed development in Fig. 129 signifies the main problem of the accumulator drive. While the hydraulic pump is capable of supplying a relatively constant oil-flow rate, which can be rendered almost fully constant by pressure compensation, the speed curve of an accumulator declines sharply with increasing pressure. Linearizing this curve requires an extensive amount of control for pressure compensation with very high losses from throttle effects. Pressure surges in fast-operating accumulator systems can be avoided only with special, smoothly switching valves. By their nature, accumulators are not suited for constant or controlled injection speeds. In conclusion, the accumulator can presently be judged as well suited for special machines or as supplementary equipment. This is particularly true if large oil-flow rates, beyond the capacity of a pump, should be available for rapid injection (Fig. 133) [69].

5.8.2 Efficiency and Energy Consumption of the Electromechanical Drive

The performance diagram of an electromechanical drive may have a similarity with that of an electrohydraulic drive (Fig. 134). So far, no advantages to a hydraulic system can be recognized in its characteristics. One can even see distinct disadvantages in the linearity in the low power range and the stability until the power declines. Of course, this drive is capable of attaining the operational points in the range of steeply declining power output, that is beyond about 50% maximum capacity, with great constancy. This cannot be done with a hydraulic drive because of its viscosity dependency.

Fig. 135 shows, by comparison, the energy demands of injection molding machines of the same size (800 kN) but with different drive systems. One can easily recognize that one has to know the drive system of a new and recommended machine version to be able to compare correctly. Since machines of the standard no. 5 in Fig. 135 are the rule today, the much discussed drives of 6 and 7 allow energy savings of another 7 to 21%. Frequently, higher savings are mentioned, but this may depend on the size of the tested machines. Anyway, savings of about 60% from 1977 until today are remarkable.

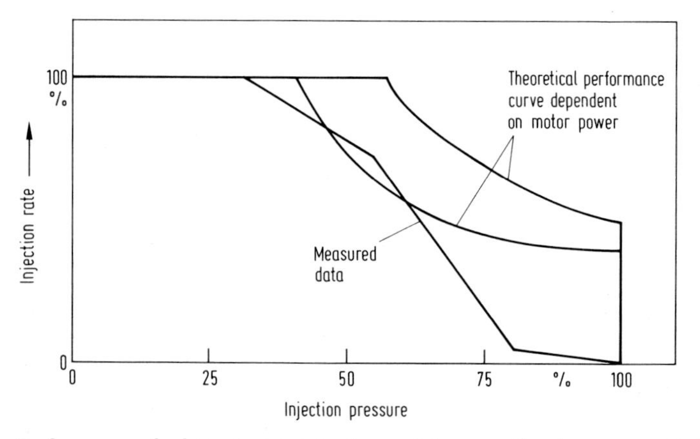

Fig. 134 Performance of a fully electric injection molding machine (compare with Fig. 122); data from one machine only

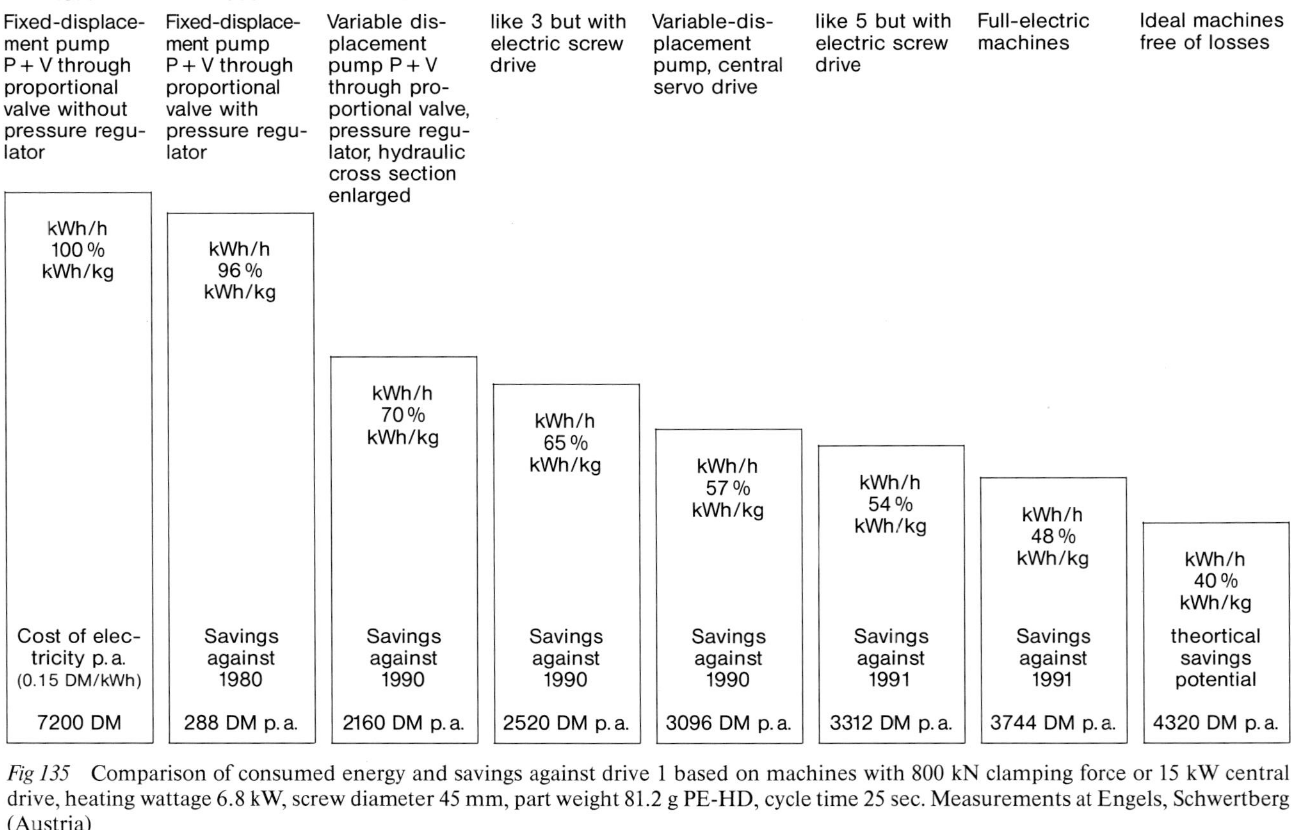

Fig 135 Comparison of consumed energy and savings against drive 1 based on machines with 800 kN clamping force or 15 kW central drive, heating wattage 6.8 kW, screw diameter 45 mm, part weight 81.2 g PE-HD, cycle time 25 sec. Measurements at Engels, Schwertberg (Austria)

6 The Control System

The control system of an injection molding machine comprises all equipment which controls oil and barrel temperatures, clamping forces, oil pressure and flow rates in such a way that they are generated and available in the required magnitude and direction at the right time during the logical sequence of one cycle or several consecutive cycles. Another most effective parameters in the molding process, the mold temperature, should not be forgotten in this context to better comprehend the entire interaction.

The quality of an injection molded part is determined almost exclusively by two processing parameters, pressure and temperature, while the injection speed (pressure dependent) exerts only a limited influence (Fig. 136). Furthermore, the design of the control system has to take into consideration the logical sequence of all principal functions, such as clamping and opening of the mold, and all secondary functions, opening and closing of shut-off nozzles, actuating cams, and so on.

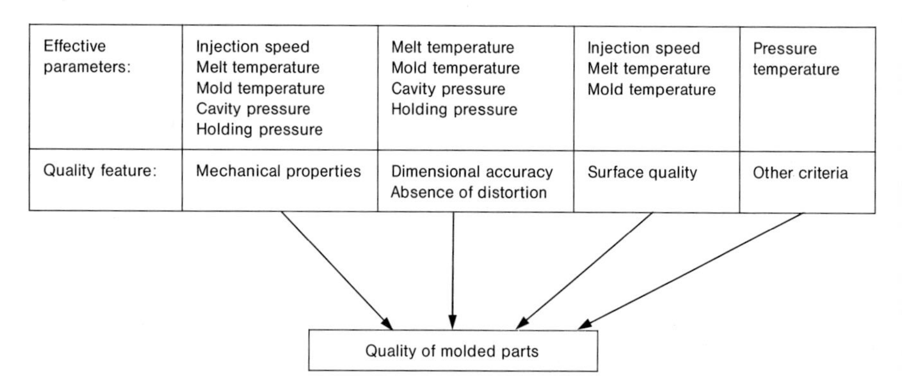

Fig. 136 Effect of controllable process parameters on the quality of injection molded parts

Meanwhile all data relevant for the distinction of the molding process are recorded and stored. This is also done to document the quality of the production. In this context, the correct selection and proper installation of recording devices is of major significance.

Consequently, the direct objective of the control system is the supervision of the process and its indirect task the assurance of the quality demands on the finished product. However, not the applied means make a great difference but the employed principles.

It is remarkable that many devices, which perform control functions, may contain an additional internal control mechanism [300, 301]. Servo and proportional valves can be mentioned in this connection. They are employed for controlling the flow rate by maintaining the spool in its position using internal pilot pressure [301].

Controlling denotes keeping a certain quantity constant during a definite period of time.

6.1 Process Control Methods

Fig. 137 shows that the term process control summarizes all conceivable methods for influencing and executing the molding process. In this sense, process control is not a modern concept but was already used with the very first injection molding machine. Today's essential methods can be classified as follows:

- manual control (mini-machines),
- control by electromechanical components after manual setting,
- control by electronic circuits and manual setting,
- control by definite programs,
- open-loop control of some important parameters (speed and pressure) with manually programmed sequence control,
- programmed open-loop control,
- closed-loop control.

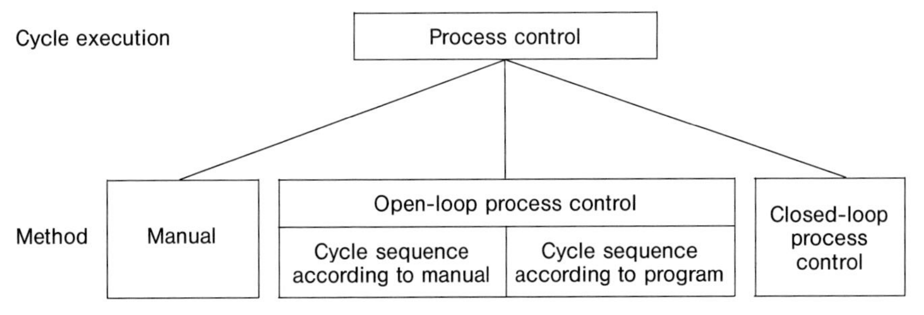

Fig. 137 Basic methods of process control for injection molding

Process control and optimizing with electronic processors, either with a small computer at each machine or a central computer supervising all machines, has been developed [56, 57, 159, 160, 228 to 299]. It has good prospects of a wider acceptance in the framework of computer integrated manufacturing (CIM).

The difference between open- and closed-loop control is demonstrated with block diagrams in the next sections.

6.1.1 Open-Loop Control

With an open-loop control, one or several input (reference) signals are modulated into an output (control) signal in accordance with the inherent interrelations of the system. It is characterized by an open sequence of actions across the individual transfer elements of the control chain [302].

Fig. 138A presents the simplified design of an open-loop control of the screw displacement as an example. The variable, which has to be kept constant, is the speed of the screw advancement x. It is controlled by the valve position y. The oil pressure z_1 as well as the oil temperature z_2 and the force z_3 acting against the screw displacement affect the speed.

Even if the oil pressure varies, the speed of the advancing screw can be kept constant by adjusting the valve accordingly with a control mechanism. The precondition for this is a known and invariable correlation between pressure and flow rate. Only then can the controller compensate pressure variations of z_1.

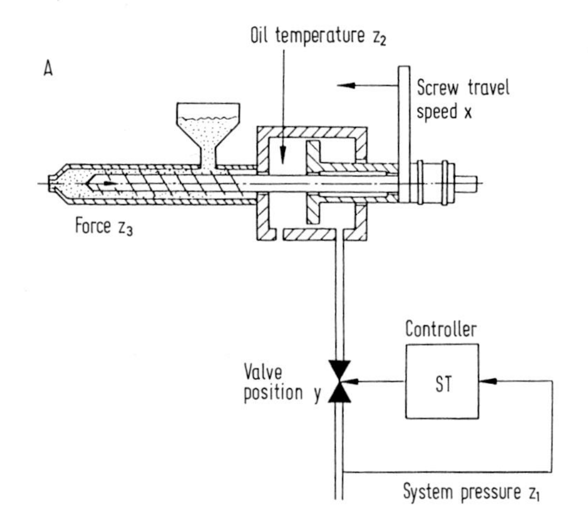

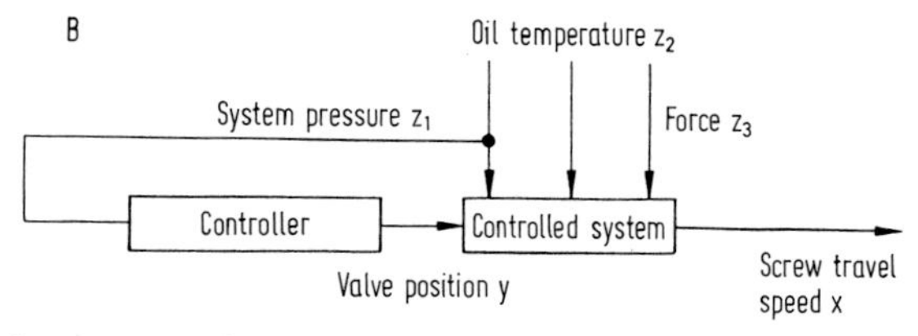

Fig. 138 Open-loop control of screw-travel speed
A: Control circuit (schematic), B: Block diagram of control [325]

Other disturbance variables, the oil temperature or the reactive force depending on the melt viscosity, cannot be recognized by the system and therefore not compensated.

Fig. 138B pictures the block diagram of the open-loop control. The course of action is indicated by arrows, and one can see that the effect of z_1 is offset by the controller. Other effects, however, influence the variable x directly.

Thus, an open-loop control is capable of compensating the effect of any interference that can be measured. The course of action is an open chain. Such a control chain can result in excellent constancy of separate process steps if distinct control circuits prevent any influence of other variable elements. Example: Injection with constant melt, oil, and mold temperature and little variations in the viscosity of the molding material.

6.1.2 Closed-Loop Control

A closed-loop system feeds back the output (controlled) signal and continuously compares its magnitude with the input (reference) signal. Any resulting deviation from the set value produced by interference is used to correct the controlled output. The course of action is a closed control circuit [302]. The drive for the axial motion of a screw shall again serve as an example for the operation of a simple closed-loop control (Fig. 139A). The variable x (actual speed of displacement) should be kept constant and is compared with the set speed value w. A controller determines the difference between set point and actual value and affects the valve position y in such a way that the difference becomes zero. Thus the influence of the oil pressure z_1, but also the other interferences z_2 and z_3 are caught and eliminated.

In closed-loop control, the output signal, which has to be controlled, is measured and the controller can compensate the effects of several disturbances, which, on their part, are not measured. The course of action is a closed circuit in contrast to the open-loop control (Fig. 139B).

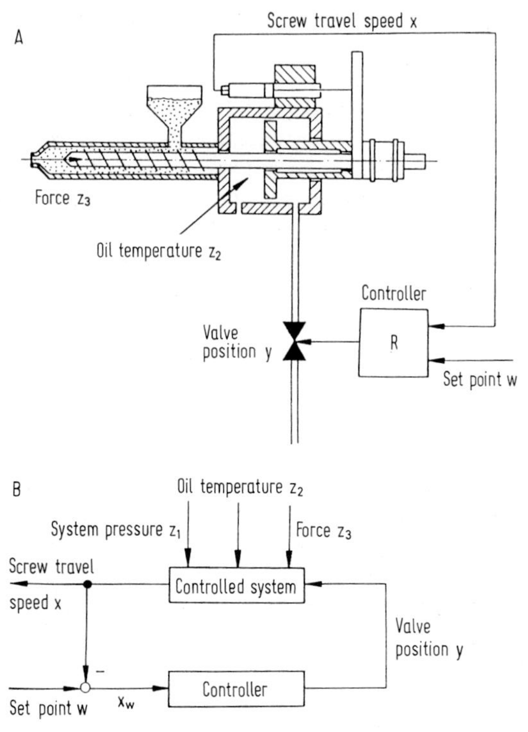

Fig. 139 Closed-loop control of screw-travel speed
A: Control circuit schematic, B: Block diagram of control [325]

6.2 Temperature Control

The material to be processed is melted and prepared in the barrel of the injection unit. The quality of this procedure is crucial to the standard of the injection process and the quality of the final product. The following temperatures are of significance:

- barrel temperature,
- melt temperature,
- temperature of the hydraulic fluid,
- temperature of the tie bars,
- mold temperature,
- temperature of the hot-runner system,
- (in some cases the ambient temperature may have an effect).

All these temperatures have to be measured in the first place and then controlled most of the time.

A uniform and constant melt temperature is an important goal although it presents, for most of the installed control circuits, an interference factor since it is affected by the initial temperature of the material, the temperature of the feed hopper, the feeding process, and the plastic material itself. Nevertheless, a considerable amount of technology is used to get this temperature as close to an optimal level as possible and to keep it as constant as needed.

Temperatures influenced by the process, shear-dependent injection temperature or demolding temperature, will not be discussed in the context of this systematism.

6.2.1 Temperature Measurements

6.2.1.1 Wall Temperatures

Temperatures of barrel and cavity walls can be measured with thermocouples and/or resistance temperature detectors (RTD). They are mostly equipped with a spring-loaded bayonet fitting to bottom the tip closely in the well. It is often overlooked that good contact is mandatory because, if the contact surface is inadequate, the measuring results may be falsified by a distinct error in thermal conduction. This is especially important for the larger resistance temperature detectors.

As far as thermocouples are concerned the ANSI Type J with iron/constantan lead wire calibration and an absolute error of about 2°C up to 300°C is mostly the choice. The hot junction is at the end of a protecting stainless-steel tube and galvanically insulated to avoid interference. Fig. 140 shows the installation of a

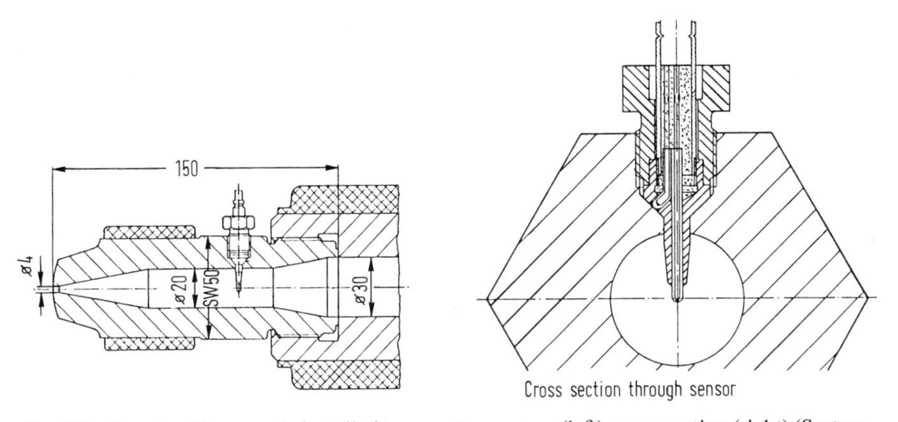

Fig. 140 Nozzle with correctly installed temperature sensor (left), cross section (right) (System Bayer) [325]

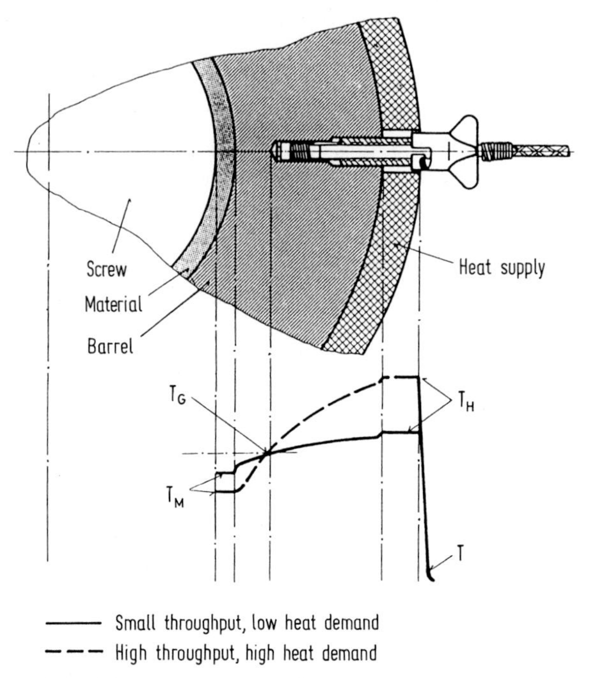

Fig. 141 Cross section of a barrel zone with profiles of radial temperature for low and high heat requirement [325]

temperature sensor in a barrel and Fig. 141 the resulting profiles of radial temperatures at different heat requirements.

Modern resistance temperature detectors are being used in an increasing number. Their sensor elements are made of high-purity platinum wire. The resistance of this wire changes rapidly with the temperature. Thus the RTD is reportedly many times as sensitive as a thermocouple. In combination with a controller it provides for freedom from voltage effects.

It turns out that only the temperature at the tip of the sensor is independent of the provided amount of heat. At the inside barrel wall and in the melt, however, temperatures are dependent on the supplied heat flux. Therefore temperature sensors in the wall should be mounted relatively close to the melt. On the other hand, the control circuit can get into uncontrollable oscillation if the distance to the heated surface is too great and the signal from a temperature rise arrives late. Then the remaining wall thickness of the barrel is overrun by the energy from the outside. The machine manufacturer knows the rules and adapts his product accordingly.

Depending on design and conditions of installation, there may be deviations of the displayed wall temperature from the real one (Fig. 142).

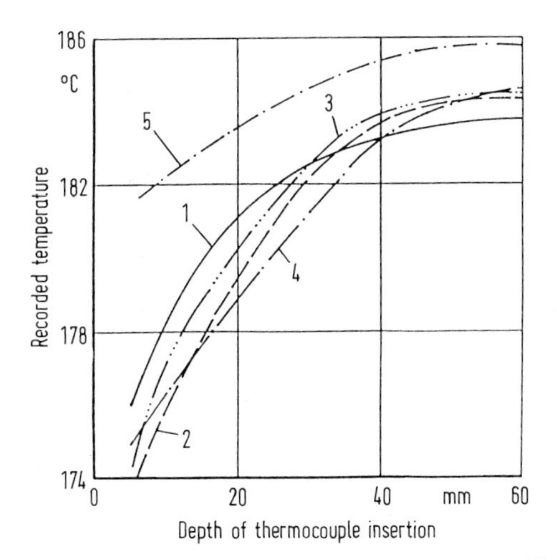

Fig. 142 Error of temperature measurement dependent on design of thermocouples and assembly conditions [325]. 1–4: Thermocouple with heat-conducting metal tubing, large error, 5: Thermocouple with heat-insulating tubing, small error

The curves 1 to 4 are the results from differently mounted sensors. One recognizes that the displayed temperature is basically dependent on the depth of the sensor well (remaining wall thickness). With adequate depth, about 40 mm in this example, further effects such as position of the sensor in the barrel (top, bottom, sideways), details of the installation, or the occurrence of draft are of secondary importance. Their influence on the display is less than 1 °C. Curve 5, however, represents a temperature that is about 2°C higher. It was determined with a thermocouple of externally equal design but with its shaft behind the hot junction made of a material with poor heat conductivity. This results in a distinctly inferior heat conduction and, with it, a larger recording error.

To measure the temperature of cavity walls, thermocouples with protecting steel tubes with a diameter of 1 mm and more are often employed. Here too, care has to be taken to minimize the heat transfer resistance by suitable mounting and the use of a heat-conducting paste. Soldering cannot be recommended because there is always the hazard of impairing the insulation between the lead wires and the tubing.

6.2.1.2 Melt Temperatures

Melt temperatures are still preferably measured with thermocouples. In the demanding environment of injection molding only permanently installed sensors are employed. Sliding probes are known for scientific purposes. They permit the investigation of several locations within a cross section (Fig. 143) [304].

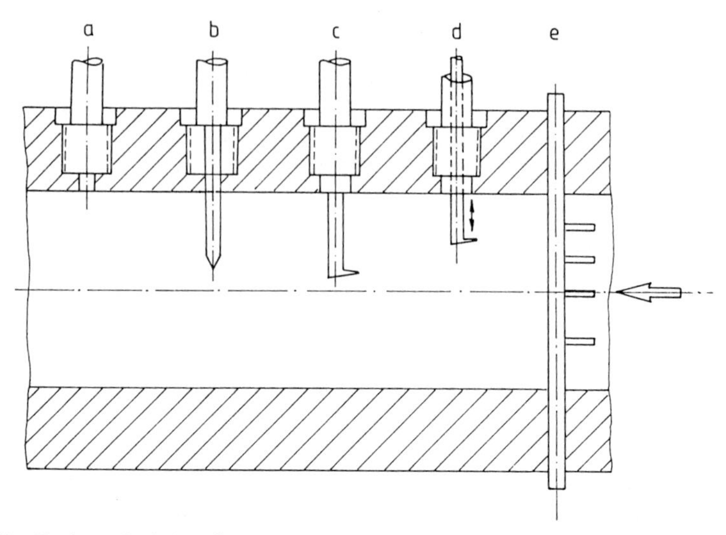

Fig. 143 Design principles of temperature sensors [325]
a: Sensor flush with barrel wall, b: Immersion sensor, c: Hook sensor, fixed, d: Hook sensor, sliding, e: Measuring bar with several sensors

The hot junction is always placed at the tip of the probe and, in the hook and cross bar design (Fig. 143 c, d, e), always directed against the flow of the material. Thus, they are placed on an isotherm in the flow. Most of the time, only too small a length-over-diameter ratio (L > 7D if possible) can be realized for reasons of strength. Thus, one should pay attention to a possible error from heat conduction.

Using such equipment, the melt temperature can be determined with adequate accuracy. Because the rough conditions of real production call for solid solutions, sensors mounted flush with the inside wall are also used in barrels. Of course, they can only measure the temperature of the melt next to the wall. The same sensors have to be employed in the area along the screw. Immersion sensors (Fig. 140 and 143) are simple and ruggedly built and can be used for production. Since the sensor is positioned transverse to the direction of flow, there is always a read-out error dependent on the radial temperature profile. Measurements and calculations imply that the best location for the tip of the sensor is at a distance of 2/3 R from the inside barrel wall. For reasons of strength, 1/3 R usually has to be chosen.

6.2.2 Temperature Controls

6.2.2.1 Introduction

In the last years, the degree of automation in the plastics processing industry has increased more and more, and temperature control has probably been the starting point of this development.

The use of temperature controls is rather manifold, e.g.:

- temperature control of the barrel,
- control of mold temperatures,
- temperature control of the hydraulic fluid in the drive system,
- control of the coolant temperature, and more.

Demands and Quality Criteria

The demands on temperature controllers, well as on their necessary accuracy as on their dynamics (response time, transient performance), differ widely depending on requirements, e.g.:

- short response time,
- minimal overshooting,
- high circuit stability even with system variations,
- transient suppression,
- sluggishness to keep temperature variations small.

Since not all demands can be met in an optimal manner at the same time, trade-offs have to be accepted in the adjustment of controllers and in the selection of suitable control elements.

Today, controllers operate in digital mode and employ microprocessors. The input signal is converted into a numerical value and mathematically manipulated. The results are summarized to obtain an output signal, which regulates the power output in such a way that the temperature is maintained at the set value.

Earlier proportional controllers did not allow the actual temperature to be at the set point. For compensation, the offset was squared and, thus, a larger deviation carried more weight than a smaller one. Regardless, the temperature always differed by some fraction of the proportional band and the offset was constantly changing. For sufficient accuracy, additional functions were needed. Since the input signal is a continuous signal, the microprocessor performs an integration and adds an averaging term which permits the temperature to be kept at the preset point from ambient temperature to full heating, even if large time lags are present. This function is called automatic reset.

Most controllers provide an additional derivative term, the rate term. This is an anticipatory characteristic which shortens the response time to changing conditions. A final useful addition is a circuit that limits overshooting.

Design and Components

In the following the temperature control of a barrel shall be examined more closely.

Fig. 144 presents a closed-loop temperature control. Its essential components are the measuring element, the controller, the power element, and the load.

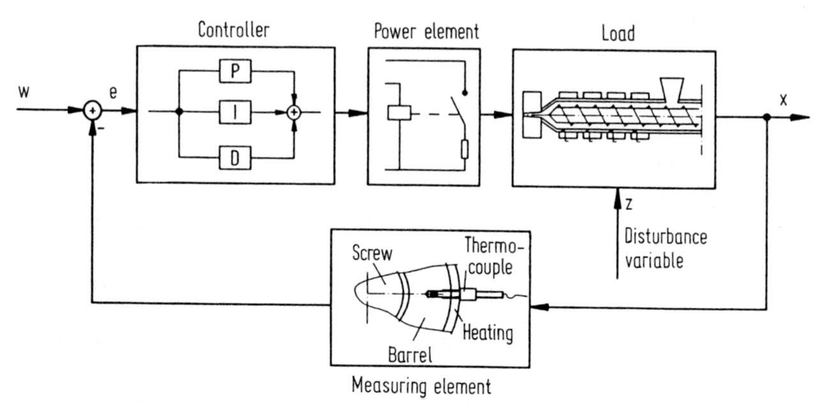

Fig. 144 Temperature control circuit [325]

Measuring Elements

Fig. 145 depicts the initial response of various thermosensor, which have been discussed before. Their behavior corresponds approximately to a system of the first order, that is a system with energy storage. The time constant of the response is determined by the mass of the sensor. If the sensor is mounted into the barrel wall, this constant can be neglected because that of the barrel wall is considerably greater. One has to see

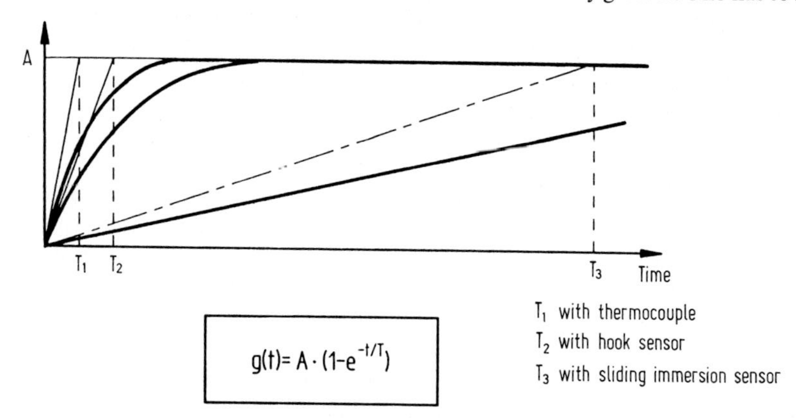

Fig. 145 Response time of various thermosensors [325]

to it that neighboring band heaters or cooling lines have only a minimal effect on the result of the measurement. The greater the influence of adjacent zones is, the greater is the degree of undesired coupling of control circuits. The control algorithms of severely coupled circuits is considerably more complicated, however, than that of loosely coupled ones. A complete neutralization is impossible because the moving melt is part of the controlled system.

6.2.2.2 Control Algorithms and Parameters

Various controller systems are available for the temperature control on injection molding machines. If the output variables of commercial controllers are examined, there are generally controllers with a steady standard output (e.g. +/– 10 V) or switching heating and cooling outputs with a linear relation between the controlling element and the controlled variable.

The control algorithms is rarely based on analog circuits anymore. Today's controllers operate in a digital mode and are engineered as microprocessors. This makes it possible to integrate several more tasks besides mere controlling. Thus, digital controllers are, at the same time, interfaces for supervisory monitoring and guidance. Their response to various malfunctions can be programmed in advance and a breakdown during production prevented. Such microprocessors are marketed as one single module with analog input for temperature sensors, analog output for control elements, digital input and output to record and control different states, and interfaces for communication. The know-how for these controllers is based on their algorithms and the procedure of computing the optimum parameters.

The design and mode of function of a digital PID controller shall now be briefly examined (Fig. 146) [323].

The input signal is in analog form. A scanner converts it into a discrete time- and value-dependent signal, The scanning speed has to be adjusted to the expected signal frequency. One has to stay within the limits of Shannon's theorem. Since the actions during temperature control are not very fast, the necessary scanning speed can always be attained. The control algorithm is derived from the analog controller differential equation. The idealized equation for a PID controller is:

$$u\,(t) = K\,[e(t) + \frac{1}{T_I} \int_0^t e(\;)\,d \;\; + T_D \frac{de(t)}{dt}\,]$$

The parameters are named:

K coefficient of gain,
T_I integration time (reset time),
T_D derivative time (rate time).

With high scanning speed (at least 10 times cutoff frequency) the differentials can be converted to differences and the integrals to sums by square integration. After

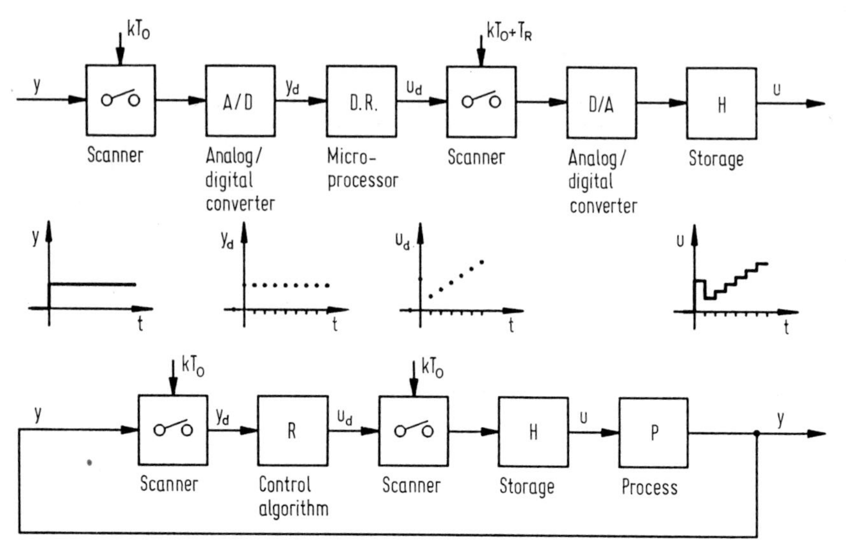

Fig. 146 Control circuit with scanning controller [325]

transposition of the equation into a recursive programmable equation, the following equations for the PID control algorithm are the result:

$$u(k-1) = K[e(k-1) + \frac{T_O}{T_I} \sum_{i=0}^{k-2} (e(i) + \frac{T_D}{T_O} (e(k-2))]$$

$$u(k) - u(k-1) = q_0\, e(k) + q_1\, e(k-1) + q_2\, e(k-2)$$

The quantities q_0, q_1, and q_2 can be derived from the parameters of controllers with steady function. Fig. 147 demonstrates a typical pattern of set point and actual value after a disturbance. The effect of P, I, and D components is clearly visible.

Modern controllers support the user in setting the optimum parameters with a trimming circuit. A learning cycle at the point of operation is activated either by the starting action or initiated by the user. The behavior of the load is determined by the input of a test function and, with this, the optimum data for the gain factor, reset time and rate time computed. The necessary function is mostly provided by the manufacturer.

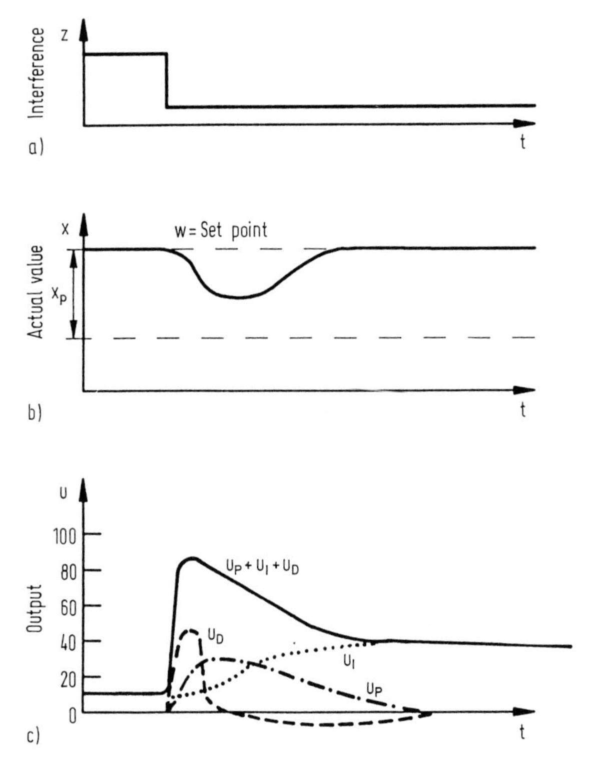

Fig. 147 Set points and actual values of a PID controller [325]

6.2.2.3 Power Elements

The output signal of a controller is either analog or digital and has to be fitted to the power element. Controlling devices for electric resistance heaters are mechanical or electronic relays or thyristors. Relay switches generally convert the signal into a square wave. This leads to good results with moderately priced and robust electronic components because of the great time constant of the load. The least expensive solution is the mechanical relay. Its disadvantage is the need for maintenance. The electronic components are maintenance-free but more expensive.

Thyristors as power elements with a steady input are not yet found often. They can adjust the heat supply very precisely and quasi-continuously, but they are considerable more expensive than the options mentioned before.

The presented control algorithms are adjusted and optimized for a certain point of operation. This is necessary because the algorithms assume a linear behavior of power elements and load. If the deviation from the point of operation is too large, the parameters are not valid anymore. In this connection, the nonlinear operation of fans should be mentioned. It is caused by a time lag between the switching on of the fan and and a stable thermal equilibrium at the cooling ribs.

6.2.2.4 Load

Generally, heating the barrel is accomplished by electrical resistance heaters. They can be made in a variety of shapes and power classes. Fans are used for cooling, if needed. Temperature control with fluids are primarily employed in rubber processing.

If the heating of the barrel of an injection molding machine is examined, then the controlled quantity is the wall temperature and the load is the barrel itself. The melt temperature, the quantity that should actually be controlled, is not picked up in each zone and appears as a disturbance if one inspects the interaction more closely. Depending on the location where the sensor is mounted, close to a band heater or farther away, one can anticipate a greater time lag and a smaller effect from variations of the melt temperature or vice versa. The behavior of the load can be adequately characterized by simple models of the first or second order. Up-to-date controllers provide, in addition, thermocouple break protection, which inhibits power supply to the load and alarms the operator. State of the art also is a power control module for monitoring the power supply to the heaters.

6.2.2.5 Design and Setup of Heaters and Controllers

With today's state of engineering, set and actual temperatures are displayed side by side on the monitor of the machine on a page reserved for temperature readings. Inadmissible deviations are made visible or release an alarm. Display of the data as numbers or bars in different colors is common.

The heat supply of a barrel has, by all means, to be divided into several independent zones. At least five individually controlled circuits should be employed in machines larger than with 500 kN clamp force. The nozzle temperature, too, has to be controlled independently in an equal manner if special elements such as heat pipes are not used. The band heaters have to have a large area of contact with the barrel. Best suited are ceramic heaters. In contrast to common practice, heaters should be used with insulating covers. This saves energy (2 to 3%) and provides for a more uniform heat supply along the barrel. Variations in the ambient temperature have less effect, and the heater bands are protected against damage [99, 136]. The displayed temperature and the attained constancy (± 1 to 2 K) should not be confused with temperature of the melt and its constancy. Radial and axial temperature heterogeneities were already discussed. They depend primarily on the design (length, flight depth) of the screw, its mode of operation (feeding travel, rotational speed), back pressure (changes along the feeding stroke), injection rate, enthalpy of the plastic material, and its residence time. Temperature deviations of ± 30 K are possible, and even in positive cases, temperature irregularities of ± 5 to ± 10 K should be taken into consideration. More elaborate systems of temperature control such as cascade control or control with the actual melt temperature as parameter have largely failed because of expenses or the inadequacy of sensors and have not found acceptation in injection molding. The necessity of controlling the temperatures of the hydraulic fluid and of the mold will be discussed in the appropriate sections. They are likewise recorded on the temperature page of the monitor.

6.2.2.6 Special Means of Controlling the Barrel Temperature

Machines for processing thermosets, rubber, and silicone, control the barrel temperature most of the time indirectly with an external heat exchanger, which is operated with a liquid heat-transfer medium (oil, brine). Controlling the temperature of the medium is done under the assumption that the barrel temperature can be affected faster. Because of the extent of all conceivable and practiced variants, they are not dealt with here [326].

6.2.2.7 Homogenizing the Melt Temperature

By varying the rotational screw speed and the back pressure a controlled change of the temperature profile along the feeding stroke can be achieved. The effect of the speed is only small [222]. However, programming the back pressure was more closely investigated. The necessity of such a move is not given for adequately long screws of 20 L/D in the same way as it is for mostly shorter thermoset screws and the higher temperature sensitivity of the processed materials.

6.3 Pressure Measurements

6.3.1 Pressure in Front of the Screw Tip

During injection molding, the pressure in front of the screw tip shows the same pattern as the hydraulic pressure (Fig. 7). In contrast to the hydraulic pressure, however, the former provides information about the material flow between nozzle and mold during the holding pressure stage. This permits a better insight into the process. Pressure in front of the screw tip is monitored, in particular, if installing a permanent pressure sensor in the mold is not feasible for lack of space or its effect on the cosmetics of the molding, or not justifiable for economic reasons.

Mostly piezoelectric transducers are employed as pressure sensors because of the rapidly occurring pressure changes. Their output is a signal which is proportional to the mechanical loading. It is subsequently amplified and converted into a corresponding voltage (Fig. 148). Only dynamic and quasi-static forces can be measured.

The charges on the faces of the crystal generated by forces are conducted to an amplifier with high input resistance. With a feedback resistor RT and the circuit capacitance CT, the amplifier input is compensated for losses, which result from the finitely high input resistance of the whole measuring circuit. The amplified voltage

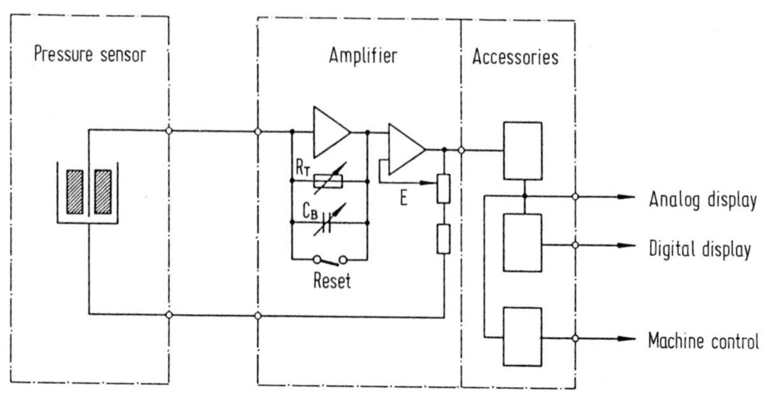

Fig. 148 Measuring system with piezoelectric material

output signal, which is proportional to the charge, is amplified a second time in such a way that the output reaches a voltage of 10 V under full load. This is accomplished with a potentiometer E. This adjustment results from the specified sensitivity of each sensor.

Charges, which exist at the input in the off-load state, can be electrically compensated with the help of a switch. This renders the system into a defined state before each measurement. It can be done automatically before injection by a signal from the machine control.

A drift of the measured signal, which is linked to the system, is better than 0.5%/min. even in the most unfavorable range. A noticeable increase is caused by humidity and soiling of the leads, which have to be kept clean by all means. For cleaning, the advice of the manufacturer should be observed [309].

For measuring the pressure in front of the screw tip, two different concepts are available. The first design, which can also be used in hot runners, has a metal cylinder as the foremost part of the sensor which is placed upon a membrane. Cylinder as well as membrane have to be in contact with the melt. This is achieved by an annular gap between sensor and bore. If plastic solidifies in this gap or becomes degraded, a correct pressure transmission to the membrane is not ensured anymore. Therefore and in accordance with the manufacturer's advice, such sensors are less suited for use under the conditions of a production run [309].

Sensors of the second design concept have the advantage of not being in direct contact with the melt. They are not part of the mold but of the machine and measure the stresses or strains respectively caused by the pressure of the melt in the nozzle body. Fig. 149 demonstrates how they are installed. The sensor is preferably mounted tangentially to the nozzle bore in the body. The output, which is proportional to the melt pressure, can be calibrated approximately by comparison with the output signal of a sensor in the mold or the hydraulic system.

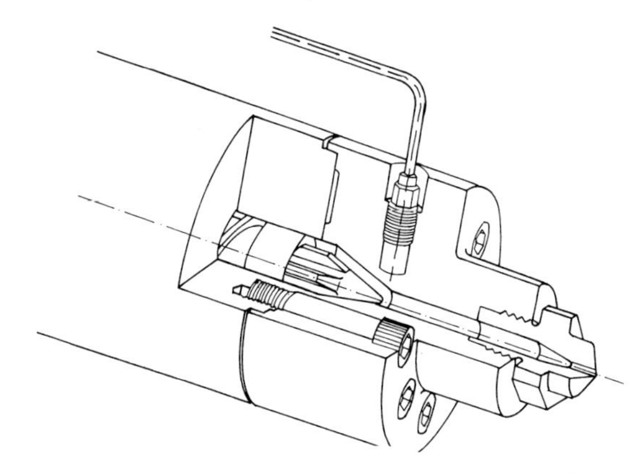

Fig. 149 Schematic display of an inserted piezoelectric stress sensor for indirectly measuring the pressure in front of the screw tip [325]

6.3.2 The Hydraulic Pressure

The pressure pattern of the hydraulics provides a picture of the flow resistance in the nozzle-gate system during injection even before a sensor mounted in the mold can respond. The pressure rises with the increasing flow resistance during the filling of the cavity. The hydraulic pressure is not well suited to judge the correctness of the holding-stage development, though.

To measure the hydraulic pressure, piezoelectric sensors are employed, too. Most of the time, however, pressure sensors on the basis of strain gages are used, which consist of a metal grid deposited by evaporation on a carrier foil (Fig. 150) [306]. The hydraulic pressure acts directly on the mensuration membrane [11, 12, 13, 248, 249, 250].

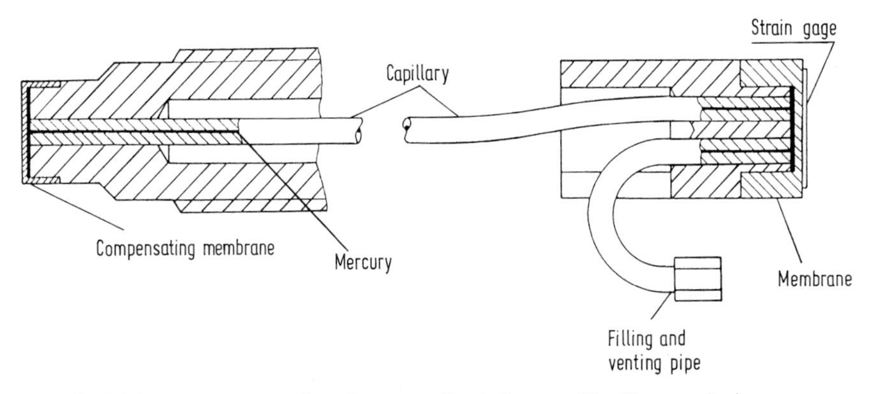

Fig. 150 Melt-pressure sensor based on use of a strain gage (Capillary version)

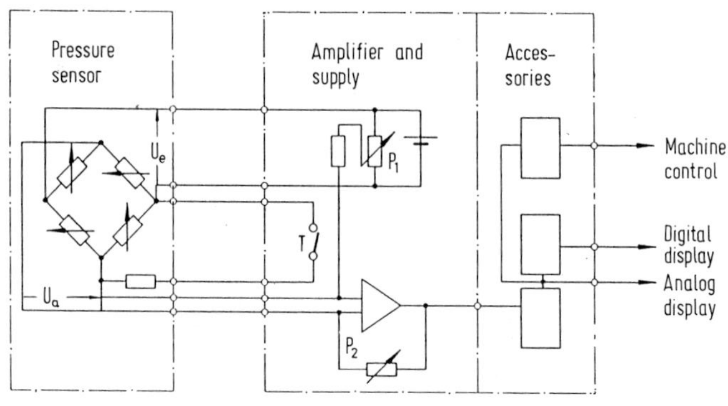

Fig. 151 Measuring system for strain gages [325]

The strain gage of the sensor is connected to a Wheatstone bridge and a power supply E_i (Fig. 151). If the gage is unstressed and the bridge balanced, the output voltage E_0 is zero. With the switch T, a resistor can be connected in parallel to one of the gages, which simulates a load and causes an imbalance of the bridge without external load. The simulated value usually corresponds with 80% of the measuring range. Thus a calibration of the circuit can be achieved in a simple way.

6.3.3 Cavity Pressure During Injection Molding

The qualification of the cavity-pressure pattern to make valid statements depends primarily on the location of the inserted sensor (close to or removed from the gate). The sensor does not record the pressure during the injection stage before the flow front has reached the location of the sensor and, during the holding-pressure stage, records it only as long as the molding does not shrink from this place and loses contact with the cavity wall. For this reason, sensors should be mounted close to the gate because the pressure pattern measured there describes best and for the longest possible time the formation of the molding. For an eventually planned control of the cavity pressure, this is the best location, too.

For monitoring the filling conditions, however, it is better to use a sensor far removed from the gate. One selects this position in molds which tend to flashing at points far away from the gate.

Because of cyclic dynamic loading, like that in front of the screw tip, piezoelectric pressure sensors are a favored option for measuring. One has the choice between two design concepts, too [309]. To directly determine the cavity pressure, the bore ends in the cavity (Fig. 152) and the face of the sensor is in contact with the melt. It contains an integrated piston, which transmits the pressure to the quartz element behind it. The face can be machined within limits to match contour and surface of the cavity. Even so,

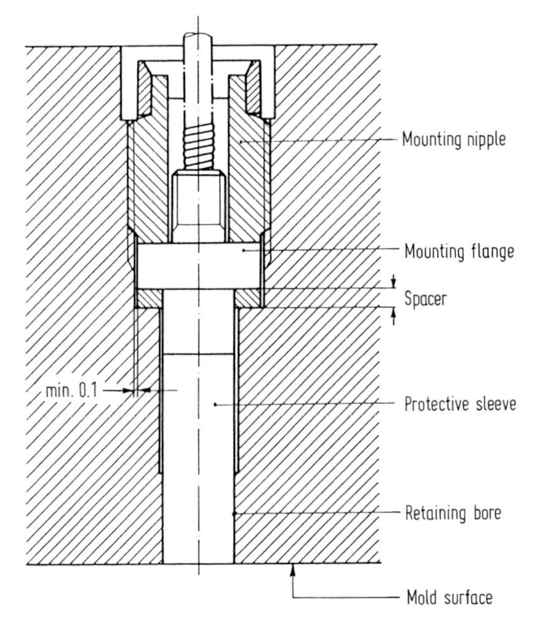

Fig. 152 Direct measuring of internal pressure with piezoelectric pressure sensor

they leave marks on the surface of the molding. This holds particularly true if sensors have to be taken out and remounted because of a mold disassembly.

Since marks on the molding surface are always undesirable if they are noticeable, one should measure in an area that does not become visible.

To avoid additional surface blemishes, so-called button mold-pressure sensors offer an easy way out. They can be mounted in the ejector plate behind an ejector pin or, in the absence of a suitable one, behind a dummy pin. The latter solution, of course, does not eliminate another mark.

A second design concept also permits an indirect measure of the melt pressure without causing additional marks. They are the so-called slide mold-pressure sensors. They are available on the basis of piezoelectric elements and strain gages with the same exterior. They are well suited to measure the cavity pressure and are likewise installed in the mold behind an ejector pin (Fig. 153), which transmits the forces from the cavity pressure to the sensor. Apart from losses due to friction, these forces are in proportion to the cavity pressure.

The pins have to be carefully fitted during mold assembly to reduce friction and be kept free from dirt to avoid unreliable transmission. This possible disadvantage is opposed by their advantage:

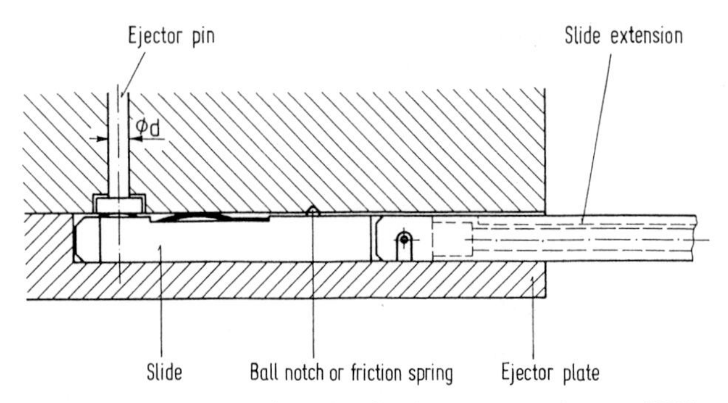

Fig. 153 Slide pressure sensor with piezoelectric element or strain gage [309]

- if an already existing ejector pin is used, no additional surface mark is generated,
- a slide sensor may be removed after a test run if a continuous control is unnecessary and replaced by a dummy slide. The slide sensor is then available for another mold.

Another method of determining a quantity proportional to the cavity pressure results from measuring the extension of the tie bars during mold filling. This procedure was initially generated to control the straining of tie bars but was further developed to convert the signal from the tie bar extension into a signal proportional to the cavity pressure [308, 312].

Fig. 154 shows the development of the clamping pressure during a molding cycle. One can recognize the rise of the clamping force, which indicates the strain on the tie bars, at the onset of injection. The lower curve represents the signal which results from

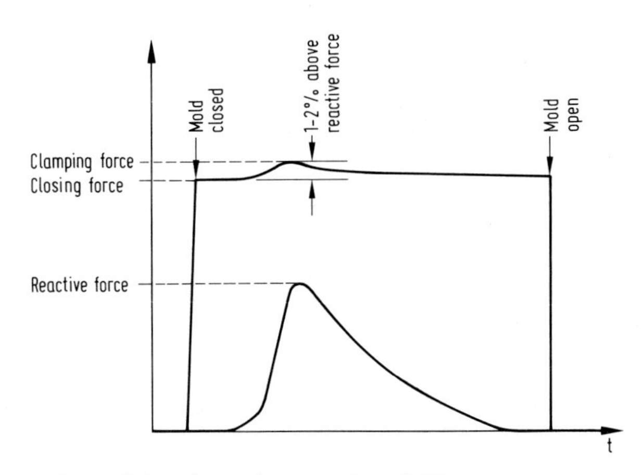

Fig. 154 Comparison of clamping and reactive force [325]

an electric compensation of the clamping force or the corresponding tie bar extension and amplification of the remainder of the signal. This curve actually presents the reactive force and is proportional to the cavity pressure. It is suited for controlling the injection process. Clamping and reactive forces form an equilibrium.

6.4 The Control Unit

6.4.1 Objects of Process Control

Fig. 155 shows quantitatively the pressure development in the cavity and discrete properties of the finished part associated with certain stages of the pressure profile.

Effects on the surface structure of a molding as well as degree of orientation and crystallinity in layers close to the surface, and thermal and mechanical strain of the melt can be linked to the filling stage.

Complete shaping of all contours in accordance with the cavity but also generation of flash and eventually resulting damage to the mold is caused by a steep pressure rise during the compression stage.

The holding pressure can be associated above all with weight and shrinkage of the finished part. Similar to the filling stage but to a lesser degree, crystallinity and orientation of the macromolecules, primarily in the interior of the molding, can be affected.

The object of process control is keeping those properties constant, which are determined by the pressure development in the cavity (Fig. 155). Appropriate control devices are used to achieve this goal.

If the optimum velocities of the flow front and the holding-pressure build-up have become known from a trial run or computer simulation, then the corresponding speed of the screw advancement can easily be calculated and the holding pressure adjusted

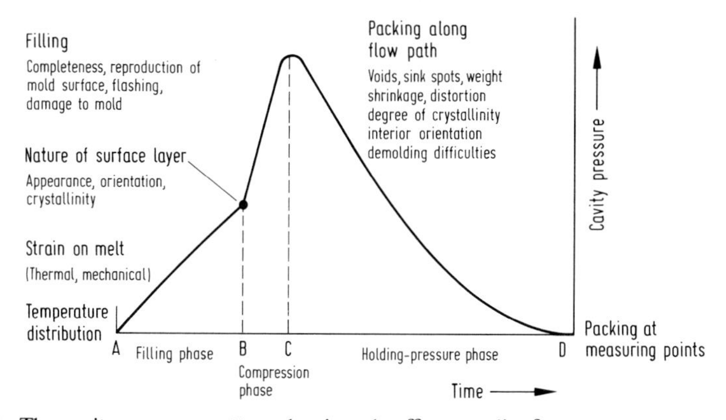

Fig. 155 The cavity-pressure pattern dominantly affects quality features

in such a way that differences in shrinkage along the flow path are minimized [313 to 315].

In older injection molding machines the stroke provides for constant injection rates and holding pressure. Modern machines usually allow for a programming with a number of set points. The more complex a molding is, the more necessary is a large number of selective points for the pressure profile.

If the position of points along the stroke can be freely selected, a point has to be set only where there is a change in velocity. In such a case 10 set points are certainly sufficient. If only an equidistant setting is possible, a larger number of settings may be required.

The transition from the injection stage to holding pressure occurs in dependence on a limit value, which can be determined as time, travel, or pressure dependent. Standard is the travel-dependent switch over to the holding-pressure stage. Usually, however, a switch over dependent on the cavity pressure results in closer tolerances of the weight of the molding because it can compensate deviations in feeding and axial screw speed [316 to 320].

A switch over dependent on pressure calls for machines with quick response, that is, with fast hydraulic elements and little time lag because otherwise the possibly very high rates of pressure increase, up to 2000 MPa/s, may lead to undesirable pressure peaks during the transition to holding pressure.

While a relatively high hydraulic pressure is needed during the injection stage to attain the desired injection rate, a lower level of pressure is called for during the subsequent holding- pressure stage. During this stage the melt in the mold cools down and reduces its volume. As much melt has to be conveyed into the cavity by the holding pressure as is necessary to compensate for this shrinkage. At the same time, cooling increases the viscosity of the melt until its flow is terminated by freezing and solidifying.

Accordingly, there is a subsiding pressure development in the mold as it is pictured schematically in Fig. 12. Here too, a pressure profile over time can be defined, although the number of selected program points is mostly smaller than that for the injection rate.

From this the demand results that axial screw speed and holding pressure have to be controllable with respect to time. This holds true regardless of the kind of control, open- or closed-loop control.

If one assumes that essential disturbances such as oil- and melt-temperature variations are compensated in advance, then no generally valid recommendation can be given whether an open-loop control of this process interval is sufficient or a more elaborate closed-loop control is required.

The accuracy and, above all, the reproducibility of the method of operation of injection molding machines depend fundamentally on the hydraulic components and the mode of their control.

6.4.2 Components of the Control Unit

As far as the electric and electronic elements of this section are concerned, the control unit is composed of input, signal-processing, and power stages.

Input stage: rotary, toggle, and preferably push-button, or touch switches, limit switches, emergency switches, decade switches, selector switches, etc.

Signal-processing stage: relays, interrupters, timers (analog and digital timing), diodes, transistors, microprocessors, and multiprocessors.

Power stage (output): solenoids, electric motors.

All these components should provide the timely and logical sequence of desired pressures, forces, and speeds during the course of a molding cycle. This requires a combined logical action of directional, pressure-, and flow-control valves in the hydraulic system or of purely electromechanical drive elements.

Appropriately, this sequence is open-loop controlled. It is done so in all injection molding machines either by a sequential logic or by programmed microprocessors, mostly multiprocessors, and so-called transputers [271, 272, 328].

Like manually operated machines of the first generation, controls of the following kind have almost disappeared from the market, too:

a) conventional cycle control with electromechanical components (relays) and pressure and speed control with manually operated valves,
b) conventional cycle control with solid-state modules and pressure and speed control with manually operated valves,
c) conventional cycle control with solid-state modules and remote control of proportional and pressure-control valves from a control cabinet,
d) like c) but with central stroke programming,
e) conventional cycle control with integrated digital control of flow-rate and pressure-control valves with or without central stroke programming with individual modules.

The logical sequence of the newest machines is carried out as follows:

f) conventional cycle control with integrated circuits for speed and pressure programming,
g) processor controlled cycle and parameters.

In addition to the components just listed, every system has control circuits for temperature control of barrel, mold, and hydraulic oil. Each of these controls can reliably accomplish individual functions within the broad range of injection molding operations.

6.4.3 Disturbances

In the following a summary of reasons will be presented which explain the need for specific actions against accidentally or systematically occurring adverse effects, so-called disturbances. Disturbances affecting injection molding are caused by:

human nature:

- unauthorized or unintentional interference with the operation of control elements,
- uncontrollable corrective actions, which result in deviations from set data (e.g. manual opening and closing),
- lack of sufficient monitoring devices (no control display, therefore subjective assessment),

environment:

- local temperature variations (during the day, seasonal),
- changes in humidity,
- contamination,

technology:

- reproducibility of mechanics, hydraulic systems, and control elements,
- mold defects,
- temperature (oil, barrel, mold), primary cause,
- from lot to lot variations of the material (comparable with melt-temperature effects,
- fluidity of material, feeding behavior.

With regard to negative effects, one can assume that human nature takes a leading position and contributes to a considerable degree to rejects or inferior product quality by its incapability of performing the necessary control manually. The following positions are usually taken by a mold of inferior design, and its temperature control (heat exchanger) resulting in temporal and local fluctuations. Frequent causes for imperfections are erroneously dimensioned runners and gates or the hot-runner manifold and its inadequate temperature regulation.

Every other disturbance is significant but takes a second place behind those mentioned here in detail. Thus, one can conclude that an important step in the direction of quality improvement, above all, quality constancy, is made by eliminating the causes for adverse effects e.g. by increasing the degree of automation, monitoring, and adequate construction before one takes recourse to the complicate methods of open- or closed-loop control.

Besides the mere sequence control, which cannot be replaced by a closed-loop method, every other control measure has the goal to reduce the effects of disturbances on quality as much as necessary and keep them within tolerable limits.

Every expense, clearly noticeable in accounting, should be measured by the achievements in this direction.

6.4.4 Methods of Process Control

One can assume that the molding sequence is largely open-loop controlled. Suitable for closed-loop control are:

- barrel temperature (standard),
- oil temperature (necessary for high-quality molding),
- mold temperature (standard),

above this, as far as needed:

- injection,
- pressure profile during compression and holding-pressure stages,
- clamping forces.

From discussions and publications, one can obtain the impression that closed-loop control by itself is superior to open-loop control. Such a conclusion is not admissible. Before a final judgment is formed, one always has to ask:

- which problem has to be solved?
- which qualitative and quantitative results are expected?

If a process has to be carried out and repeated (reproduced) which is very constant by its nature, then a stable open-loop control can be recommended as a general principle. If a process is subjected to unsystematic disturbances, the use of closed-loop circuits for controlling individual disturbances (individual parameter control e.g. temperature control) or for the whole process may be meaningful or even necessary.

The question of the frequency of the cycle sequences must be posed, too. Control operations of the magnitude of fractions of a second can hardly be used successfully in injection molding.

Thus, a clear-cut position in the sense of “pro” or “con”, open- or closed-loop control, can hardly be taken in injection molding.

The accuracy and, above all, the reproducibility of the procedure of operating an injection molding machine depends essentially on the hydraulic components and the methods of their control.

Digital control of the hydraulic system shows less hysteresis and a better reproducibility than proportional valves with equal response time. Only by applying servo valves in closed-loop circuits a reproducibility of $< 0.5\%$ is attained as it is with digital control. It is obvious, therefore, to equip open-loop controlled machines with a digital hydraulic system and closed-loop controlled ones with servo valves.

6.4.4.1 Cycle Control with Electric or Electronic Timers and Manual Control of Pressure-, Flow-Control, and Directional Valves

Such controls carry out solely the logical timing of a sequence of functions of one or several consecutive cycles automatically after manual presetting has been done. This sequence control can be a wired-in relay control, a wired-in programmed solid-state control, or a control with a programmable microprocessor with memory [250 to 261, 264, 320, 321, 322]. Pressure, speed, and stroke are not centrally but manually adjusted at the valves or with limit switches. Injection rate, injection pressure, holding pressure, back pressure, and screw travel are determined by the functional mode and the reproducibility of electric, electronic, and hydraulic elements. Changes in the material (viscosity changes from temperature variations) constantly influence pressure and temperature during injection and holding-pressure stages and, with this, the quality of the molded part. Nevertheless, if the quality is still in acceptable limits, then these control systems serve their purpose even today.

6.4.4.2 Control with Central Control Setting

A central control system is characterized by a central setting of all instructions at the control cabinet. The read-out usually appears on the screen of a monitor, frequently in color today. A function status can even be displayed with the so-called window technique. This permits an especially effective observation [274]. The instructions are manually given from general experience or taken from a data sheet. The results can be directly observed on the screen. This ensures a high degree of supervision and reliability in machine setting. Repetitive actions become reproducible. The instructions are converted into an executable code and stored in memory to be used by a solid-state control or a microprocessor [236, 239, 244, 250, 251]. Most of the time, set and actual values of vital parameters are displayed on the screen side by side. A connected printer can deliver a print-out for documentation. A comfortable method may allows the print-out to present only those data, independent from what is displayed on the screen, which are relevant for describing the machine's setting to ensure the required quality level. However, the accuracy of the machine's operation still depends primarily on the used analog proportional and pressure-control valves [207, 250, 252, 254, 257]. Pressure-control as well as flow-rate valves in proportional technique have different characteristics and exhibit a drift during their time of use [230, 252].

6.4.4.3 Control System with Central Control Setting and Digital Flow- and Pressure-Control Valves

While logical functions, sequence control, and display on the screen are processed in digital mode, the hydraulic execution of the instruction is performed by analog and digital systems. Since a number of years, digital hydraulic elements can be found in well-known injection molding machines [191, 192]. This control uses analog/digital converters on the hydraulic side as an alternative to analog servo valves for pressure and flow-rate control. The desired values for pressure and flow-rate are selected as usual from the control cabinet. With respect to efficiency and quality, there is no significant difference from the analog control of the hydraulic system. The available production results suffice almost all high demands of injection molding [207, 225]. With the common gradation of 8 (8^2–1 steps), they can have a resolution of 60 MPa [192]. If this gradation is insufficient, one is faced with an adaptation problem. The reproducibility is better than with proportional valves, though [191, 192].

6.4.4.4 Control of Injection and Holding-Pressure Stages

To control the injection rate, the position of the screw is determined in analog or digital mode and integrated over the time and the result fed back to the controller for comparing set and actual values (Fig. 156).

Presently, the controllers still in use are controllers with PI or PID characteristics and fixed parameters [321].

The output of a flow-rate controller causes a corresponding adjustment of the servo valve. The flow-rate control remains in effect until the switch over to the holding pressure occurs. As already mentioned, this switch over can be done in dependence on time, travel, or pressure. Here the pressure-dependent switch over is presented. Fig. 156 shows a control concept, in contrast to presently offered controls, which does not directly affect the holding pressure but a control of the pressure increase. Thus the transition to holding pressure is accomplished gradually and without pressure peaks.

Input of this controller is the differentiated signal from the cavity pressure (increase in cavity pressure) and a set value of zero. The controller affects the servo

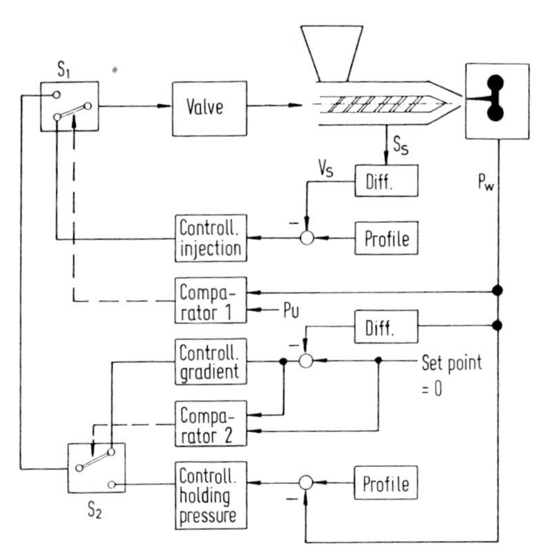

Fig. 156 Diagram of injection-rate and holding-pressure control by controlling the rate gradient

valve in such a manner that the rate of pressure increase becomes zero. When this is accomplished, the transition to holding pressure follows [264].

Since injection and holding-pressure stage differ largely in their control functions (amplification, response time, a range from solid to melt), for each phase a controller with fixed parameters, adjusted to this phase, is associated with the presently used analog controllers [322]. Another option is the use of controllers with parameters, which cannot be adapted optimally by necessity and have an sufficiently stable (slow) response. One can expect, however, machines to be marketed in the near future which have adaptive digital controllers with self-optimizing parameters [263].

With this adaptation of control parameters, an adjustment can be achieved in correspondence with the changing behavior of controlled systems during the individual stages and in dependence on the viscosity of the material

By knowing a suitable strategy [56, 57, 229, 238, 249, 250, 253, 254, 264, 265, 266] and with the availability of sufficiently fast servo valves, it is possible to largely eliminate disturbances. Thus, an excellent and constant quality of molded parts can be usually ensured.

6.4.4.5 Optimizing the Holding-Pressure Stage with the P-V-T Diagram

Always the same degree of orientation, residual stresses and shrinkage, and with those, a constant quality should be the goal of isochoric process control. Basis for this is the P-V-T diagram, which describes the dependence of the specific volume on melt temperature and holding pressure.

The connections shall be illustrated with Fig. 157. Injection with constant temperature is presented (line I). At point II injection pressure is switched over to holding pressure. At this point the holding-pressure control sets in. The pressure is kept constant for some time (line III). Beginning at point IV, a phase of constant volume is maintained (isochoric phase). This isochoric phase is especially important because, with it, one strives for a minimum of orientation and residual stresses and, thus, a minimum of distortion. This phase is decisive for the dimensional accuracy of a molding. There are different opinions and different theories and methods about a methodical approach of point IV. This is permissible because they are unimportant for the dimensional accuracy and affect, at most, the degree of orientation, the surface quality, and others. Reaching point VI on the 100 MPa line (ambient pressure) in a uniform way is decisive for the constancy of weight and dimensions of the molding. After point VI, the molding cannot be influenced anymore. It shrinks unaffected along the 100 Mpa line usually down to ambient temperature.

From the requirements of

- isochoric process control during the holding-pressure stage
- and constancy of the specific volume after arriving at the 100 MPa line during each cycle

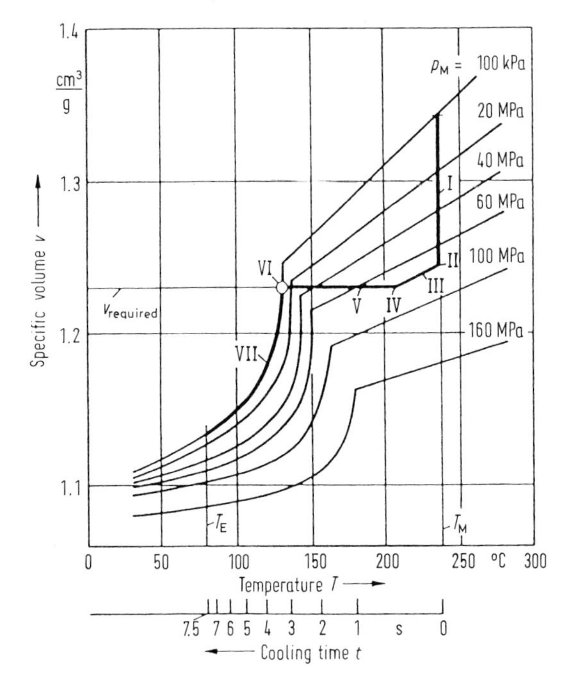

Fig. 157 Procedure during the holding-pressure time presented in a P-V-T diagram [268]
p_M: Melt pressure, T_E: Ejection temperature, T_M: Melt temperature

an optimizing strategy was developed [266, 268]. It renders the automatic finding of a point of operation feasible by using measured melt and mold temperatures and a computation of the cooling process (Fig. 158).

Machine control on the basis of microprocessors do not only permit to optimize individual control parameters but whole process segments. Thus, an optimum point in the P-V-T diagram can be accurately targeted independently from variations in melt or mold temperature and parts with constant weight are produced in this manner.

During setup, maximum holding pressure and holding-pressure time are provided as input. With these data, the microprocessor computes a holding pressure profile based on experience and suggests it as a starting point. Thereafter, the mean demolding temperature, the average wall thickness of the molding, and the effective thermal diffusivity of the material are put in. Then the cooling time is calculated by the system taking melt and mold temperature into consideration. Now the setup personnel has the option to change the proposed holding-pressure profile in such a manner that parts are produced, which conform with the demands.

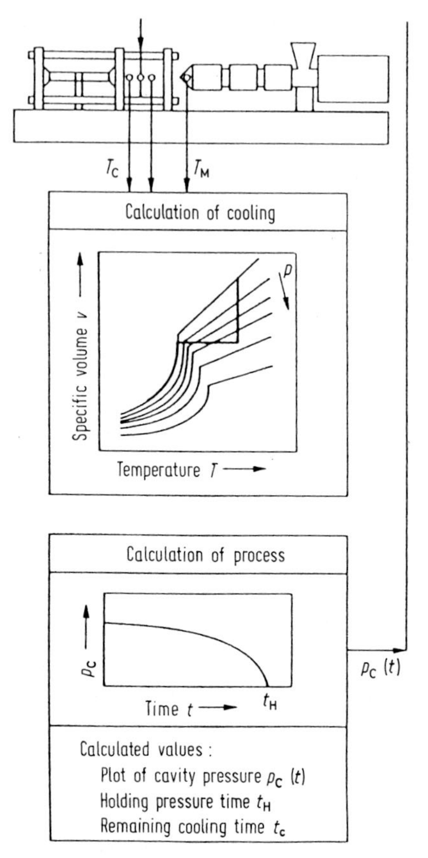

Fig. 158 Basic method of optimizing the holding pressure with the aid of a P-V-T diagram [268]
Measured values: T_C average cavity temperature T_M melt temperature before injection

After this phase has been terminated, automatic optimizing of the holding pressure is turned on. If melt and mold temperature change, holding-pressure profile and cooling time are adjusted by the microprocessor that always the same target in the P-V-T diagram is reached at the end of the holding pressure or at demolding time. Thus, molded parts with constant specific volume can be produced within limits, in spite of varying melt and mold temperatures [268].

Fig. 159 presents the weight variations during start up. Start up without optimizing is shown in the top section, with optimization in the lower section.

All this facilitates the setup process considerably and most machine manufacturers offer this option now. However, such controls have not yet been able to make a breakthrough.

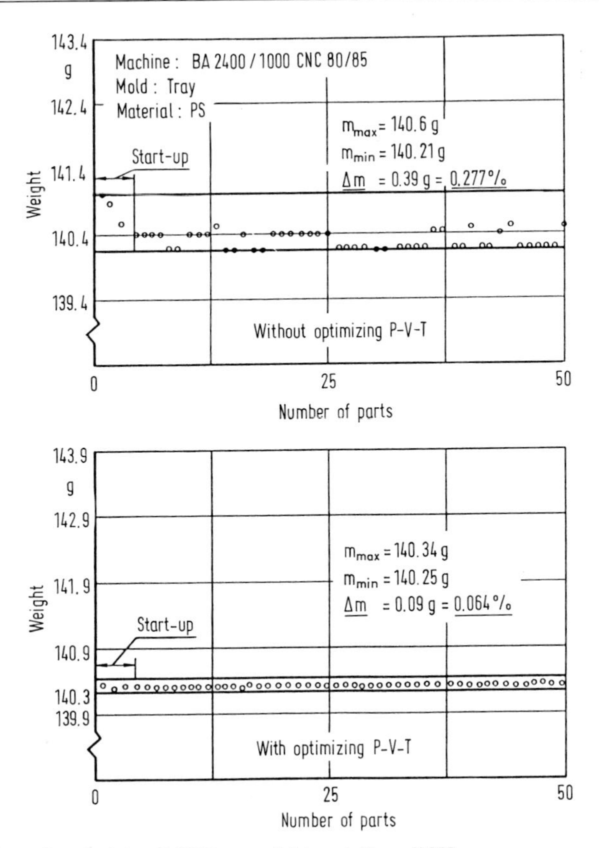

Fig. 159 Effect of optimizing P-V-T on weight variations [322]

6.4.4.6 The P-m-T Control

The P-m-T control has been developed from the P-V-T control. It calculates an ideal holding-pressure profile under the assumption that a mold cavity, which is considered constant, containing a constant amount of material (mass) produces constant moldings. The specific volume has to be determined with the weight of the molding. An acceptance of this strategy for a machine control has not yet become known. If someone is interested in this concept, he should refer to [268, 273, 280, 327].

6.4.4.7 Flow-Number Control

In [283] a product-adapted control is explained which uses the so-called flow number. This flow number is measured from a pressure-travel integral during the injection stage. It is generally known as filling index [59 to 64]. The flow number correlates with the viscosity. Viscosity affects the switch-over point, and its variations can be corrected by determining the flow number. Any application of this strategy is not known.

6.4.4.8 Integrated Control of Injection and Holding-Pressure Stages

In [268] a control is pointed out, with which it is supposedly possible to achieve equal orientation of molecules and crystallites, equal dimensions and residual stresses with one and the same mold on different machines even with different screw diameter. This method determines an optimum injection time, which remains constant during the process. Injection remains invariable within small variations in temperature or viscosity. The injection energy is measured in front of the screw. It is tied to a holding-pressure control. The selected algorithm takes care of an automatic adjustment with a gradient to achieve the desired dimensions of the molding. Viscosity variations can be compensated.

6.4.5 Microprocessor Control Units

The concentration of the logical circuits of a microprocessor onto a single silicon chip accomplished by Intel Corporation in 1969 has since led to a situation where almost all marketed injection molding machines use this universal logic element. In conventional logic, all sequential functions were programmed by means of a wiring matrix. In a microprocessor control system, this matrix is replaced by software, which contains the instructions telling the processor how it has to process the signals from the operator panel and link them with other process parameters stored in electronic memory modules. In other words, the microprocessor takes over the entire logic of the machine run. Since the processing speed of such a vast amount of data is not arbitrarily high, the sequence logic is divided into individual independent blocks with separate microprocessors running their programs. Therefore it usually takes several processors for the entire logic. This is called multiprocessor technique in modular design. The latest version are so-called transputers, which may become the successor of the multiprocessor technique.

6.4.5.1 Design of a Microprocessor

Fig. 160 presents a block diagram of the design of a microprocessor. The so-called data bus is the primary path between memory and processor. A 16-bit-wide data bus allows a full word (16 bits) of memory to be moved into the processor in a single operation. Fetching the data by their addresses and returning them to the correct address is done with the address bus. All operations are controlled via the control bus.

The heart of the microprocessor is the central processing unit (CPU), which contains the control unit for controlling and monitoring the data exchange, and the arithmetic-logic unit, which performs basic operations. The Read Only Memory

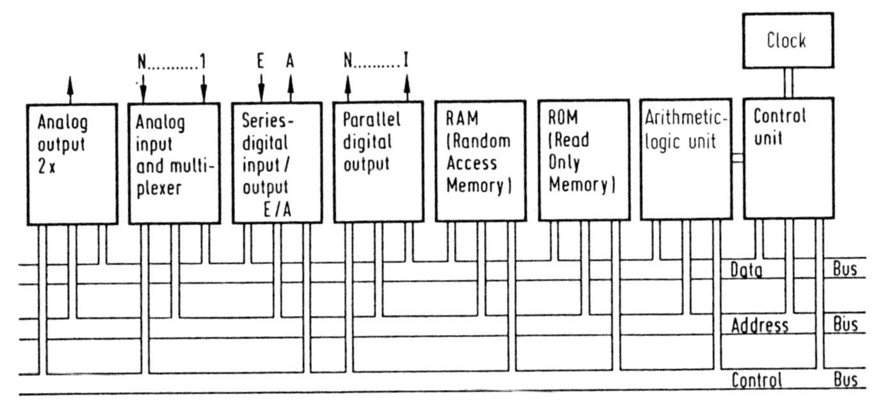

Fig. 160 Structure of a microcomputer

(ROM) contains fundamental data and instructions, while the Random Access Memory (RAM) stores all additional data generated in the data-processing operation and the necessary instructions. From there they can be retrieved anytime.

Input and output modules connect the processor with a machine, the injection molding machine in this case. Analog signals, such as temperatures, pressures, displacements, and more are converted to digital signals. So are the instructions from the control panel for activating hydraulic valves.

The first generation of microprocessor controls mostly employed a single-processor system, that is, all instructions during a cycle were executed in sequence. Because response times were usually too long for injection molding, with the exception of temperature control, this system became unsuitable, especially if a connection with peripherals was considered.

Consequently, one proceeded rapidly to a decentralized system, which connects several microprocessors with one another (multiple-processor system). Such systems can execute several instructions at the same time (multi-tasking). With this a sufficient processing speed for every cycle sequence can be attained. It can also handle extensive programs for automated devices around the machine with adequate speed.

The latest improvement, the transputer technique, was developed by the British company INMOS about 7 years ago. It reportedly can execute more data in less time by built-in communication-link interfaces [271]. This favors an easy expansion of automation because their inter-unit communication capability increases with each addition and does not require more storage space. A presentation of processor-controlled systems is made with Fig. 161 [271, 272, 328].

The processor use in an injection molding machine is often equated with process control. However, control of injection or holding-pressure stage are still the exception today. There are good reasons for this. In most cases, an open-loop constant-parameter

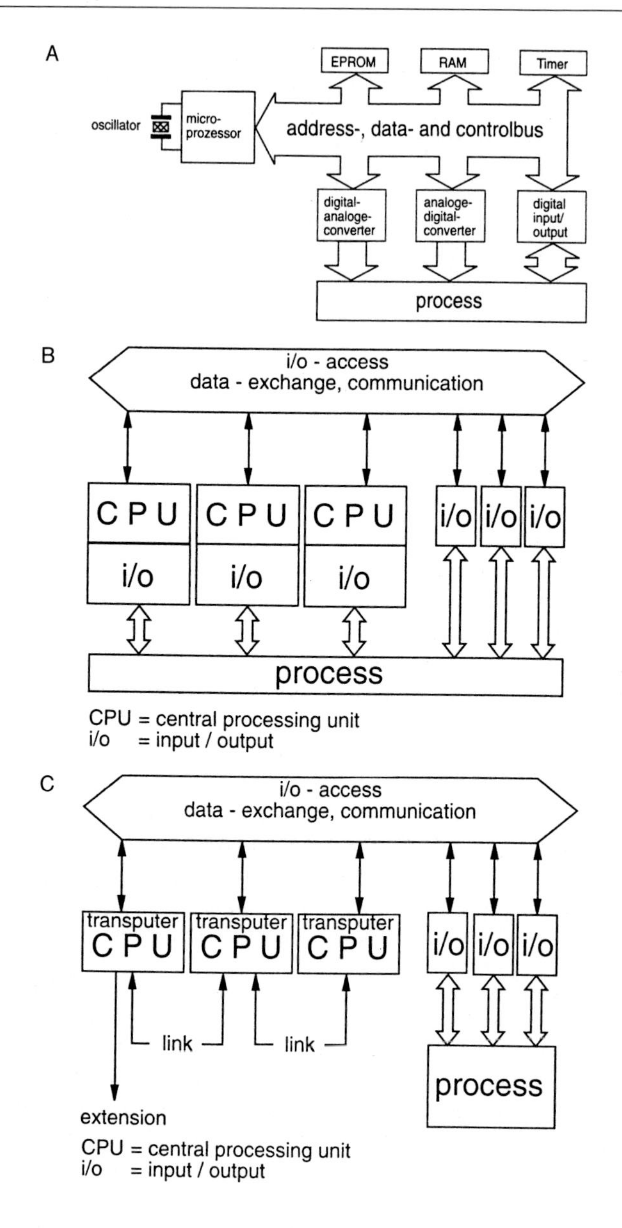

Fig. 161 Processor systems (Courtesy Battenfeld)
A: Single-Processor system, B: Multi-processor system, C: Transputer system

controlled machine produces parts which are of equal quality with those produced by a closed-loop controlled machine. On the other hand, it is reasonable to employ the immense capacity of a low-cost computer to quickly process large amounts of data in designing controls which can be adapted to the molding process or even to the product.

A microprocessor or multi-processor system has to carry out the following control and monitoring functions:

a) Standard functions:
- Sequence control,
- Timer functions,
- Displacement-actuator function,
- Alphanumeric data display,
- Limit-value monitoring,
- Malfunction indication.

b) Monitoring functions:
- Self-diagnosis of malfunctions,
- Screen display, dialog program,
- Control of setup procedure by means of operator guidance,
- Calculation of operating data,
- Input of data from a data carrier,
- Connection to a central monitoring system and evaluation of data via central interface,
- Corrections via a central monitoring station.

c) Control functions:
- Control of temperatures of barrel, mold, and hydraulic system,
- Control (process control) of speed and holding pressure,
- Control by using the filling energy (so-called filling index control) [283],
- Control of switch-over point from injection to holding pressure,
- Optimizing the process by self-adaptation [340 to 344].

As long as the microprocessor assumes only the functions listed under a), its application is not different from conventional machine control. The only exception is the ease with which the program can be altered. Since this has to be usually carried out by the manufacturer, no advantage is obtained in most cases. The advantages of a processor for direct machine control can only be fully utilized if it carries out all monitoring and control functions. Help in malfunction diagnosis and in setup operations can provide welcome rationalization in numerous workshop situations. Above all, controlled setup operations, monitored on the screen, avoid subjective errors and promote standardization of the process strategy. Determining operating points by trial and error is inevitably eliminated. Calculations of indirect operating data, such as shot volume derived from the screw displacement, melt pressure from the hydraulic pressure, and clamping force from the clamping pressure, can be performed automatically [289].

Processor systems offer interfaces for connecting additional peripherals (documentation) [285, 290, 291, 293].

If frequent mold changes are involved, data input by means of a data carrier or from a central storage area is an extremely rational method, leading rapidly and reliably to reproducible operating conditions. It is relatively easy to connect microprocessors to a central monitoring system by means of a standard interface. This improves and facilitates supervision of operations. For reasons of compatibility as an essential condition, the user of injection molding machines must make it a requirement that the manufacturers use uniform interfaces. Multi-processor technology and the resulting simple control technique became the basis for the automation of injection molding beginning about 10 years ago [250, 339 to 344].

On the other hand, it must be borne in mind that microprocessor- control systems with either open or closed loops do not readily ensure the reproduction of good processing conditions. A whole series of factors, which my not at all, or only with difficulties, be handled by a processor, affect the reproductive behavior of an injection molding machine. Some of these factors are:

- Reproducibility of valve positions or control of their characteristic by integrated readjustment,
- Constancy of clamping force,
- Wear resistance of plasticating unit, function of nonreturn valve,
- Constancy of melt temperature,
- Constancy of initial material temperature,
- Uniformity of material,
- Reproducibility of mold sealing along parting lines, mold temperature, and deformation (effect of clamping force and hot-runner system),
- Constancy of ambient conditions, especially the temperature.

Appropriately, one uses single parameter control for the temperatures. Closed-loop control of the actual molding process is confined to the injection and holding-pressure stages.

6.4.6 Electronic Data Processing in Injection Molding

The injection molding machine is frequently a component of an integrated production system. In this connection, a considerable amount of data transpires. Essential tasks of planning, including part and mold design, of production control, and of the administration of operational data have to be handled. Since these data by themselves do not represent any value, it is necessary to perform their processing as economical as possible. A computer can take over such routine work from humans [281].

If an integrated system from ordering goods over production to the delivery of the finished product is generated and administrated, the accumulation of data in a molding shop is very large. Fig. 162 provides a view at some components of a computer integrated production system in immediate vicinity of the injection molding machine.

Beyond the machine area, one has to form a network of many stations (Fig. 163). Such systems are the prerequisite for the so-called Computer Integrated Manufacturing (CIM) [289, 291, 347 to 352 and for more references 395]. Today, all machine manufacturers offer systems of this kind, which can be incorporated into the CIM needs of a shop.

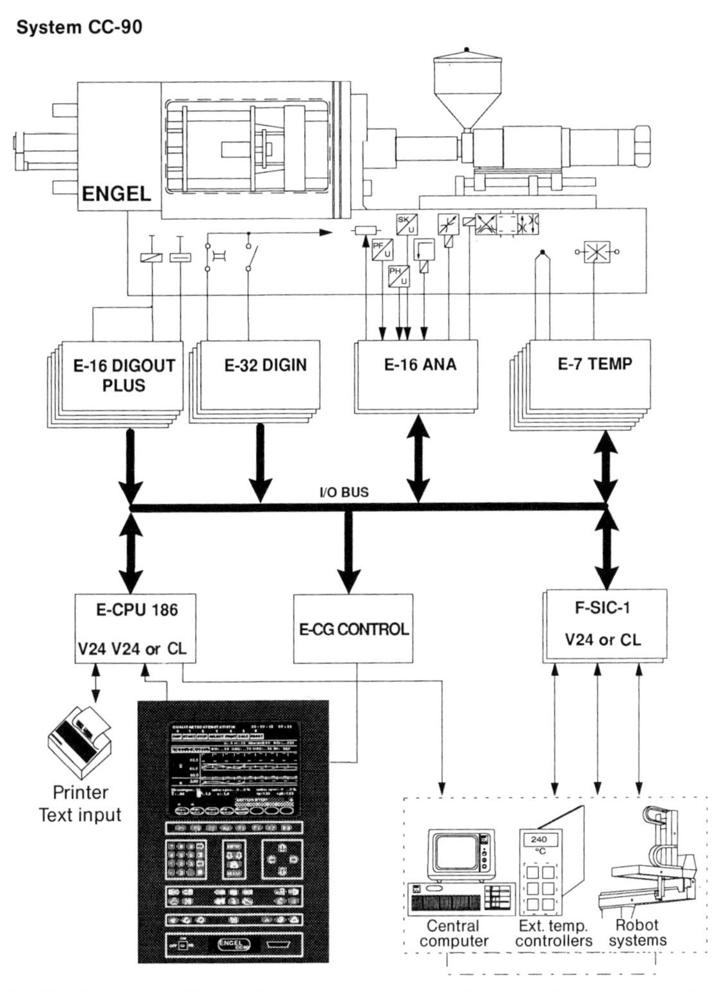

Fig. 162 Configuration of a multi-processor control with connection to a central computer, external temperature control and control of robot systems (System Engel)

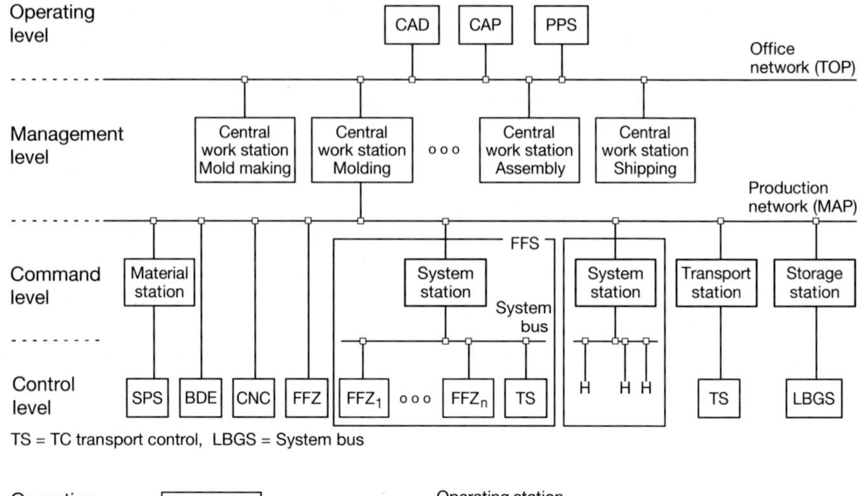

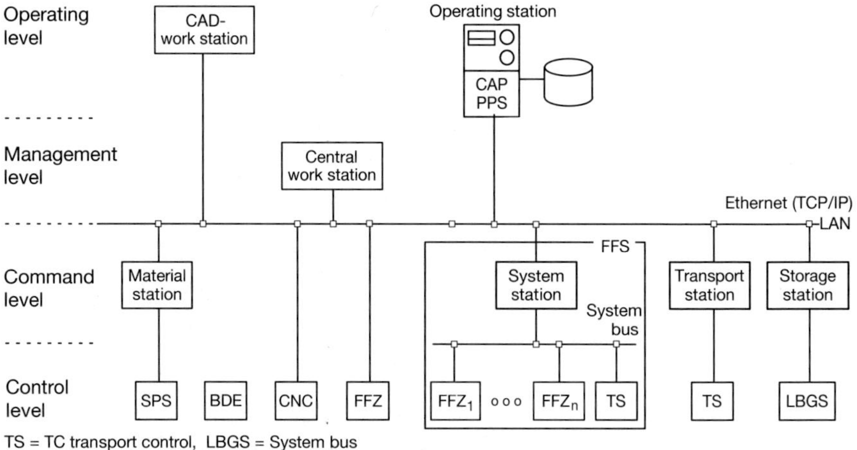

Fig. 163 Concept of a processor system for various organizational levels of a future molding shop (top); presently realizable processor network operation for medium-sized molding shops (bottom) [289]

6.4.7 Computer-Controlled Automatic Injection Molding Installation

Of the 2 to 5 hours of changeover time, 20 to 30% are allotted to preparations, 25 to 35% to taking down and setting up the mold, about 6% to mold repair and adjustment, and 20 to 30% to restarting production. This time can be reduced to several minutes by various measures:

- Work preparations before the machine is stopped for the changeover,
- Mold change and eventual exchange of the plasticating unit with devices operating fully automatically,
- Automation of machine setup, control circuits,
- Master computer station for automatic operation of the installation.

A conveyer system for molds and plasticating units is of particular interest for the automation of a molding shop. Working automatic injection molding installations were exhibited for the first time in Europe at the K'83 in Duesseldorf (Fig. 164). A master control computer supervises the whole production sequence and selects a new order after completion of the previous one. It takes care that the old mold is exchanged with the correct new one, that the plasticating unit is exchanged or provided with another material as needed, and new processing data are available for the selected machine.

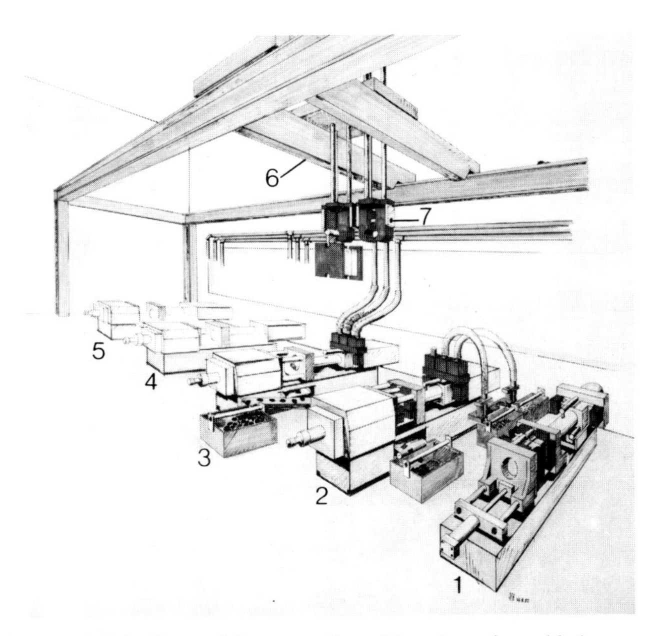

Fig. 164 Automatic injection-molding operation with automatic mold change, exchange of plasticating units and automatic material supply (Courtesy Netstal)
1: Preheating station for mold and plasticating unit, 2 to 5: Molding machines, 6: Automatic hoist, 7: Grab with or without mold to be exchanged

Such an installation can automatically execute and monitor the following steps:

On the clamping side:

- preheating of molds,
- automatic transport of the mold from the preheating station to the clamping unit,
- automatic mounting of mold,
- automatic connection with oil lines for hydraulic core removal, with cooling lines for mold cooling, and electric lines for temperature and pressure sensors in the mold,
- automatic daylight adjustment,
- automatic adjustment of clamping force,
- automatic adjustment of opening stroke.

On the injection side:

- preheating of plasticating unit,
- automatic mounting of plasticating unit,
- automatic coupling of screw,
- automatic connection to energy supply for plasticating unit (Fig. 165),
- automatic adjustment of barrel retraction,
- automatic device for supply with different material.

Moreover, such systems can be combined with automated storage areas for storing molds and/or plasticating units.

The computer can be provided with all necessary data for the desired order and stores and processes them. A transportation system, equipped with grabs (Fig. 166) carries out all desired exchange functions. In case of a breakdown, the computer turns of the respective equipment in a systematic way. It records time, number of produced

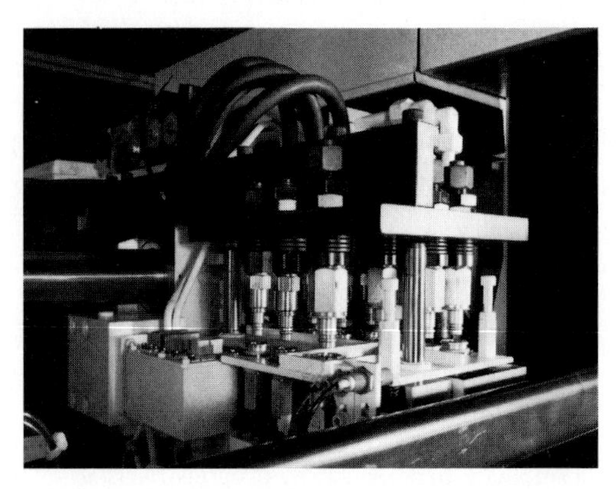

Fig. 165 Connectors for energy supply of an injection unit during automatic exchange (Courtesy Netstal)

Fig. 166 Automatic hoist during exchange of an injection unit (Courtesy Netstal)

acceptable parts and rejects. Of course, these machines have to be closed-loop controlled for optimizing. This is a must for an orderly operation.

Some installations of this or a similar kind (floor transportation systems) have already been realized. They are the result of strategic decisions. Reliable analyses of their economy are not yet available [289]. Considerable uncertainty exists in determining the time and expenses for planning as well as startup and the optimum procedure for mold exchange in dependence on a changing market situation among others.

7 Types of Injection Molding Machines

The fundamental design of common injection and clamping units has been discussed in the appropriate chapters. Only special types will be presented here. Machines with a rotating injection or clamping unit, which, in rectangular position, inject into the parting line, are not treated separately here since they have become standard during many years of use. The are standard machines with options for additional variations.

The systematism of separating processing technique and machine design with the latter being the main interest of this book is deliberately left here. To present machines of a special design partly in connection with mold design and processing technique independent of the pertaining course of processing and its particularities does not appear to be possible and meaningful for a full understanding. This is the reason for placing emphasis on describing processes in the following sections of this chapter.

7.1 Machines for Off-Center Molding

Some machines offer the option of off-center injection by either pivoting the injection unit through a relatively small angle with respect to the horizontal axis or laterally moving the injection carriage on the base. The only additional change required is an enlargement of the opening in the stationary machine platen by an amount determined by the displacement (Fig. 167).

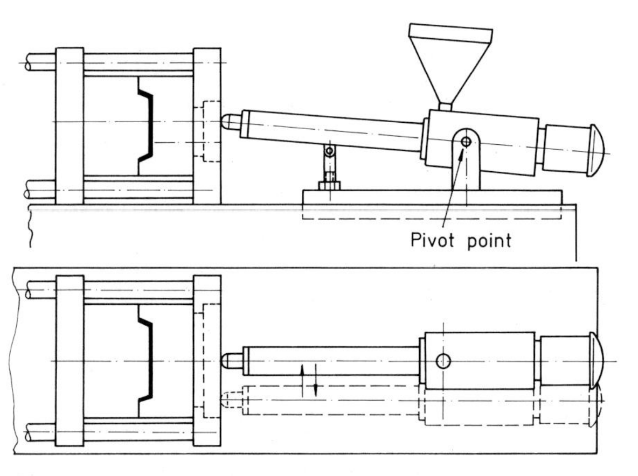

Fig. 167 Tilting and sliding injection units with enlarged opening in the stationary platen

7.2 Multi-Component Injection Molding Techniques

At first, multicolor injection molding was employed to produce keys for typewriters and cash registers. After the original in-house manufacturing of special machines needed for this procedure, an important market sector for such machines has developed, which was stimulated by the dcmand for multicolored taillights in the automotive industry. Multi-component injection molding received another impulse by the production of moldings which consist of two or several different materials.

Today, plastics are combined as follows:

- materials of two or more colors,
- rigid-soft composites,
- inseparable but livc connections,
- sandwich molding, among others with foam in two- or three-component technique.

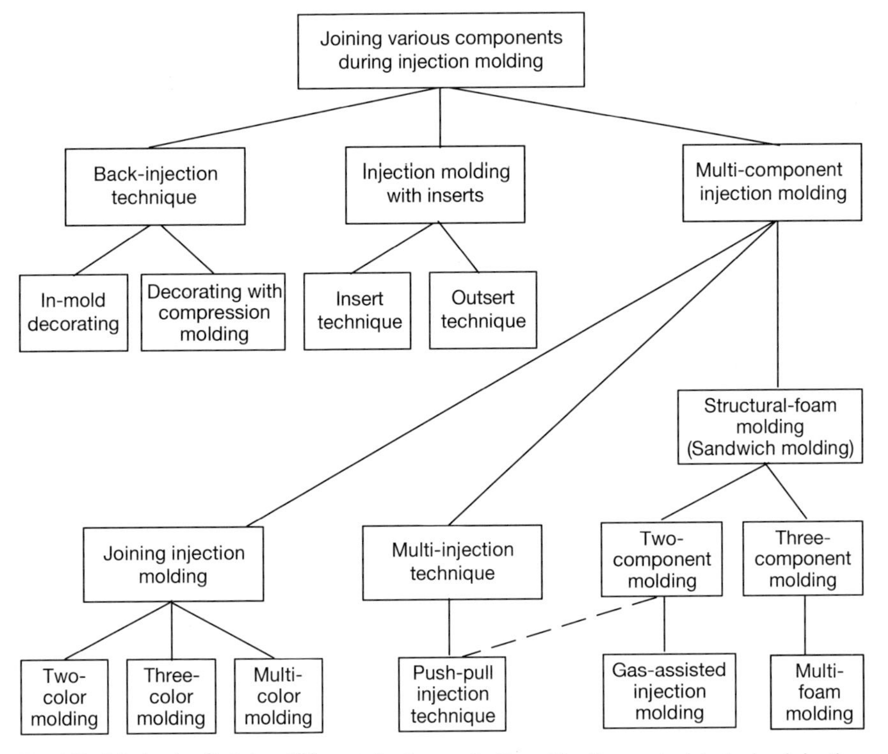

Fig. 168 Methods of joining different plastics or plastics with other materials during injection molding

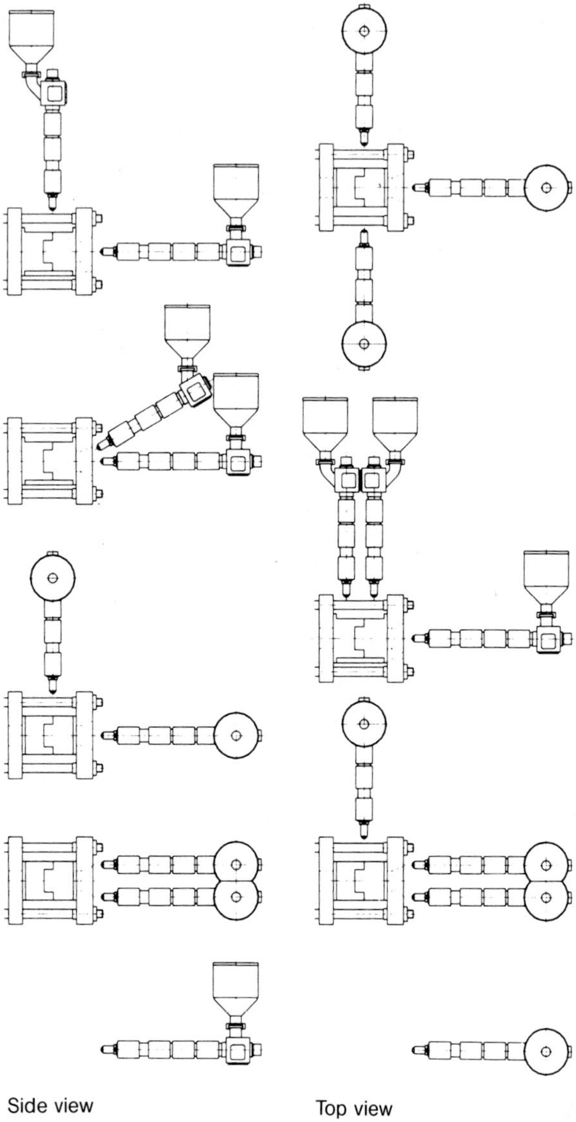

Side view

Top view

Fig. 169 Variations of injection molding machines for multi-combonent molding (Source: Kloeckner Ferromatik Desma GmbH, Malterdingen)

The whole range of possibilities is summarized with Fig. 168 [222]. This presentation clearly explains the significance of injection molding in the area of plastic composite parts. The corresponding machines can be classified into the following categories:

- Horizontal design with several injection units in parallel or angular position to one another.
- Vertical design with vertical clamping unit and lateral injection unit,
- Machines with two injection units operating through one single nozzle, which allows the injection of two components with one another, after one another, or in intermittent sequence.

Fig. 169 shows a variety of combinations for the multi-component technique.

7.2.1 Machines in Horizontal Design

The characteristic of this machines is the arrangement of two or three injection units, either parallel to the axis or at an angle with one another. They are usually adapted from standard machines as complete units with separate axial and rotatory screw drive systems and can be of the same or different size. They are independently fed, controlled, and heated. In each unit, the stationary platen is separately holed through for injection (Fig. 170).

While the mold or a mold half in conventional molding moves only in the axial direction of the clamp unit, multi-component molding requires the transport of the molding from one injection station to the next one by rotating part of the mold. This is done only when the mold is opened. The function is usually provided by a revolving component of the mold. It is even feasible to mount an additional rotating plate to the stationary machine platen. Multi-component machines of the design shown in Fig. 170 can also be used to produce special color effects by mixing the melt with a nozzle that joins the individually colored materials. Parts of composite materials, too, can be coinjection molded with coaxial nozzles (Sect. 7.2.3).

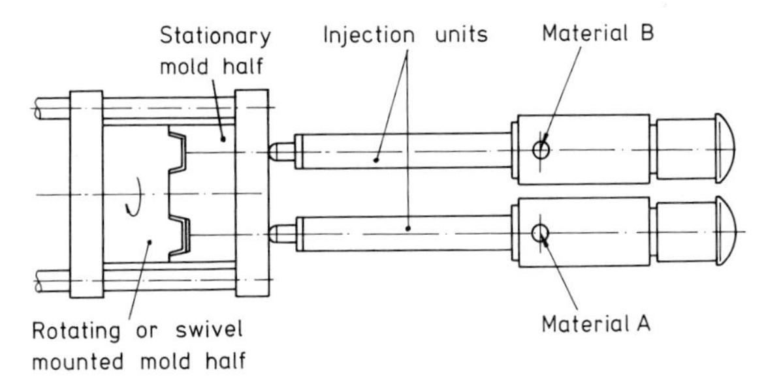

Fig. 170 Injection molding machine with two injection units for two-color molding

7.2.2 Machines in Vertical Design

A machine of the type shown in Fig. 171 has a vertically operating central clamping unit with three or four injection units grouped around it. It is utilized primarily for three- or four-color molding. The melt is generally injected into the parting line. One mold half, usually the top half, can rotate around a vertical axis (rates of rotation: 180°, 120°, or 90°) and moves the part molded at station 1 subsequently to the stations 2 and 3. At each of these stations new material is injected into the free space that is left after closing and reclamping. Since the fourth side has to be left unoccupied for ejection, two injection units are placed at one side in a four-color molding arrangement. For an automotive taillight, for example, the color (material) sequence can be transparent natural – amber – red side by side in the adding-on process or the entire part is molded in transparent natural with amber and red subsequently molded on top of it, which is called the adding-up technique (Fig. 172). Both cases represent a three-stage process. The same equipment also permits a two-stage technique (Fig. 172). In the first stage, a transparent natural part is molded. After a 180° rotation, the colors amber and red are injected simultaneously through separate runner systems (Fig. 172).

The application of such machines can even be extended further. During the last years, a drastic development in rational injection molding has taken place by using multi-component molding for integrated assembly [338]. Incompatible materials can be used for items with living joints, inseparable though, produced by injection

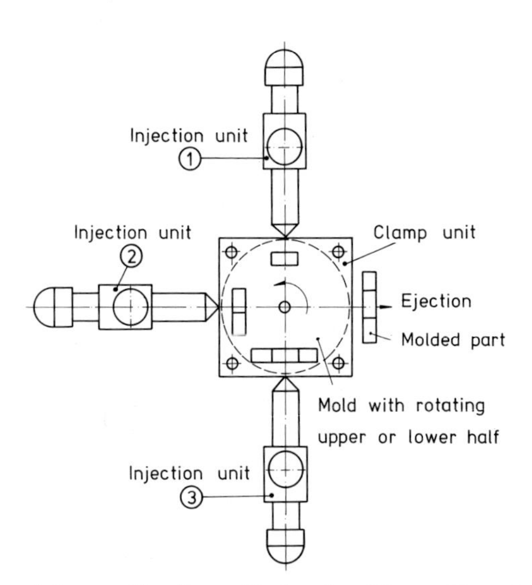

Fig. 171 Injection molding machine for multicolored moldings with vertical clamp unit, rotating mold half, and three injection units [337]

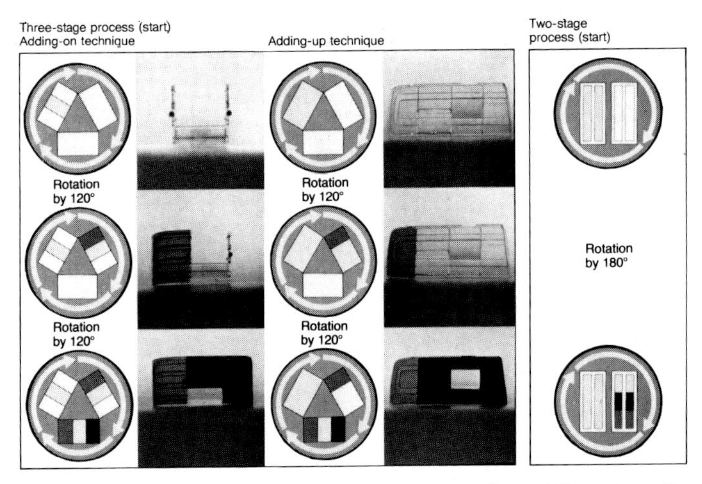

Fig. 172 Comparison of a three-stage with a two-stage process for molding automative tailights. Three-stage process: Simultaneous injection of three different colors into three different cavities and subsequent rotation of the mold by 120°. Two-stage process: Simultaneous injection of two different colors (components) in one mold position and subsequent rotation by 180° [337]

molding [360]. A multitude of combinative parts, ready to use, are produced in one working step, from toy figures, with which it all began, to automotive grills.

7.2.3 Coinjection Molding Machines

While until more recently, multi-component molding primarily employed the same basic material although in different colors to obtain optical effects, coinjection molding machines open up entirely new techniques and applications. With this, economical and technological characteristics have moved into the foreground.

Derived from the structural-foam machine which produced moldings with a compact outer skin and a foam core, the two-component coinjection process was developed (Fig. 173). How the two components are brought together, is pictured in Fig. 174. In the center of this figure, the ideal distribution of the two components has been accomplished. If the configuration is a complicated one, this uniform distribution is rarely attainable. Most of the time, a filling like that in the lower part is sufficient and is achieved with a liberal amount of leading skin material. A short shot of insufficient leading skin material, as pictured at the top, results in a rupture of the skin caused by the core material. This has to be avoided, because the core component should not become visible on the surface.

With this method a cosmetically impeccable surface can be attained enclosing a rigid, strong, and heat-resistant core if the same basic plastic is used but a glass fiber-reinforced grade for the core. With metal-filled core materials, electromagnetic (EMI) shielding can be accomplished [345, 355].

The principles of nozzle connections can be viewed in Fig. 175. It is necessary to provide appropriate elements to allow or block the flow of material in the desired order (Fig. 176). A large diversity of combinations have already been tested (Fig. 177). A good adhesion between components is not always of major importance.

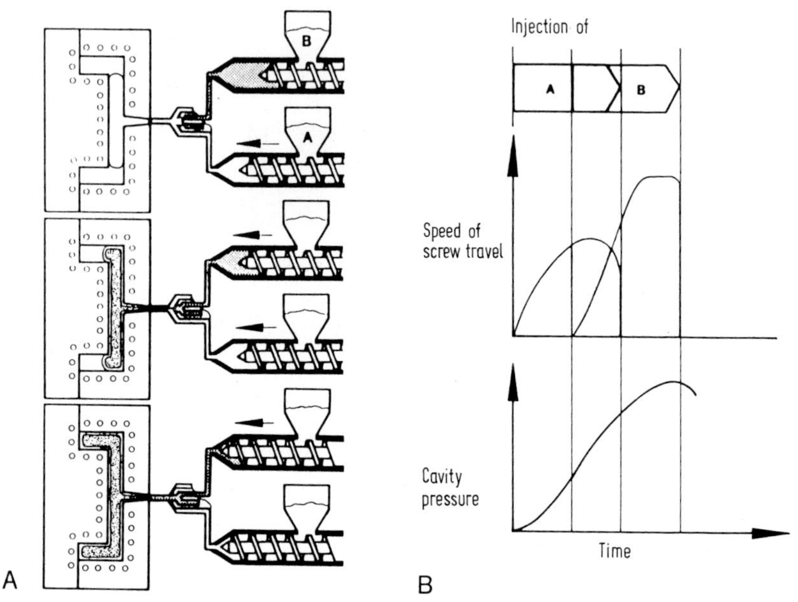

Fig. 173 Two-component injection molding (Courtesy Battenfeld)
A: Injection unit with two separate injection elements for the components A and B; presentation of a basic filling image. B: Cavity pressure and screw travel over time with typical simultaneity phase

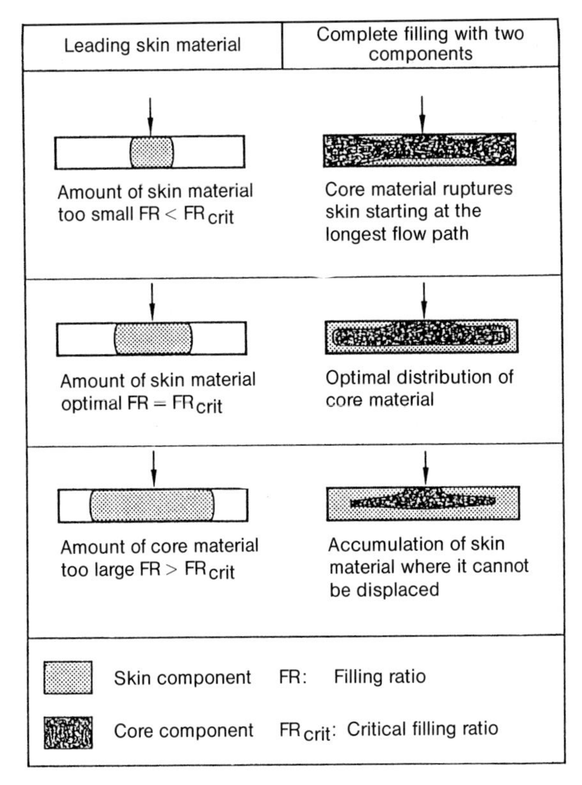

Fig. 174 Definition of critical filling ratio

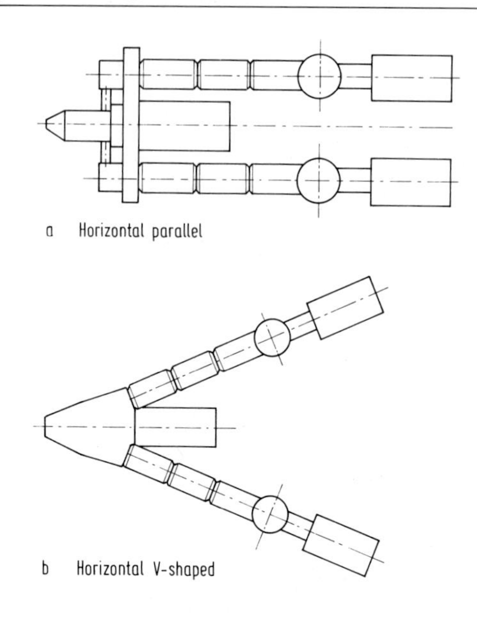

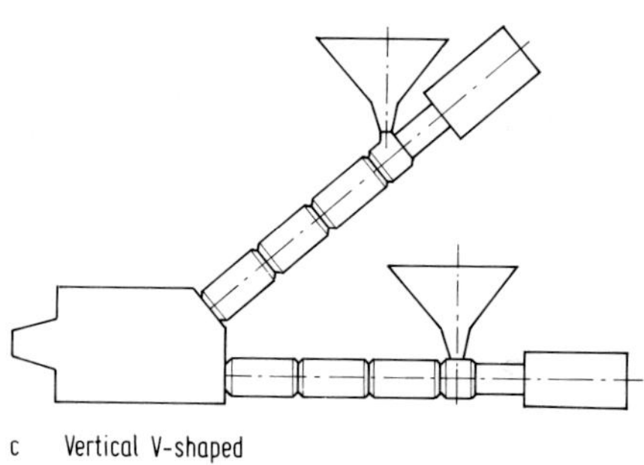

Fig. 175 Various arrangements of injection units for two-component co-injection

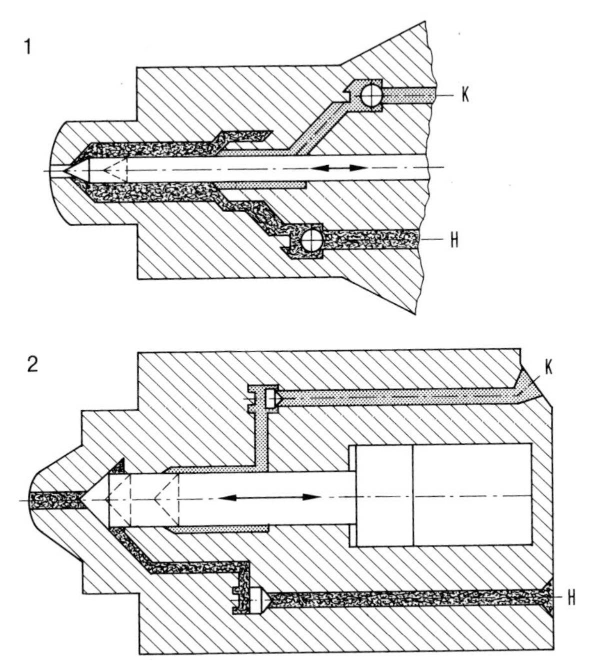

Fig. 176a Nozzle design with the option of freely programmable simultaneity phase, types 1 and 2. Side-by-side melt channels in nozzle head, coaxial joining of components before nozzle outlet

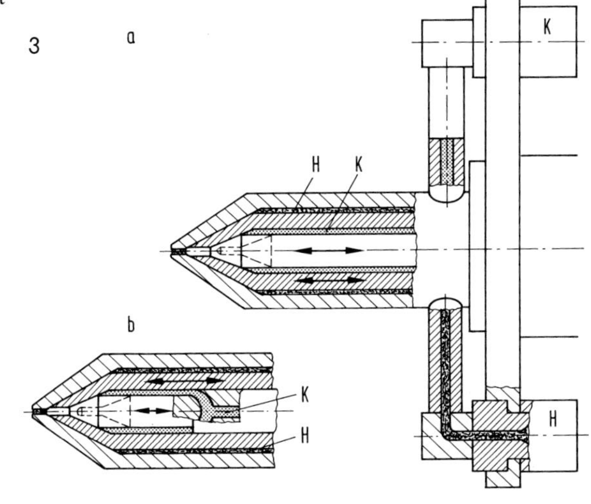

Fig. 176b Nozzle design with the option of freely programmable simultaneity phase, types 3a und 3b. Coaxial melt channels in nozzle head, coaxial joining of components before nozzle outlet

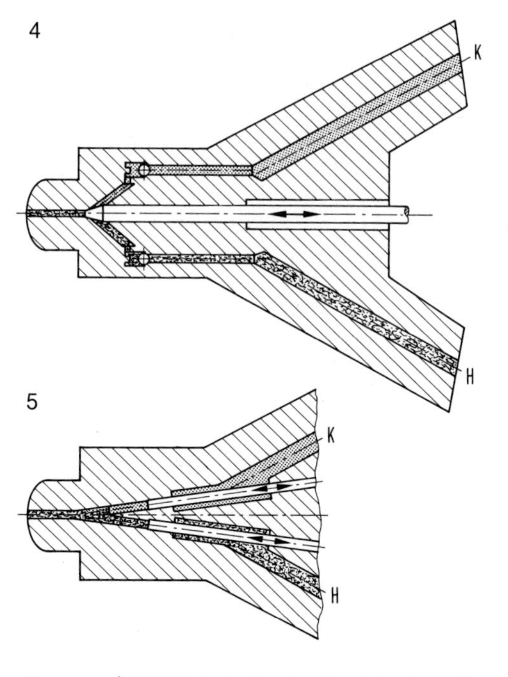

Fig. 176c

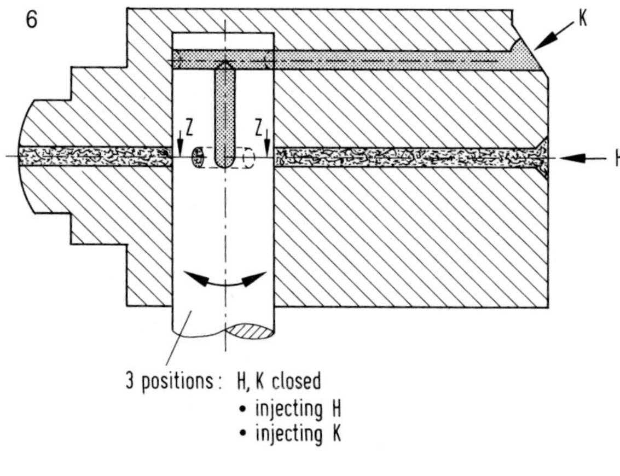

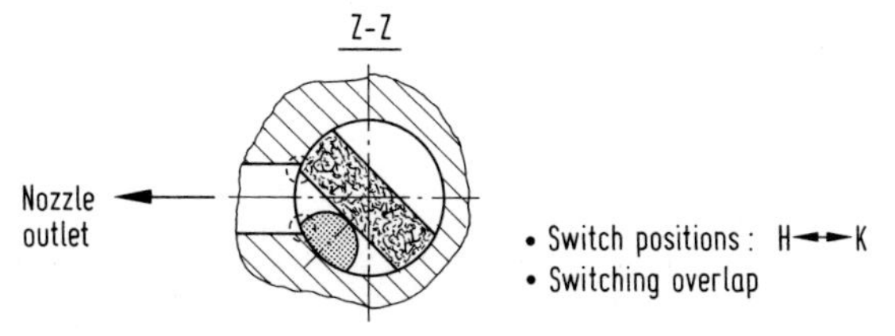

Fig. 176d

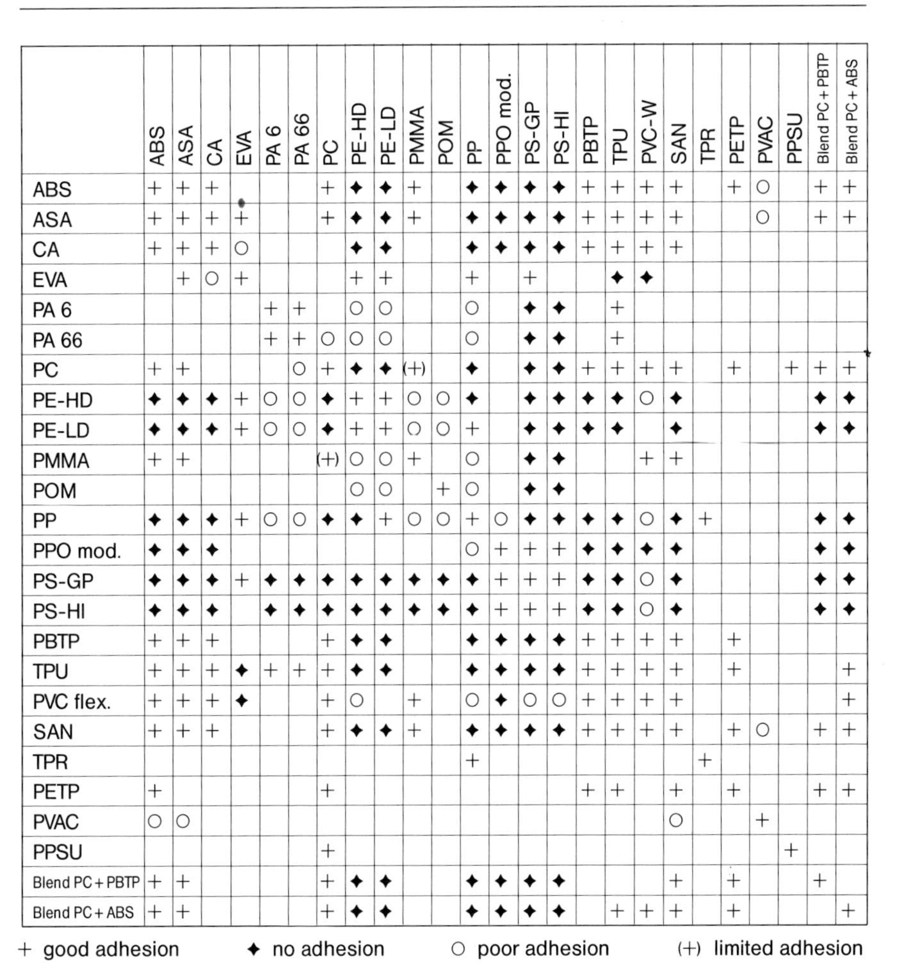

	ABS	ASA	CA	EVA	PA 6	PA 66	PC	PE-HD	PE-LD	PMMA	POM	PP	PPO mod.	PS-GP	PS-HI	PBTP	TPU	PVC-W	SAN	TPR	PETP	PVAC	PPSU	Blend PC + PBTP	Blend PC + ABS
ABS	+	+	+				+	✦	✦	+		✦	✦	✦	✦	+	+	+	+		+	○		+	+
ASA	+	+	+	+			+	✦	✦	+		✦	✦	✦	✦	+	+	+	+			○		+	+
CA	+	+	+	○				✦	✦			✦	✦	✦	✦	+	+	+	+						
EVA		+	○	+				+	+			+		+			✦	✦							
PA 6					+	+		○	○			○		✦	✦		+								
PA 66					+	+	○	○	○			○		✦	✦		+								
PC	+	+				○	+	✦	✦	(+)		✦		✦	✦	+	+	+	+		+		+	+	+
PE-HD	✦	✦	✦	+	○	○	✦	+	+	○	○	✦		✦	✦	✦	✦	○	✦					✦	✦
PE-LD	✦	✦	✦	+	○	○	✦	+	+	○	○	+		✦	✦	✦	✦		✦					✦	✦
PMMA	+	+					(+)	○	○	+		○		✦	✦			+	+						
POM								○	○		+	○		✦	✦										
PP	✦	✦	✦	+	○	○	✦	✦	+	○	○	+	○	✦	✦	✦	✦	○	✦	+				✦	✦
PPO mod.	✦	✦	✦									○	+	+	+	✦	✦	✦	✦					✦	✦
PS-GP	✦	✦	✦	+	✦	✦	✦	✦	✦	✦	✦	✦	+	+	+	✦	✦	○	✦					✦	✦
PS-HI	✦	✦	✦		✦	✦	✦	✦	✦	✦	✦	✦	+	+	+	✦	✦	○	✦					✦	✦
PBTP	+	+	+				+	✦	✦			✦	✦	✦	✦	+	+	+	+		+				
TPU	+	+	+	✦	+	+	+	✦	✦			✦	✦	✦	✦	+	+	+	+		+				+
PVC flex.	+	+	+	✦			+	○		+		○	✦	○	○	+	+	+	+						+
SAN	+	+	+				+	✦	✦	+		✦	✦	✦	✦	+	+	+	+		+	○		+	+
TPR												+								+					
PETP	+						+									+	+		+		+			+	+
PVAC	○	○																	○			+			
PPSU							+																+		
Blend PC + PBTP	+	+					+	✦	✦			✦	✦	✦	✦				+		+			+	
Blend PC + ABS	+	+					+	✦	✦			✦	✦	✦	✦		+	+	+		+				+

\+ good adhesion ✦ no adhesion ○ poor adhesion (+) limited adhesion

Fig. 177 Matrix for evaluating boundary adhesion between co-injected materials

◀ *Fig. 176c* Nozzle design with the option of freely programmable simultaneity phase, side-by-side melt channels in nozzle head, side-by-side joining of components before nozzle outlet

Fig. 176d Nozzle design without the option of freely programmable simultaneity phase, side-by-side melt channels in nozzle head and separate exit of components

7.2.4 Fi-Fo Molding Machines

Another method, which is aimed at a modern application, is the injection molding of two components, one after the other. An appropriate amount of material is prepared for the following injection process as it is known from blow molding as Fi-Fo Process (first-in, first-out) [70, 365]. This means that the component, which has been fed first into the space in front of the plunger or screw, is injected first, followed by the second component. In blow molding, it was the purpose of this method to achieve equal residence times for the stored materials. The same goal was pursued in rubber molding to prevent unequal curing. In injection molding, sandwich molding can be done without the usual two injection units.

The injection process is a sequential one [466, 471]. First an extrusion screw (Fig. 178) feeds the desired volume of the main component into the space in front of the injection screw. After completion, the injection screw feeds the necessary volume of core material into the barrel while it retracts. Then the switch valve is actuated and injection initiated. To the surprise of any skeptical viewer, the desired operation occurs reliably. Both components are placed in the molding in an ideal distribution. The core is fully enclosed by the primary material. Mixing can be avoided if the ratio of both volumes is controlled and kept in accordance with that of conventional sandwich molding.

The interesting part of this process is the ease of supplementing an already existing injection molding machine. The expenditures are considerably smaller than those for a conventional sandwich machine with two injection units.

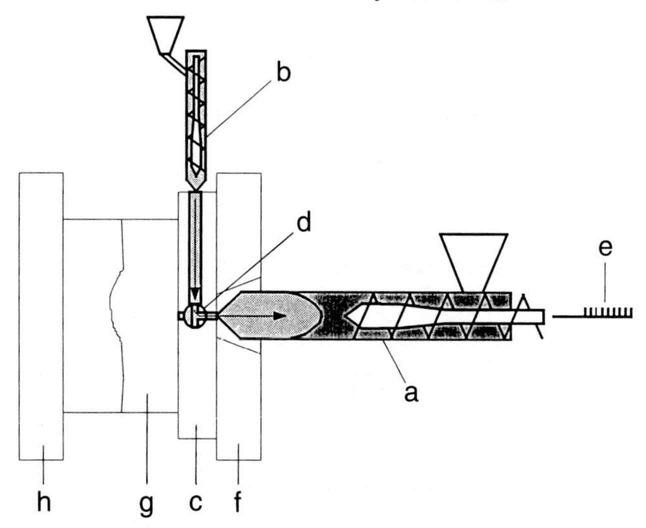

Fig. 178 Principle and function of a sandwich injection-molding mono-system with two plasticating units (Fi-Fo process).
a: Principle injection unit. b: Secondary extruder. c: Hot-runner manifold. d: Switch valve. e: Displacement transcuder. f: Stationary machine platen. g: Injection mold. h: Movable machine platen

This procedure is particularly suited for using recycled material as component b. The inventor replies to objections that there will be mixed regions and color impairment with the acceptable comment that only material of the same kind is primarily used.

One may assume that the success of this technique depends largely on the interest of the market in recycled materials [471]. Therefore its prospects seem to be good.

7.3 Gas-Assisted Injection Molding

Gas-assisted injection molding enables the molder to produce parts with a combination of thick and thin sections [366 to 374, 452]. Instead of a second component like in coinjection, an inert gas is used. There are already reports about the use of an easily vaporizing liquid, too.

A short and defined shot of melt is first injected into the cavity. Then the gas is injected into the already present amount of melt (Fig. 179). Gas injection can be part of the injection or the holding-pressure stage. Today the gas is nitrogen for reasons of low cost and safety. Air is not suited because explosive mixtures may develop [369].

Presently, there are two different design concepts for the gas supply with pressures up to 30 MPa maximum. Fig. 180 presents a piston unit for gas injection. The amount of gas can be well metered. However the piston cannot hold pace with the expansion of

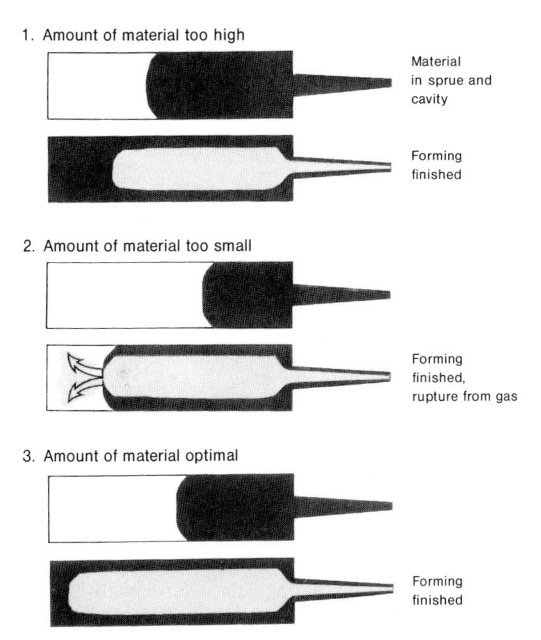

Fig. 179 Gas-bubble formation during a gas-assisted injection process [370], three different amounts of injected material

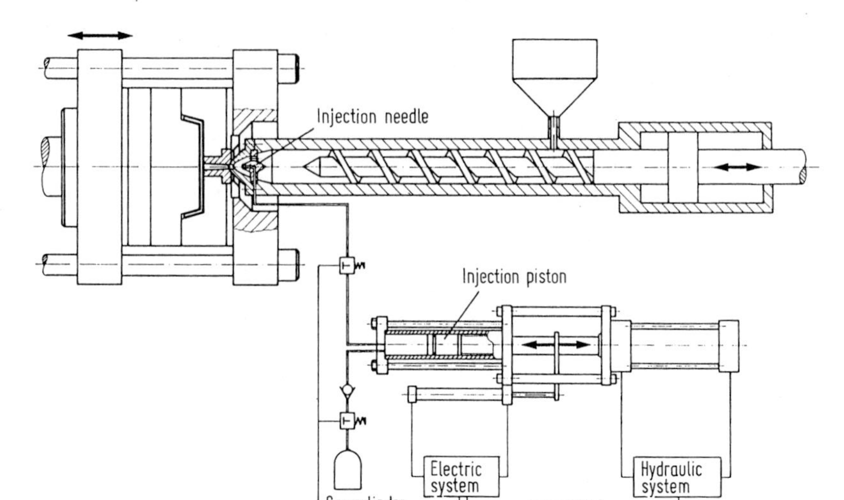

Fig. 180 Gas-assisted injection unit with piston injection of gas, Cinpres - method (System Peerless)

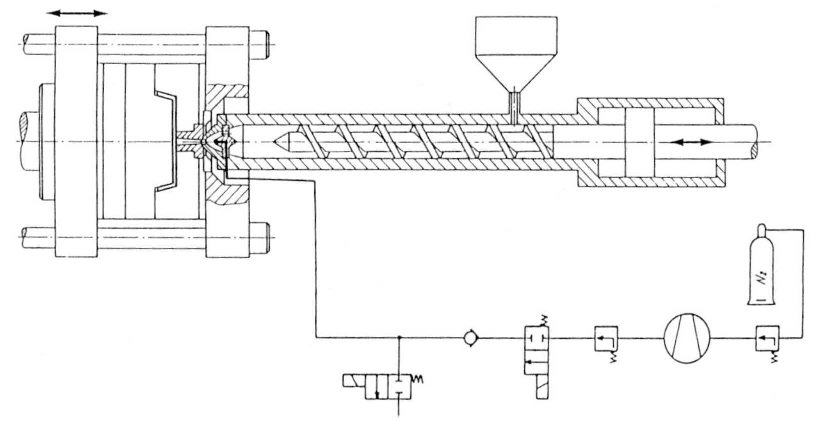

Fig. 181 Unit for gas-assisted injection with pump injection, 'Airmold principle' (System Battenfeld)

the gas and a heavy pressure loss has to be anticipated when the injection needle is opened. The gas filling is out of control because the gas expands faster than the piston can act. Therefore any effort to control this stage is useless.

The gas-injection unit shown with Fig. 181 operates directly with pump pressure. This pressure can supply the necessary holding pressure up to its maximum value. On top of this, one pump can supply several machines simultaneously.

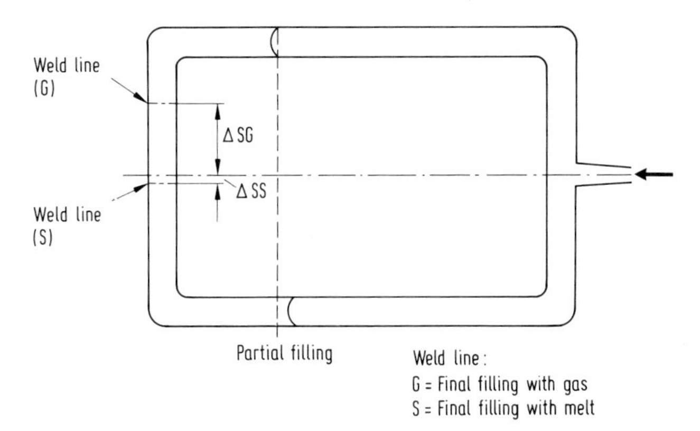

Fig. 182 Shifting of weld line from two flow fronts [388]

Fig. 182 displays the characteristics of melt and gas and illustrates best the entire process. In regular molding, if there is even a slight imbalance between two parallel channels at any time during the filling stage, then the weld line is moved in accordance with this imbalance. The weld line occurs in this case below the center line as it is predicted by all rheological simulation procedures. If, however, the cavity is filled by the gas-injection process, then the weld line is clearly moved into the opposite direction because the gas takes the path of lowest resistance. Parallel channels should be avoided if the same amount of gas is required in each channel.

A simultaneous filling with melt and gas should be prevented in any case since the gas is always in advance of the melt front and moldings of no use would be produced. This is not the case if gas injected with the melt cannot break through the melt front anymore because the latter has reached the end of the cavity. Then, however, a simultaneous technique does not make sense. The injection of gas can be achieved through the runner system or with a separate injection needle immediately into the cavity (Fig. 183). The gas displaces the melt from the inside by some kind of swell flow as it is known from flowing melt. This creates internal cavities, primarily in heavy sections. The gas pressure extends to the farthest end almost without any drop. This results in an economic production of moldings without sinks.

Although the increasing velocity of the flowing gas reduces the wall thickness but the thickness of the remaining section is almost independent of melt viscosity and process parameters, such as pressure, size of the first short shot, mold temperature, and timing of gas injection. However, the kind of material and its type and amount of filler have a crucial effect on the wall thickness [371, 388]. This contradicts in part existing theories [384, 385]. By now it is clear that a description of the process which uses models based solely on shear viscosity does not suffice. It seems that the ratio between shear and strain viscosity is of eminent significance [383].

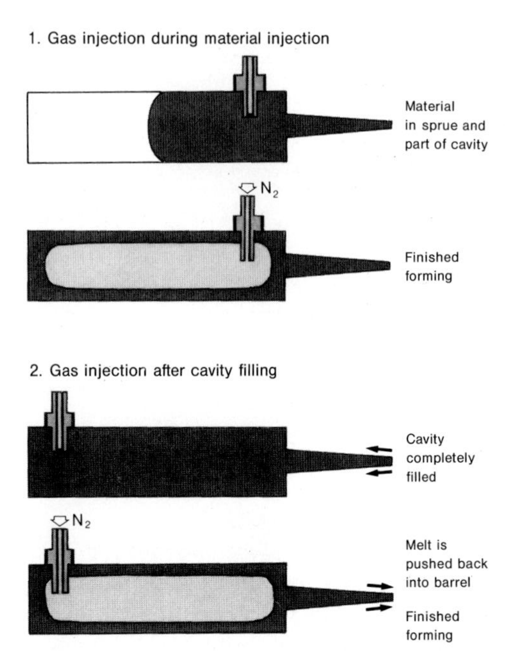

Fig. 183 Variations of the gas-assisted injection process [371, 388]

A particular degree of efficiency is accomplished by a reduced cooling time because of less material concentration and the resulting shorter cycle. At the same time, the cavity pressure is reduced and, with it, the need for high clamping forces.

Advantages of gas-assisted injection molding:

- reduction of parts weight (especially of parts with heavy sections),
- reduction of cooling time (compared to solid parts),
- improved holding pressure, less sinks,
- less distortion (compared to solid parts due to less orientation and lower residual stresses).

Disadvantages:

- surface blemishes from jetting in large cross sections,
- surface marks from longer standstill of the flow front during switch over,
- poor reproduction of the cavity surface because of nearly pressureless injection.

Special nozzles are employed for gas-assisted injection molding, which allow the injection of gas but with the gas channel protected against the entering of melt. Since in this process the wall thickness of the molding can only be affected by the speed of

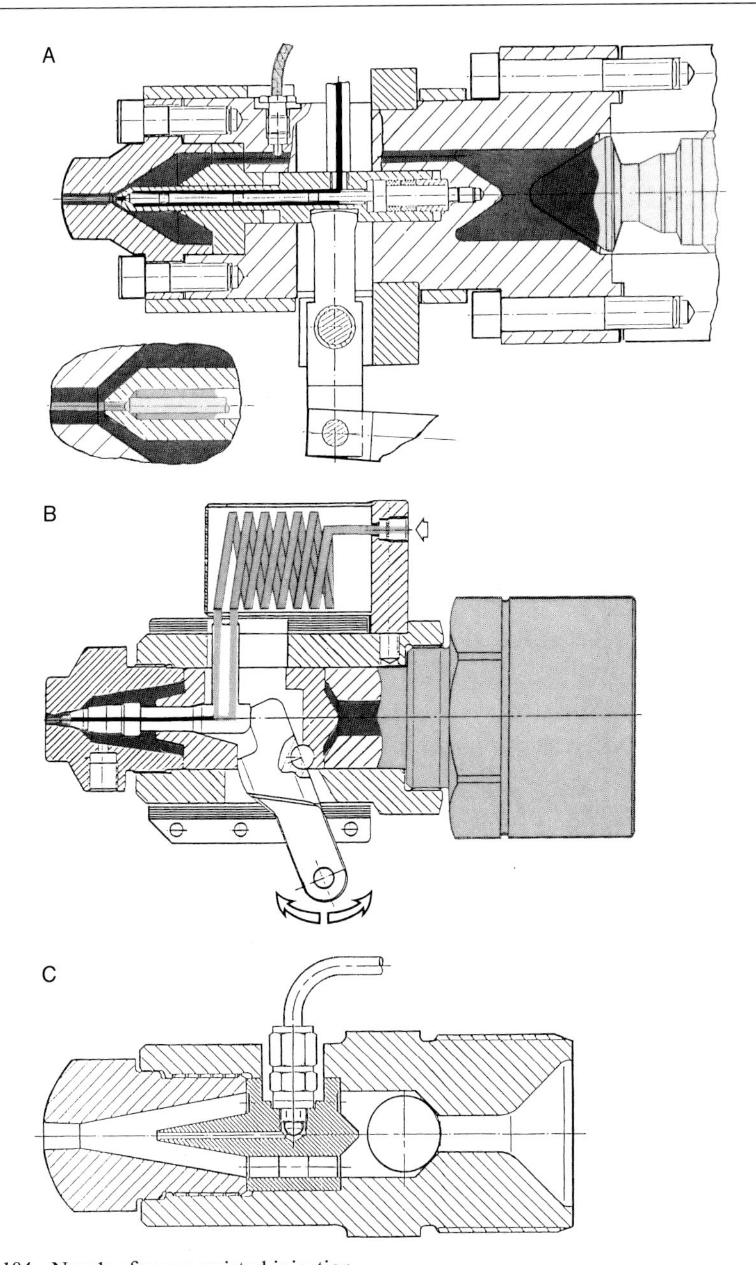

Fig. 184 Nozzles for gas-assisted injection
A: Gas-melt nozzle (System Engel), B: Gas-injection nozzle (System Krauss Maffei), C: Airmold nozzle (System Battenfeld)

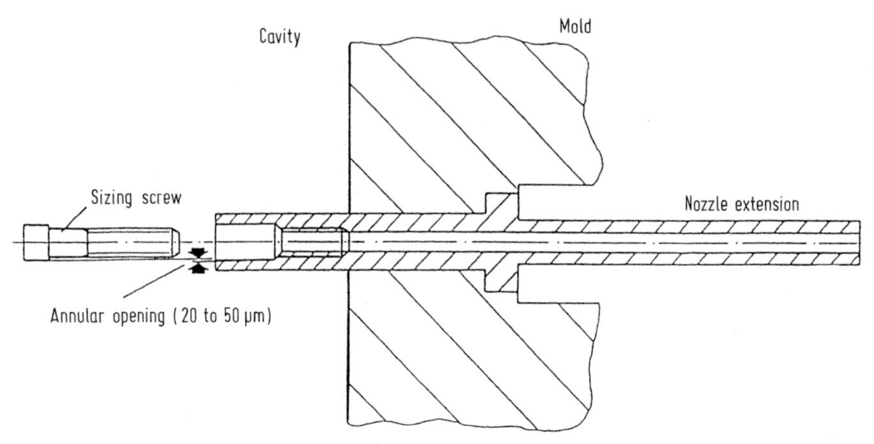

Fig. 185 Gas-injection module [388]

gas expansion, large cross sections for the gas flow have to be maintained in spite of problems with the design. Fig. 184 depicts some nozzles for gas-assisted injection. For direct injection into the cavity a nozzle according to Fig. 185 with a defined narrow annular gap is particularly suited [388]. It is well secured against entering melt by the narrow passage of its gap.

The following procedures and designations are known today:

GAIN process	Gas-Assisted INjection molding through the nozzle. Patent rights by Michael Ladney, Gain Technologies (former Friedrich patent of the Roehm Co.)
Cinpress II process	Gas injection directly into the runner system or mold (Peerless Co.) [372]
Airpress process	Gas injection directly into the mold (Kloeckner-Ferromatik)
Airmould process	Gas injection arbitrary but directly by pump, eventual for several machines (Battenfeld)
Gasmelt process	Gas injection arbitrary with piston (Engel)
GID	abbreviation of the German term for "Internal Gas Pressure", same technique as above (various manufacturers)
GIP	abbreviation of the German term for "Internal Gas Pressure Process", Arbitrary gas injection with piston
GIT-Gegenstrom process	gas-assisted injection – counter flow (melt return) process [375] (Kloeckner-Ferromatik)

The US Patent No. 4943407 [376] defines the last process. Gas is pumped into the filled cavity and pushes part of the melt out of the cavity either back into the plasticating unit or into some kind of accumulator. For reasons of practicality, the injection nozzle for the gas is located opposite the nozzle for the melt (Fig. 183, second process). The resulting wall thickness is mostly 1 mm more than with nozzle gas injection. The surface quality of such moldings is about the same as that of conventional moldings.

The Gas-Assisted Structural-Foam Injection Process

This process is a variant of gas-assisted injection molding. Plastic material is utilized which is capable of dissolving the injected gas under pressure and at temperatures common for the gas-assisted injection technique. Only a relatively little depth of the plastic around the gas bubble is penetrated by the gas. When the pressure is relieved, a foam is spontaneously generated, which fills the space of the gas bubble entirely or at least to some degree.

The advantage of this process is the creation of an undisturbed surface with a very rigid core. To practice this technique, a coinjection molding machine with gas-injection equipment is needed [367].

A first successful trial of thermoset processing in combination with gas injection is reported in [389]. Precondition for a success is a swell flow of the plastic in the cavity. This confirms an observation made with gas-injection molding of thermoplastics. It states that the gas recognizes immediately the path of lowest resistance (smallest generation of entropy), flows into gaps in the flow front and breaks through in an irregular pattern. An effect of fillers can be observed with thermosets, too. Organic fillers promote the swell flow and ensure the success. The technique for machine, mold, and runner system differs little from that for thermoplastics.

Structural Web Process

This process [378] may be considered the predecessor of the gas-assisted injection technique since it has all the characteristics of this method. The only difference is that a molding with a large surface area with heavy ribs, a web, are filled with flat gas bubbles. This technique is impractical, though, because the formation of gas bubbles is arbitrary and irregular from one part to another. Thin sections occur and the walls of the parts are not stable. The process is mentioned here for reasons of completeness only.

7.4 Injection Molding with Pulsating Melt

Conspicuously weak weld lines were noticed in injection molding of liquid-crystal polymers (LCP), since macromolecules with a strong tendency to orientation do not form loops along the weld line [377]. Similarly weak weld lines are known from processing fiber-filled plastics. These fibers do not overlap or form loops either. Standard injection molding can exercise little effect on the strength of weld lines. It is known, however, that one-sided holding pressure in the opposite direction can break through the weld line [368, 374].

The major effects of this counter-flow technique are:

- improvement of weld-line strength,
- improvement of fiber orientation in the molding,
- more effective holding-pressure control in the cavity, less sink marks,
- positive effect on the structure of amorphous plastics.

An investigation has demonstrated that a strong, positive effect can be achieved if holding pressure is initially applied only through one flow channel immediately after the two flow fronts have joined and the melt entering from one side penetrates deeply into the opposite melt front. The speed of impingement is very important for this effect [368]. It has not become known whether the equipment portrayed later permits this method.

Tests at the Institute for Plastics Processing (IKV, Aachen) with special machines (Braas machine) had already demonstrated at the end of the sixties that the melt in runners can be kept fluid longer if the material between screw and cavity is caused to oscillate [379]. This method, however, could not be rendered practical in an economical way.

Multi-Life-Feed Injection Molding (MLFM process)

This technique is based on the justified assumption that the orientation of macromolecules in the cavity can be influenced by a displacement of portions of the melt [380]. An attempt to meet this requirement is made by dividing the flow of melt in the nozzle (UK-PS 85 31374). A hydraulically actuated piston is integrated into each of the two melt channels, and the pistons exert holding pressure alternately. This moves the entire melt back and forth. Since the plastic in the cavity solidifies uniformly from the outside to the interior, overlapping layers are produced at the weld line with each pressure surge. Tests have demonstrated that a distinct product improvement at certain weld lines could be achieved.

Fig. 186 shows the equipment in principle. It can be adapted to every injection molding machine. The arrows indicate three different options of sequences, which can be used. According to [377], they are:

- Piston a and b pulsate at the same time but in opposed phases.
- Piston a and b pulsate in equal phases.
- Piston a and b apply equal holding pressure at the same time.

The division of the melt flow, which produces a weld line already in the intermediate feed block is of some disadvantage. The feed block contains a considerable volume of material under processing temperature. This may lead to molecular degradation. The whole equipment is space-consuming and calls for additional controls. There is not yet anything known about an application of this technique in mass production.

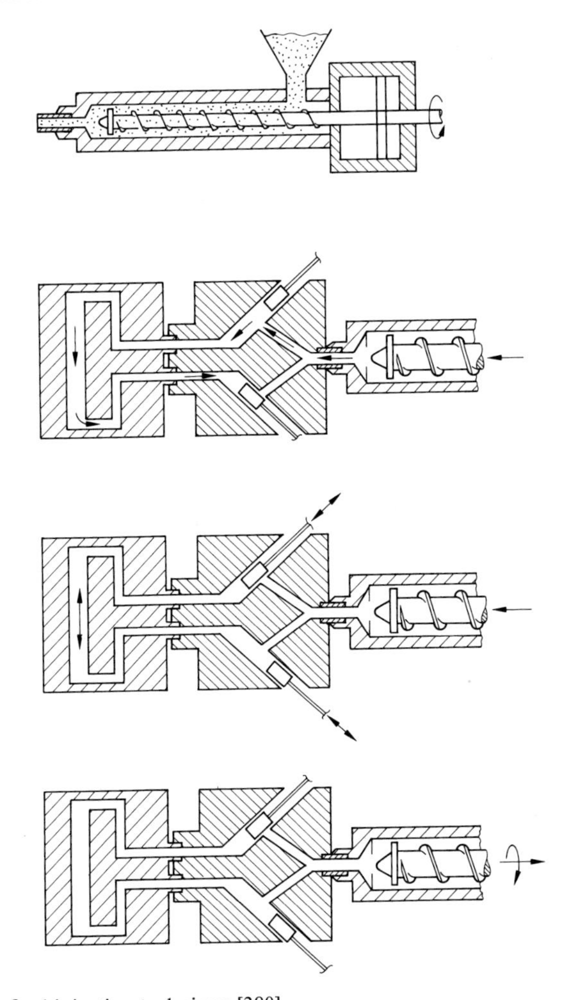

Fig. 186 Multi-life-feed injection technique [380]

Push-Pull Injection Molding

The push-pull process [377, 381, 382, 390] provides the option of pulsating injection with two or three injection units. The barrels are connected with the mold cavity by at least two sprues (Fig. 187). The so-called leading unit fills the cavity first. Then the second unit applies pressure from the opposite direction on the still molten part in the cavity. The leading unit is depressurized. Thus, still fluid melt is pushed back by fresh melt from the second unit. After some time, the process can be reversed.

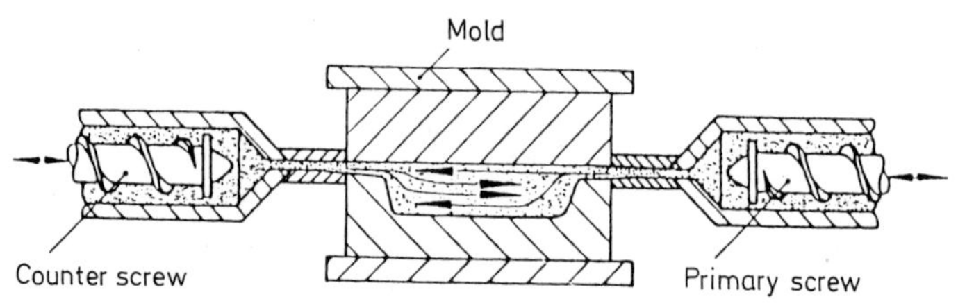

Fig. 187 Push-pull injection molding (principle) [382]

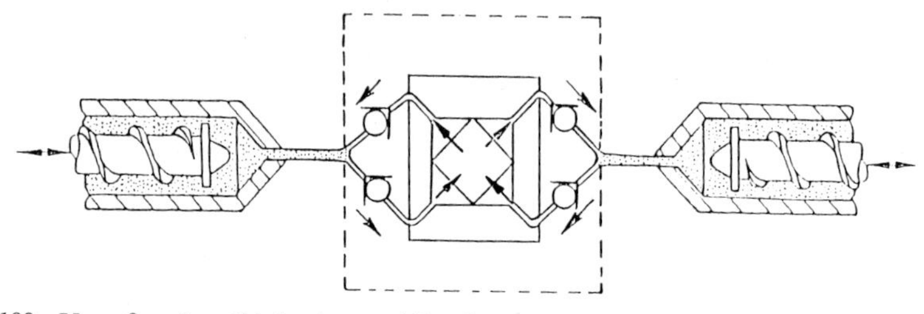

Fig. 188 Use of push-pull injection molding for parts with large areas [382]

The fluidity of the material is maintained by heat generated by shear. This makes some repetitions of the procedure possible. Pushing in alternating directions has a strong effect on the orientation, which is always parallel to the direction of flow. Especially if the weld line is broken through several times, the weakness of the weld line can be considerably lessened or, in favorable cases, completely removed. At least, considerable increase in strength has been noticed with molded tensile bars of LCP compared with those made by conventional molding [381]. The method has proved successful down to section thicknesses of 1 mm. More effort is needed for parts with large surface areas. Fig. 188 shows a proposal for diagonal gating of a rectangular plate. This calls for additional valves to switch from one corner to the other one.

7.5 Fusible-Core Technique

Parts with such a complex internal configuration that they cannot be made with conventional cores, which are pulled before or during mold opening, can be molded by using a so-called "lost core". Such a core is removed from the molding after demolding by dissolving it or, more often, melting it out. A major application is in the molding of automotive air-intake manifolds. Best suited is a technique which works with metal cores of materials with low melting points [393]. This process needs equipment for core casting and melting but no special injection molding machine. A machine with horizontal machine platen, however, offers some advantage. In the meantime, one manufacturer is marketing a machine which is almost identical with the one presented with Fig. 192.

The fusible-core technique is in direct competition with the shell technique. The latter is preferred if the molding permits a clear parting line be positioned in the article. As a crude distinction, one can say that most air-intake manifolds for four-cylinder engincs can be made by the shell technique while manifolds for engines with six cylinders and more should be molded by the fusible-core method.

7.6 Low-Pressure Injection Molding for Injection Stamping and In-Mold Decorating

Low-pressure injection molding is understood as a conventional molding process which uses all possibilities to limit the pressure level. Fig. 189 provides a rough classification of pressure levels for various techniques. A closer view reveals that the ranges from conventional molding and low-pressure molding overlap. The economical significance of low cavity pressure, if practical, is apparent with Fig. 190. Conventional injection molding requires high clamping forces against the reactive forces. Such forces become impractical with projected areas of more than 1 m^2. If the generated cavity pressure remains below 10 MPa, parts with areas of more than 3 m^2 can be produced.

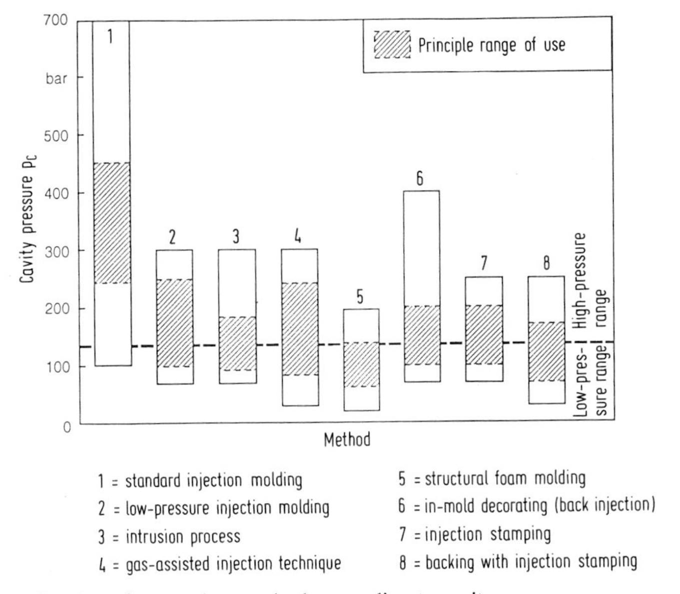

Fig. 189 Classification of processing methods according to cavity pressure

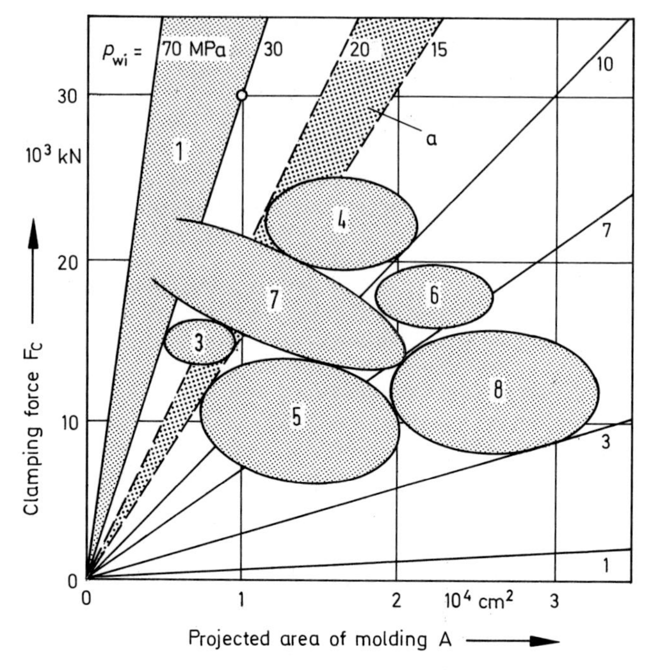

Fig. 190 Range of application of various processing methods dependent on clamping force and projected area with the cavity pressure p_c as parameter. a: Boundary region between high- and low-pressure process at 15 to 20 MPa. Symbols like Fig. 189 [407]

A low-pressure technique, which has been known for some time and is routinely utilized, is injection stamping. On can assume that the usual cavity-pressure level can be cut in half by injection stamping [409] (Sect. 7.6.1).

7.6.1 Injection Stamping

With injection stamping, the final part of cavity filling, after partial prefilling, is accomplished by the pressure from closing the mold halves and spreading the resin throughout the cavity, or by moving one or several cores into the cavity. It is the goal of this technique to generate less residual stresses in the molding or to save clamping force. The reduction in necessary clamping force is between 25 and 70%. The absolute section thickness and the ratio between flow length and wall thickness are decisive factors for the magnitude of the required stamping force. It should be mentioned, not to omit anything, that several methods of injection stamping are known:

- injection stamping with open mold,
  - with minimal opening of a few tenths of a millimeter,
  - with a mold opened by several millimeters,
  - same as before but with a movable core in the mold
- injection stamping by opening the mold against a low clamp force (reactive force higher than clamping force),
  - with minimal opening of a few tenths of a millimeter,
  - with a mold opened by several millimeters (rarely used).

In all cases, the opening movement is wholly or partially reversed by rising the clamping force. This generates pressure on the still fluid or solidifying resin in the cavity. A stamping pressure of only part of the molding surface can be created by one or several pressurized pistons in the mold. By the methods mentioned above, a considerable part of the molding (long opening stroke) or a smaller part (small opening stroke) is shaped. Long opening strokes call for for shut-off faces.

If a very short but exact stamping stroke has to be maintained, and this is mostly the case, then separate stamping cylinders and displacement controls are necessary for otherwise conventional machines. Usually, injection stamping can be done more easily with hydraulic clamp units than with toggle machines.

7.6.2 In-Mold Decorating

Injection stamping is the basis for in-mold decorating by back-injection. Both techniques, injection stamping and securing fabrics or foils in the mold before injection have been known since the sixties. The advantages of injection stamping did not become fully effective before a suitable and precise control system permitted a reproducible uniform displacement. This is the precondition for a constant quality of the moldings. It also explains, why in-mold decorating has become employed only for the last few years [397].

All substrates, which are used for in-mold decorating have to have adequate suitability [397, 410]. They have to withstand shearing, deformations from holding pressure, and high temperature. These oads have varying effects dependent on part geometry and mold surface.

The substrate, the so-called cover (foil, textiles, velour, or carpet) can be fed in a timed procedure from an endless roll (Fig. 191A) into the parting line and is back-injected. This bonds the resin to the cover. Then the molding is punched out and the remainder rolled up. This technique is preferred for small parts and multi-cavity molds. Its disadvantage is a rather high amount of wasted cover.

With another technique, the cover is loaded as a blank (Fig. 191B). A stack of covers is prepared outside the machine. The covers are picked up and placed into the mold by a needle or suction grab and kept there. After molding the finished part is taken out by another grab. This method is suited for simple flat parts with little demand on cover position with respect to part contour. Excess cover material pro-

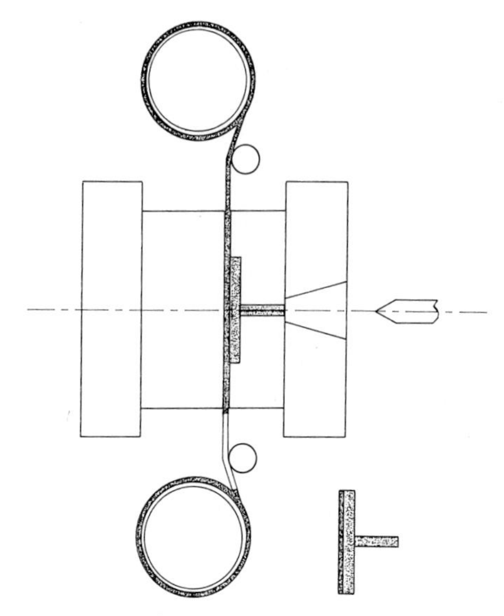

Fig. 191A In-mold decorating: cover from endless roll

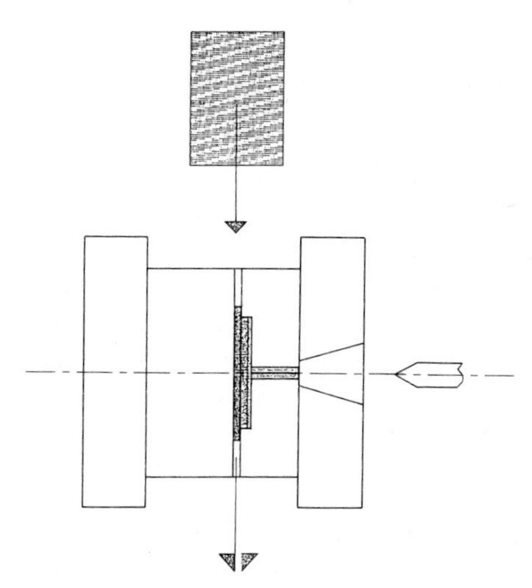

Fig. 191B In-mold decorating: cover as blank

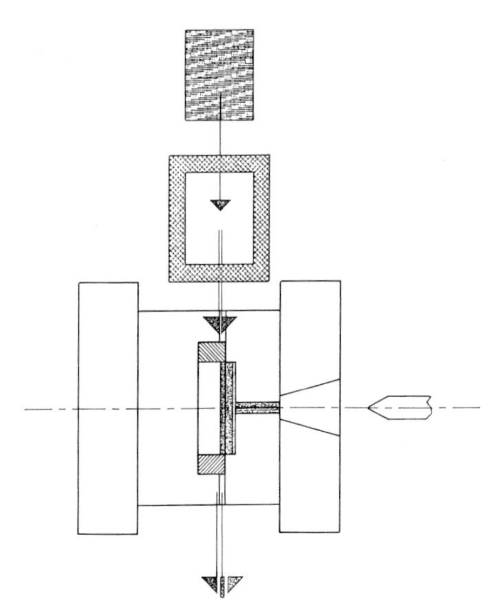

Fig. 191C In-mold decorating: blank kept in frame

trudes the edges of the molding and can be used for a folding-over technique. The amount of wasted cover material is insignificant.

A third method (Fig. 191C) employs a tenter-frame for keeping the cover in place. The blank is attached to the tenter outside the machine and placed into and taken out of the mold by a robot. The tenter may be curved and allows more complex, three-dimensional moldings. The cover can be correctly positioned with respect to the outline of the molding. Time for loading and unloading the mold can be kept short and, thus, benefiting the cycle time. Excess material can again be used for the fold-over technique. The amount of waste is larger than with the second method. Any modern, well controlled injection molding machine can be employed for either process. The only precondition is an advanced control with a complete display of all process parameters. After this, the success depends largely on a low holding-pressure level, which in turn is dictated by gating and part design [405 to 407]. There are covers, off course, made of materials which are less sensitive to pressure and temperature.

For parts with large surface areas, a horizontal working area is suggested for cover and finished article. The machine in Fig. 192 is custom-built especially for this purpose. It is similar to multi-stage rotary machines made for a different purpose (Sect.7.13).

A peculiarity is the interlining molding process, which directs the melt into the space between two substrates. For this and the fold-over technique suitable steps in mold design have to be taken for guiding the melt in the mold [408].

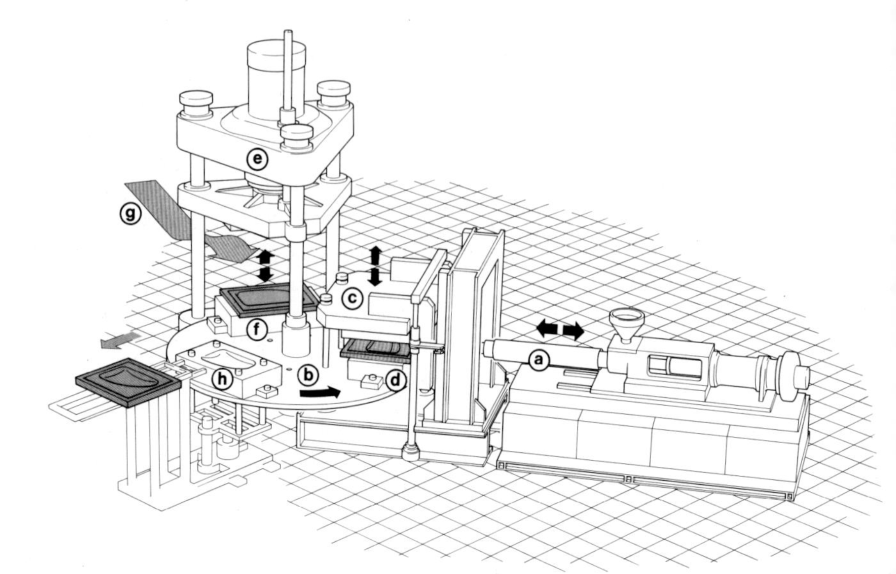

Fig. 192 Schematic presentation of installation for producing in-mold decorated parts in large numbers
a: Injection unit of a molding machine, b: Rotary table, c: Preform clamping unit (station 1), d: Preform mold, e: Hydraulic press (station 2), f: Die, g: Decorative foil feed, h: Pick-up station (station 3)

7.7 Machines for Intrusion

The intrusion process is applied if parts are to be molded the weight of which exceeds the available shot weight of the machine. Other essential preconditions for the applicability of this method are a large cross section of the sprue gate and heavy sections of the molding. The section thickness should commonly be more than 6 mm. If these conditions are met, one proceeds as follows. With the screw in the foremost position, the cavity is first filled by the rotating screw. After the volumetric filling of the cavity, pressure is built up, which returns the screw against a controlled back pressure. When an adjusted length of travel is reached, high pressure is turned on, which is the holding pressure for compensating the volume contraction. With regard to the machine, only an intrusion control has to be added.

7.8 Injection Molding of Polyester Resins

With the following terminology and the commonly used abbreviations materials are often designated as well as processing methods. To clarify the matter, the term process is applied if a processing method is discussed.

With the BMC process, polyester resins (Bulk Molding Compounds) are processed, which are prepared by blending resins, catalysts, reinforcing fibers, powdered mineral fillers, pigments, lubricants, and other additives. They are unsaturated polyester resins with about 12 to 25% glass fibers (UP-GF) and about 40% mineral fillers. In spite of their high filler content, they flow easily and are also called "wet polyesters". There are plenty of formulations which are characterized by their speed of reaction, content of resin and filler, and hardener systems. To reduce shrinkage, up to 25% of thermoplastics on the basis of PVA, PS, PMMA, or CB are added. The mostly high-molecular thermoplastics in these low-profile resins (LP systems) swell at the processing temperatures and partly compensate for the volume shrinkage of the thermosets. A smoother surface is created than with polyester alone. Even a "Class A" finish can be achieved.

BMC resins are supplied as cylindrical strands, e.g. in form of granules or as sticky, pasty-fibrous compound and can be pressed onto a screw (Fig. 193) or fed by a screw from a rotating hopper to a plunger (Fig. 194). In both cases, one has to keep the grinding effect acting on the glass fibers as low as possible. To accomplish this, one employs screws with low pitch for preparation and injection or screw preplastication and plunger injection (ZMC process), which conserves the fibers best.

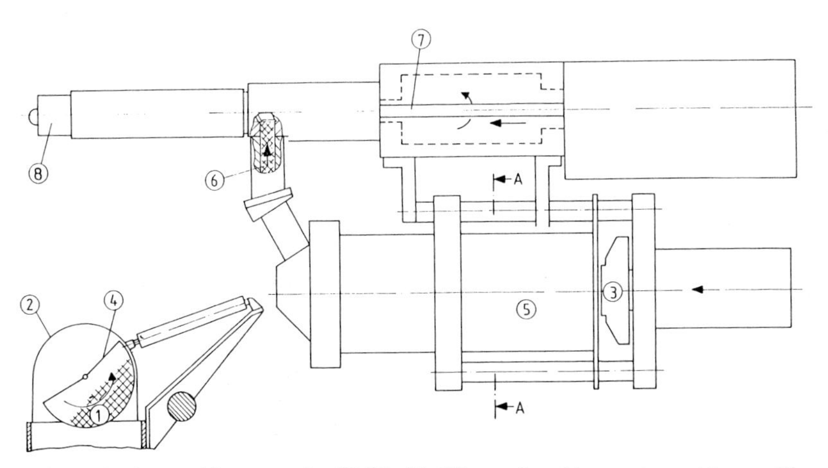

Fig. 193 Injection unti for processing BMC with filling and packing equipment low positioned in front (System Fahr-Bucher)
1: Polyester supply, 2: Resin container, 3: Packing piston, 4: Shut-off device, 5: Packing cylinder, 6: Connecting pipe, 7: Screw shaft, 8: Barrel and nozzle

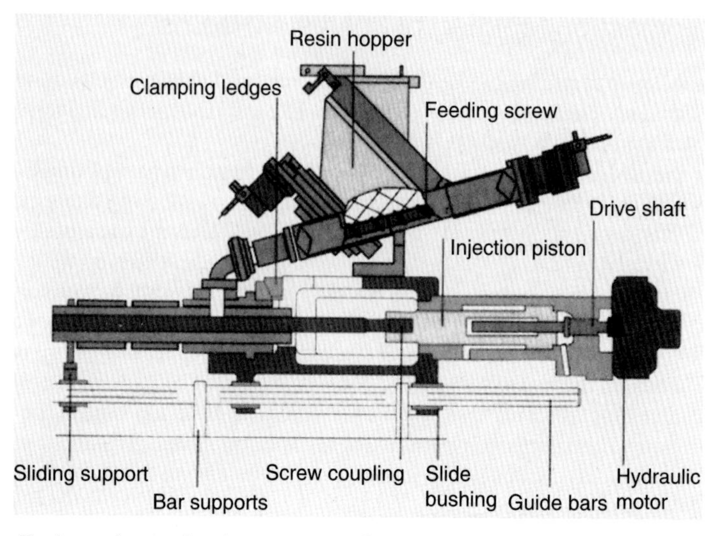

Fig. 194 Design and operational components of an automatic supply unit for BMC processing (System Kraus-Maffei)

Moldings have good toughness due to their long glass fibers. Weak points are the weld lines, which should be avoided by clever part design or gating, or be placed at locations of low stresses. Molding of polyester resins has taken an upswing after successful production of large automotive parts with the BMC process. One show piece is the rear door of the Citroen BX, which is composed of two bonded sections. Lately more automotive parts have come into production such as hoods, fenders, front ends, roofs, and seats. Another field of application is the manufacturing of electric appliances. More recently however, this process has come under pressure from competitive injection molding of thermoplastics.

The TMC (Thick Molding Compound) Process

This process deals with such BMC compounds which are high-viscous, with a dough-like consistency from high filler loading, therefore also called DMC (Dough Molding Compound). They exhibit improved mechanical properties. Otherwise they are processed like BMC compounds.

The ZMC Process

The Institute for Plastics Processing (IKV) in Aachen (Germany) has developed a ZMC process, which features a plasticating screw coaxially installed in a ram (Fig. 195). This design permits an especially careful treatment of resins with long glass

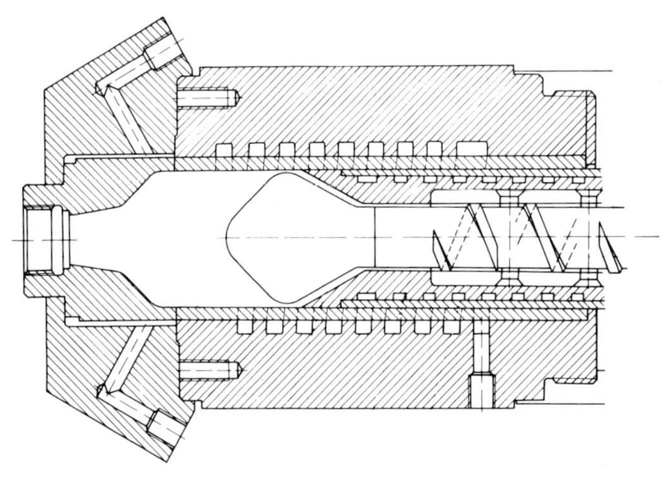

Fig. 195 Coaxial screw-piston plasticating and injection system for BMC processing developed by IKV, Aachen, similar to the ZMC machine (System SMTP-Billion)

fibers. The rotating screw plasticates the material and conveys it into the space in front of the ram, which injects the material into the mold with as little damage to the fibers as possible. Resins with a fiber length of 0.5 to 2.5 mm are employed.

The NMC (Nodular Molding Compound) Process

The NMZ process uses granulated polyester resins. For this reason one can anticipate shorter glass fibers in the finished product and reduced toughness because the fibers are shortened during plastication.

7.9 Machines and Equipment for Processing Liquid Silicones

A liquid silicone compound is pumped into a screw-injection unit with a piston (Fig. 196). The following process is similar to rubber processing. The barrel has to be cooled and the mold heated.

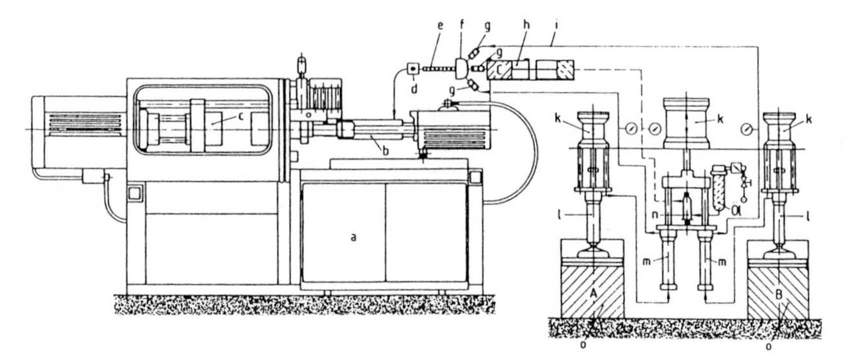

Fig. 196 Injection molding equipment for processing liquid silicones
a: Molding machine, b: Water-cooled barrel with hydraulic shut-off nozzle and special screw, c: Mold, d: Ball valve pneumatically actuated, e: Static mixer, f: Mixing chamber for components, g: Check valve, h: Unit for rapid exchange of colorant cases, i: High-pressure hose, k: Pneumatic drive cylinder, l: Feed pump, m: Metering pump, n: Metering pump for oil supply, o: Storage container for silicone components A and B

7.10 Tandem Injection Molding Machines

These machines are capable of accepting two molds in tandem, one behind the other. This makes for a certain similarity with the stack-mold technique. With such a design (Fig. 197) the economics of injection molding can be improved with machines of more than 5000 kN clamping force and cycle times of more than about 20 sec.

The tandem machine has a movable center machine platen. Injection is done laterally through this center platen, which carries a female mold half on each side. The projected areas of the two parts should be approximately the same. The advantage results particularly from that stage during which the molding in cavity 'B' is cooled, while the molding in cavity 'A' is demolded. This means a high degree of utilization with efficient use of space. Equally shaped parts offer an additional advantage. Their history of production is the same; they are molded in one operating stage of the screw.

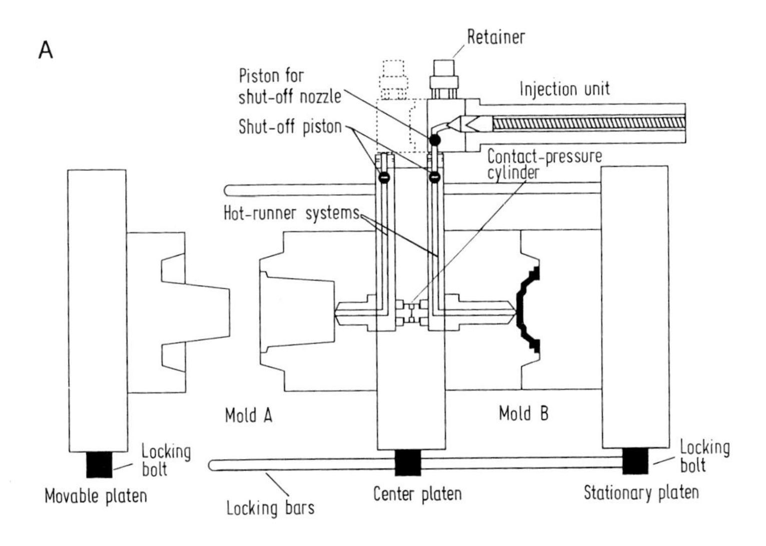
A
Retainer
Piston for shut-off nozzle
Injection unit
Shut-off piston
Contact-pressure cylinder
Hot-runner systems
Mold A
Mold B
Locking bolt
Locking bolt
Movable platen
Locking bars
Center platen
Stationary platen

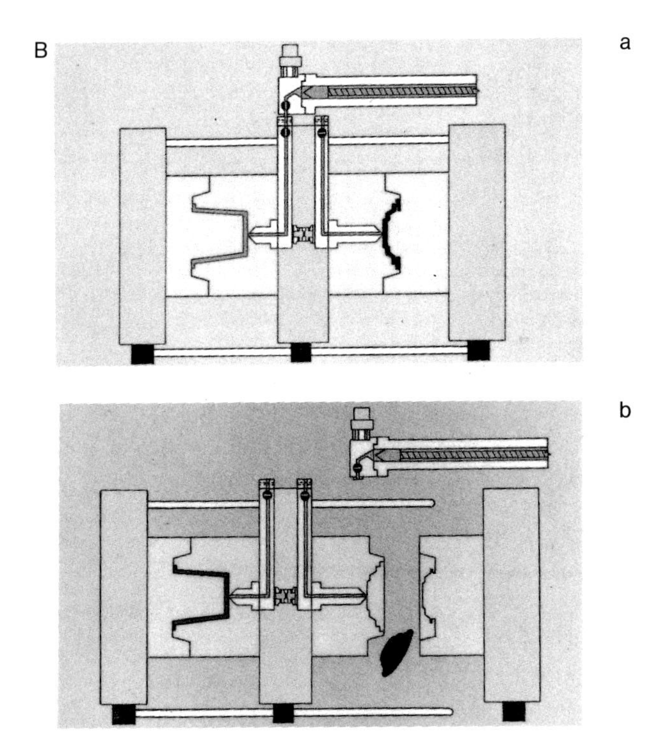
B
a
b

Fig. 197

c

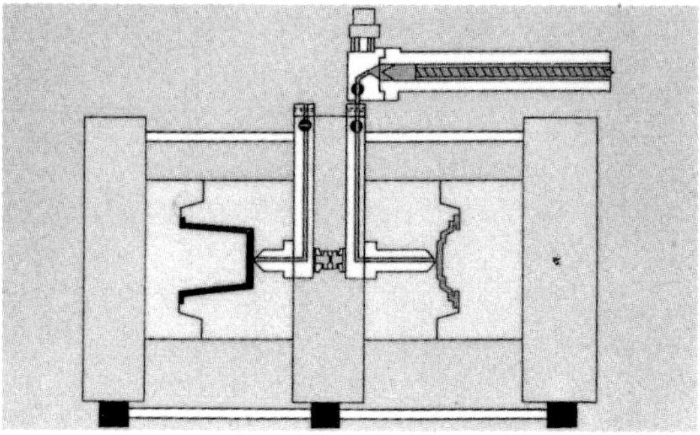

d

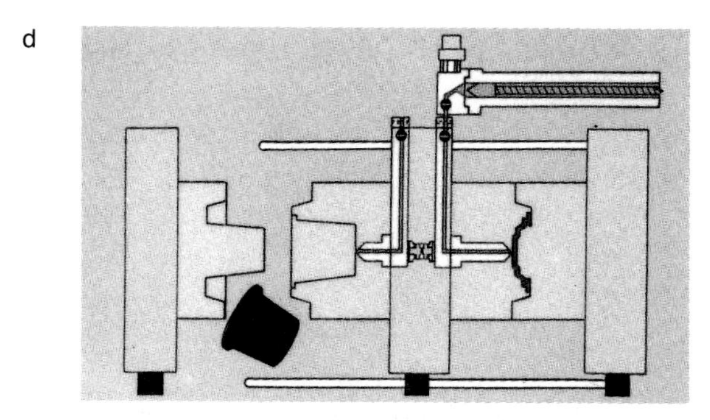

Fig. 197 Injection molding machine with clamping unit in tandem arrangement (System Husky)
A: Center platen with hot-runner system, B: Process sequence
a: Both molds are closed, injection into mold B, cooling of mold A, b: After injection and holding-pressure time for mold B, mold A is opened and part ejected, mold B remains closed. The injection unit moves to mold A. c: Mold A is closed and injection starts, mold B is cooled, d: After filling mold A, mold B is opened and mold A is cooled

7.11 Plunger-Type Injection Molding Machines

The plunger-type injection molding machine was the standard machine until 1955. A cylindrical barrel is filled with plastic pellets. They are melted by heating from the outside. The cylinder houses a plunger or ram, which builds up injection pressure by axial advancement just like a screw and pushes the melt into the cavity of the mold. Mini-machines with a plunger diameter up to about 20 mm still operate according to this principle (Fig. 198).

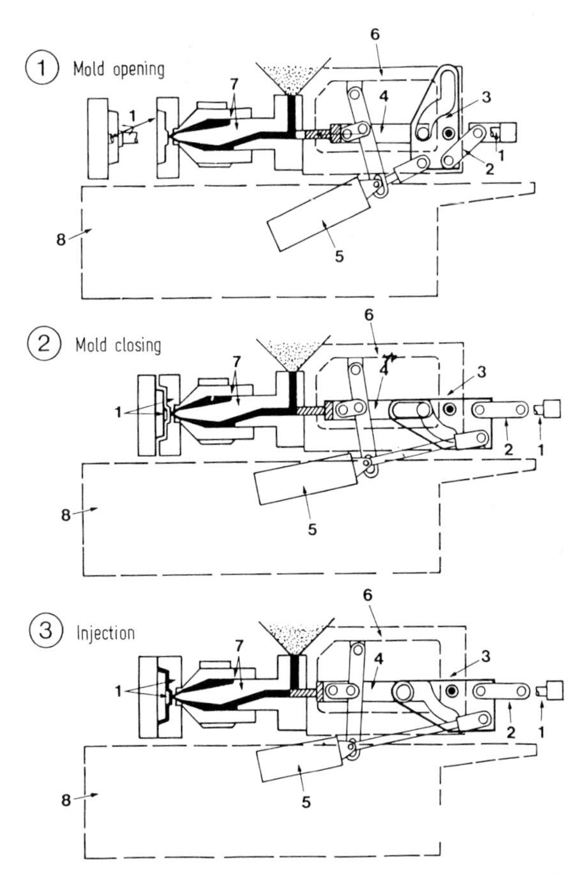

Fig. 198 Small injection molding machine for shot weights up to 2.5 gm and 30 kN clamping force with double-acting toggle clamp system which clamps the mold and actuates injection plunger with one stroke (System HEK)
1: Mold and machine platens, 2: Latch, 3: Link, 4: Thrust rod, 5: Air cylinder, 6: Frame, 7: Barrel, 8: Machine base

7.12 Two-Stage Injection Molding Machines with Screw Plastication and Plunger Injection

This design is older than the in-line reciprocating-screw machine. For a time it was almost totally displaced from the market by the standard screw machine. In recent years, however, it has experienced a comeback. It has found a meaningful field of application for extreme requirements in terms of plasticating capacity and shot size. Typical applications for this machine is the production of structural-foam moldings and of very small parts. Years ago, the high plasticating rate of a comparatively large, continuously rotating screw was utilized to manufacture packaging articles with such a machine. Injection is always accomplished with an accumulator.

Fig. 199 pictures an injection unit for mini-parts. The first mentioned application needs a rather modest plasticating rate. Structural-foam parts have a long cooling period. Therefore a small screw can feed the space in front of the plunger with enough melt for another shot. Fig. 200 and 202 show other design concepts, while Fig. 201 presents an injection unit according to the first-in, first-out principle. This machine can, after all, produce an injection pressure of about 280 MPa. A machine for molding structural foam parts is shown with Fig. 203. This low-pressure machine is characterized by a low clamping force.

Compared with a conventional machine, the ratio between clamping force and shot size is smaller by a factor of 2 to 6 and the injection pressure only half as high. For low-pressure molding of structural foam, there is a multitude of variants on the market. They are primarily aimed at achieving an improved surface quality without post-treatment. The following systems have not survived in the market place:

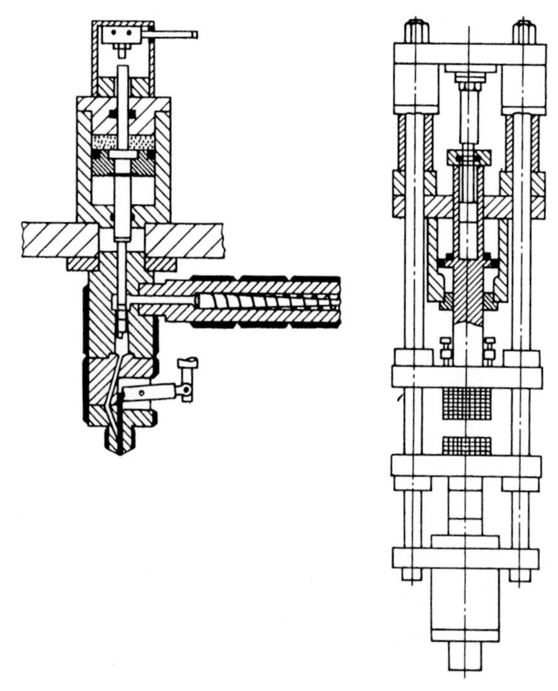

Fig. 199 Injection unit (left) and clamping unit (right) of a special machine for molding micro precision parts (schematic)

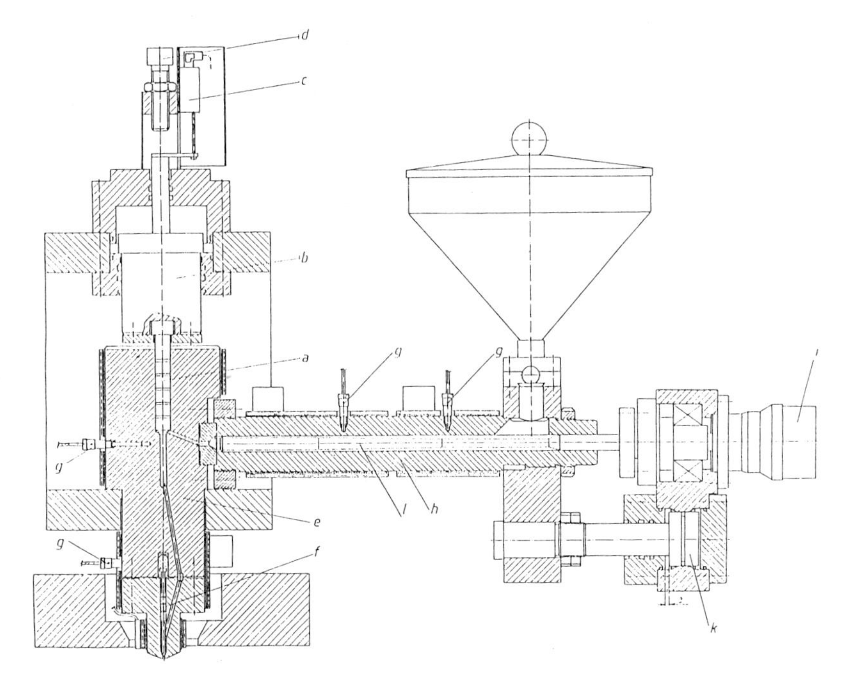

Fig. 200 Plunger-injection unit with screw preplastication for molding small parts (System Engel)
a: Injection plunger, b: Hydraulic cylinder, c: Stroke control, d: Stop, e: Cylinder block, f: Nozzle, g: Thermo-couples, h: Screw barrel, i: Screw drive, k: Hydraulic cylinder, l: Special screw

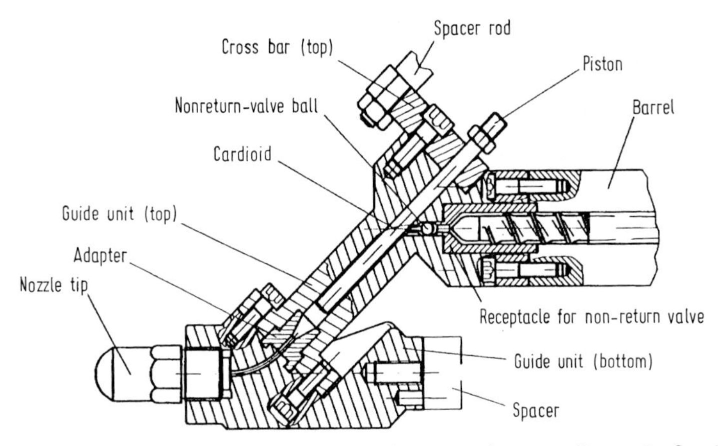

Fig. 201 Sketch of a mini injection molding machine, operating according to the first-in, first-out method (System Kloeckner Ferromatik/IKV)

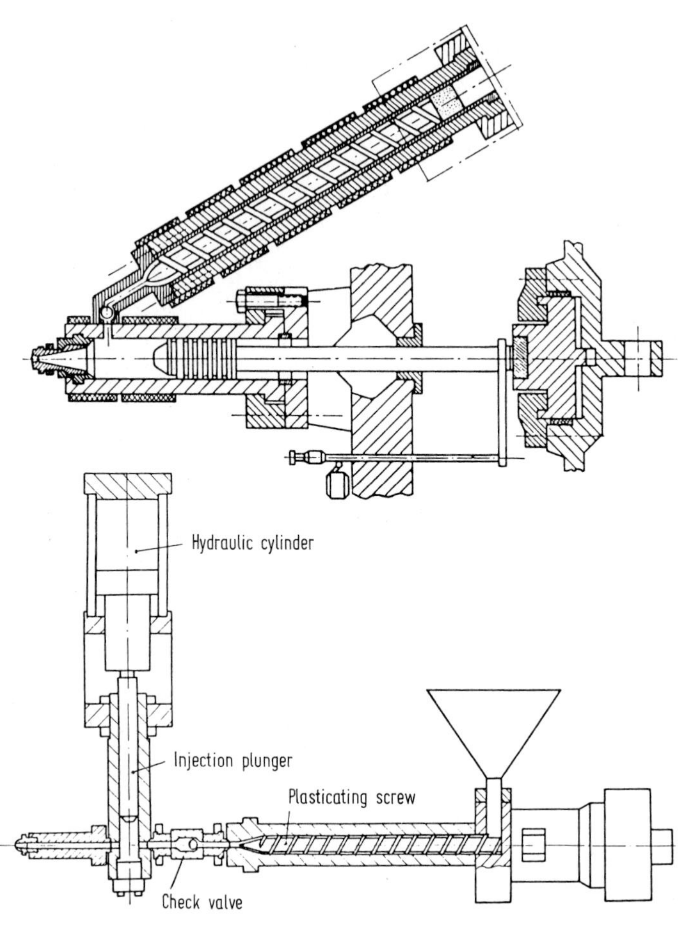

Fig. 202 Two-stage injection unit with screw plastication and plunger injection (schematic) Top: Plunger unit in machine axis (System Battenfeld) Bottom: Plasticating screw in machine axis (System Billion)

- Variotherm Process,
- Toshiba-Asahi Foam Process (TAF process),
- Thermoplastic Foam Process by Allied Chemical (TFM)
- Mixed-Injection Process (IKV Aachen),
- Thermal Insulating-Layer Process,
- New Structural-Foam Process.

This leaves only three methods to produce structural-foam parts. They are the normal low-pressure molding process, the gas counter-pressure process, and the two-compo-

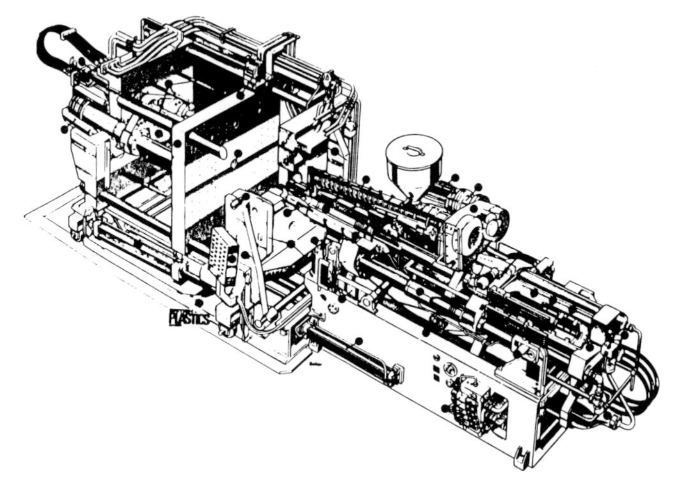

Fig. 203 Special structural foam injection molding machine for thermoplastics with blowing agent. Screw plastication and plunger injection with gas counter pressure

nent injection process with a solid skin and a foam core. The merits of the last method is an improved surface quality without sinks, comparable to or sometimes even better than solid moldings. Their level of physical characteristics is very high. This process is of great importance for a multitude of applications as furnitures, housings, and containers.

Machines with screw plastication and plunger injection are also employed for processing rubber. Besides a design similar to a structural-foam machine (Fig. 204), the

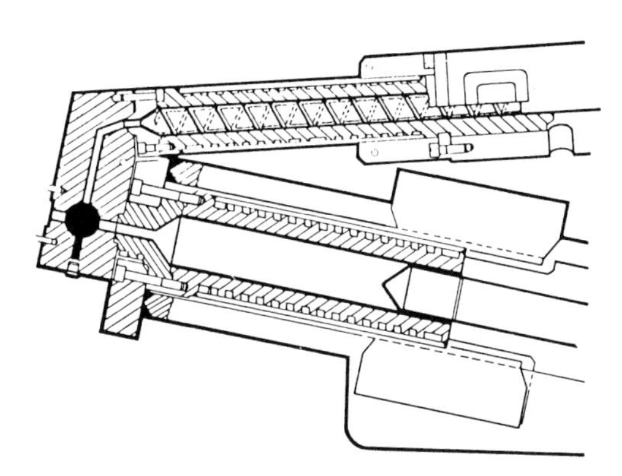

Fig. 204 Section of an injection unit with screw plastication and plunger injection for rubber processing (System Kloeckner-Ferromatik-Desma)

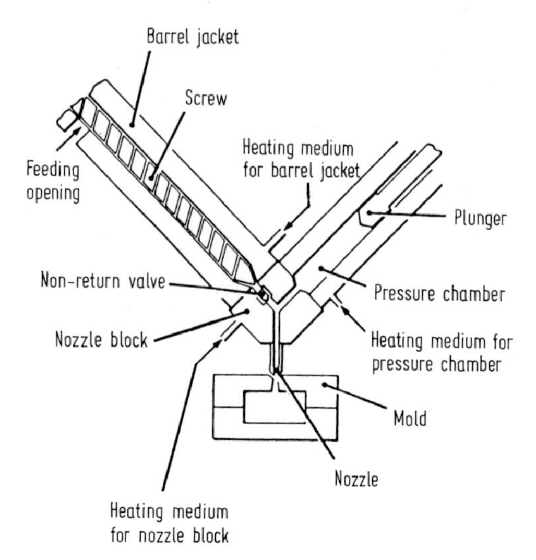

Fig. 205 Injection unit for rubber molding

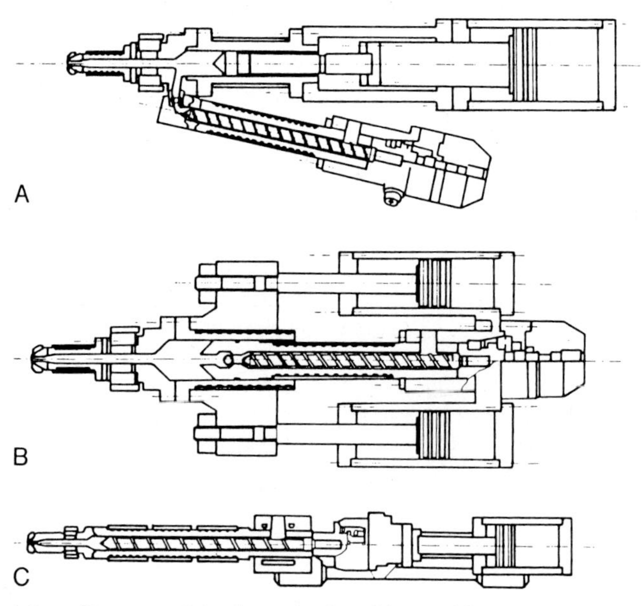

Fig. 206 Sketches of two-stage injection units for rubber molding
A: Stationary screw feeds plunger unit, B: Screw is positioned inside plunger, C: In-line reciprocating screw machine

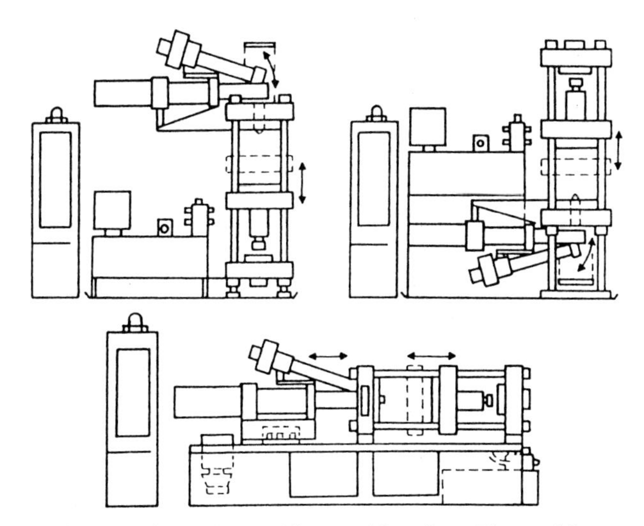

Fig. 207 Design features of injection molding machines for rubber molding

concept of Fig. 205 is successfully utilized. In this case, the plasticating and injection units can be positioned above or below the clamping unit. Similar injection units avoid the problem of premature curing of that material which, in the standard version, is fed first into the space in front of the ram and leaves it last (Fig. 206, version B). The complete machine design is presented with Fig. 207.

7.13 Multi-Station Machines

Injection molding machines with several mold stations are especially well suited for the production of large quantities of similar or dissimilar parts, if their cooling time in the mold leaves part of the injection capacity unused. A popular multi-station machine for thermoplastics is the revolving machine, which operates with a rotating clamping unit (turret) like a capstan lathe. The turret brings one mold after the other to the injection unit in a circular motion (Fig. 211). Other multi-station machines are portrayed in Figs. 208 to 213.

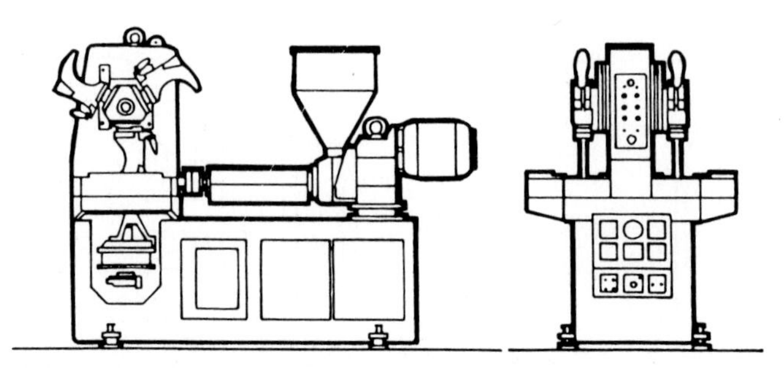

Fig. 208 Injection molding machine for molding soles onto boots with double rotating core carrier and split cavity (System Kloeckner-Ferromatik-Desma)

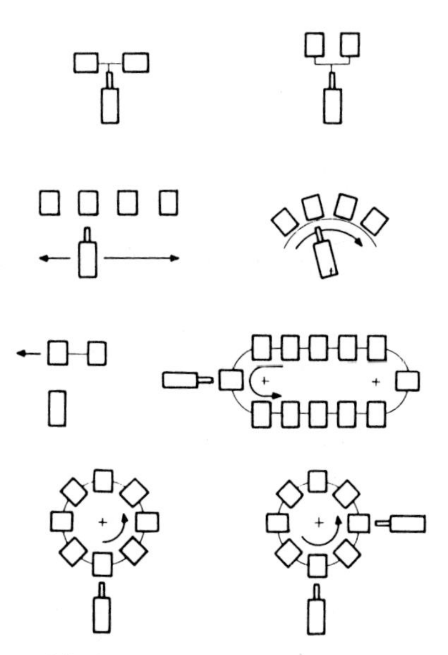

Fig. 209 Set-up of rubber molding machines (schematic)

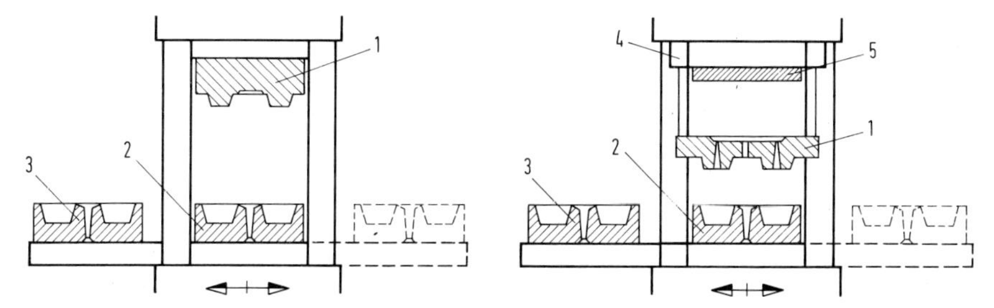

Fig. 210 Vertical clamp unit with one top mold half and two bottom halves in a sliding carrier for shuttle operation. While one complete mold is in operation, the other mold half is ready for ejection or placement of inserts (System Kloeckner-Ferromatik-Desma)
A: Two-plate mold in clamp unit with sliding carrier
1: Top mold half, 2, 3: Bottom mold halves, Lateral gating in parting line or from below
B: Three-plate mold in clamp unit with sliding carrier
1: Top mold half, 2, 3: Bottom mold halves, 4: Mechanical lifter, 5: Runner plate

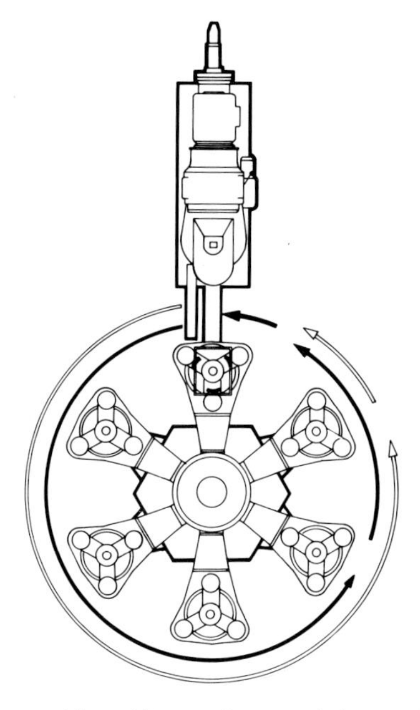

Fig. 211 Injection molding machine with turret for several clamp units in three-column design (System Kloeckner-Ferromatik-Desma)
Clamp units rotate with turret and are brought to injection unit one after the other or in arbitrary sequence. Ejection in one place. Suitable for long cooling cycles, preferably for rubber and eventually structural foam

Fig. 212 Injection molding machine with several clamping locations with particularly low clamping pressure suited for low-pressure molding (System Hettinga)

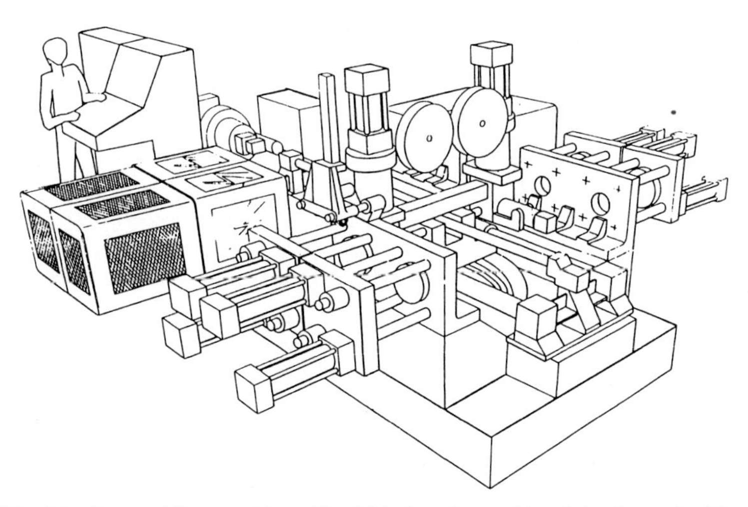

Fig. 213 Injection molding machine with eight clamping and two injection units (Two-stage operation with screw preplastication) (System EMS-France)

7.14 Special Types without Commercial Significance

There has never been any lack of effort to replace the screw as the plasticating element in an injection molding machine by other techniques. A relative recent try was a practical application of an old principle. In Belgium, a Maxwell extruder operating in accordance with the Weissenberg principle was combined with an injection molding machine. In the early 70s, Th. Engel proposed an injection unit with an axially oscillating plunger (teledynamic principle), which heats plastic materials primarily by shearing under high pressure. Pressure variations prevent the sprue from freezing. In particular, parts with heavy sections can be produced without voids. A "high pressure plastication" of the Institute for Plastics Processing (IKV, Aachen, Germany) employs a variation of the teledynamic principle. Plastication is achieved with pressures between 500 and 1000 MPa primarily by shearing and heat transfer in an especially narrow aperture.

None of these techniques has gained any significance. They cannot match the reciprocating screw with its all-purpose features, and for special applications, the technical and economic preconditions are still lacking.

8 Machine Sizes and Performance Data

8.1 Sizes

The field of injection molding is characterized by a large variety of items made from all kinds of plastics. These items include packaging, everyday commodities, equipment housings with technical and stylish functions, building elements, and large numbers of components for engineering applications. The wrist watch provides an example of the smallest technical parts. Some molded gears are so small that 10 000 of them can fit into a thimble. The hull of a sailing boat is the opposite extreme, with a surface area of more then 3.5 m^2 and a part weight of 19 to 20 kg.

This could lead to the conclusion that there are no limits to injection molding in either direction. This is not correct, however. Special equipment is needed for parts as small as the watch gears mentioned above, but molding electronic components between 1 and 10 g shot weight already is a job for standard production machines. Plunger machines are well suited for these dimensions. Reciprocating-screw machines are available from 15 mm screw diameter upward with corresponding shot weights of about 4 to 12 g. The clamping force of such machines is 100 to 250 kN. Fig. 214 illustrates the proportions of a machine with 50 kN in comparison with one with 100 000 kN clamping force. The heaviest injection molding machine has a clamping force of 100 000 kN with a shot weight of about 70 kg. It can hardly be called a standard machine, anymore than the one for molding boat hulls. Thus the technological limits are between 1 g minimum and 25 kg maximum shot weight. The upper limit appears to be 50 000 kN clamping force. Machines of larger size are singularities.

8.2 Performance Data

The attempt has been made here to provide a summary of performance data of injection molding machines, but the reader should bear in mind that part of this listing is data published by machine manufacturers [418]. At this point, only a general assessment can be made. In special cases, any evaluation must consider the required function. The inferences may be very different. Thus, an injection molding machine can be perfectly suitable for processing thermosets and yet be unsuitable, even with the use of conversion parts, for thermoplastics and vice versa. Performance data may reveal this.

Internationally, it is customary to specify clamping force and maximum shot capacity to characterize a machine. One also has to understand that overall performance is the result of an interaction of a number of individual characteristics. Data such as dry cycle time, shot capacity, plasticating capacity, injection pressure, injection rate, available injection power, clamping force, opening and ejection force, closed and open daylight, and platen size. They all provide basic information about performance.

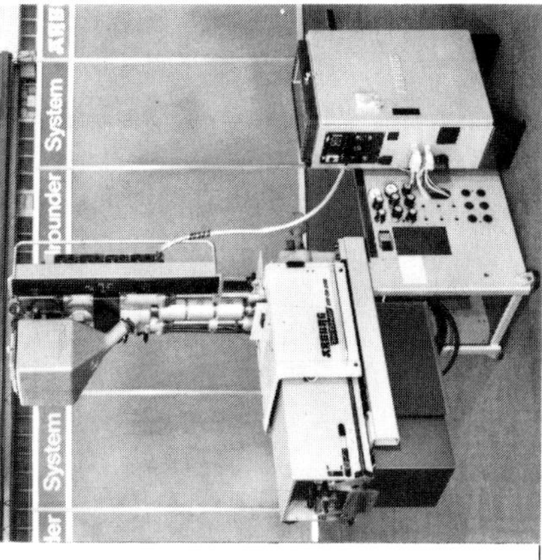

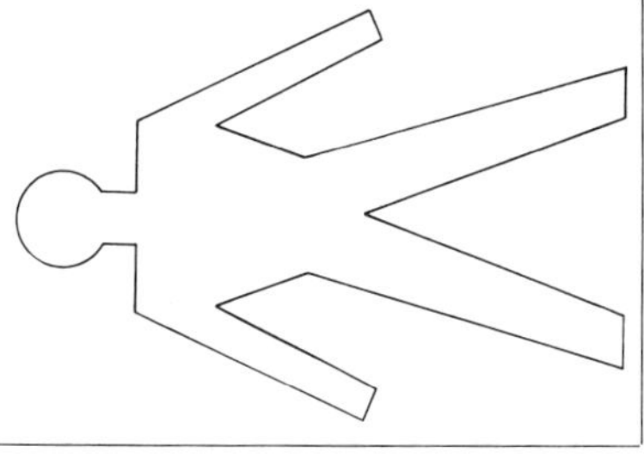

Fig. 214 Size of an injection molding machine of 500 kN clamping force compared with one of 10 000 kN clamping force

The following details and graphic presentations reveal a surprisingly broad spread of performance data for various machines. A desirable standardization is still far away, even for those data for which one should take it for granted. Demands on ease of handling, susceptibility to breakdowns, ease of setup, floor space, and accessibility are decisive factors for judging the suitability for a particular task. Interfaces for the connection of peripherals has become standardized. With respect to controls, machines built in German-speaking countries are recognized for their high standard and considered exemplary world-wide.

A number of measurable demands must be met either completely or for the most part, if a particular molding problem is to be resolved, Table 25. Machine evaluation also includes consideration of qualitative criteria that usually carry varying weights. Such criteria comprise quality of construction, method of process control, convenience of operation, cost of maintenance, price of machine, reputation of manufacturer, and others [75, 421].

Table 25 Measurable Minimum Requirements for Injection Molding Machines

Feature	Unit
Dry cycle	s
Recovery rate	g/s
Injection rate	cm^3/s
Available injection power	kW
Injection or shot capacity	cm^3 or g
Torque of screw	Nm
Injection pressure	MPa
Clamping force	kN
Opening force	kN
Platen/size (height × width)	mm × mm
Distance between tie bars (horizontal × vertical)	mm × mm
Max./min. daylight	mm
Max. stroke	mm
Drop chute	m^2
Floor space	m^2
Weight	kg

8.2.1 Performance Data of the Injection Unit

The most important data relate to maximum injection or shot capacity, injection rate, available injection power, and plasticating rate.

Shot Capacity

While in Europe shot or injection capacity is expressed as volume (cm^3) and, therefore, reflects the maximum screw displacement, in the United States, the weight (oz) of of the displaced material (polystyrene) is the preferred unit.

Hence, the shot capacity is a calculated dimension derived from the maximum screw displacement, which is the product of the maximum screw stroke and the effective cross-sectional area of the screw, multiplied by the density of polystyrene. The screw displacement should be about 10% larger than the volume of a molding if injection is not done with a rotating screw. The maximum possible part weight can be approximated as follows:

$$W_P = k \times \textit{shot capacity (oz)} \times \textit{density of material} / 0.9$$

Ounces may be converted into grams by multiplying by the factor 28.35. 'k' is a correction factor with a value between 0.7 to 0.8 indicating that the maximum possible part weight should usually not be fully utilized. A factor lower than 0.2 to 0.3 should be used in exceptional cases only. In this context, machines are considered which have a feeding stroke of 4 x D (Sect. 2.4.1).

The machine manufacturers offer a broad range of shot capacities (Fig. 213). The dashed line presents the mean value of commercially available machines, depending on their clamping force. The simplest variation results from a change in screw stroke, which is usually three to four times the screw diameter. Since 1985 more and more often the stroke is increased to 5 to 7 x D. This is a questionable undertaking because a long feeding stroke beyond 3 x D results in taking in air and produces an heterogeneous melt. High-quality molding usually does not allow a feeding stroke of more than 3 x D [75, 77, 88, 89].

Whereas in the United States, whole injection units of different shot capacities but the same injection pressure are generally combined with one or several clamping units, it is more customary in Europe to vary only screw and barrel diameter of an injection unit, which changes not only its shot capacity but also the maximum injection pressure. Three different diameters are commonly offered for each machine. This system makes it necessary to introduce the term "working capacity" for reference purposes. Working capacity is the calculated screw displacement related to 100 MPa injection pressure. Such injection units again can be joined with several clamping units.

The diagram of Fig. 215 alone does not provide information about the usable screw displacement or shot capacity in dependence on the clamping force. This is made possible with Fig. 216. This presentation is based on the several assumption:

- Using a feeding stroke of less than 1 x D and more than 3 x D is not permitted.
- Besides the standard screw, two more different sizes are available.
- The minimum injection pressure is 180 to 200 MPa.

According to Fig. 216, three screw diameters are available. Rounding off, a diameter of 60 mm presents the "standard screw" for a 2500 kN machine. The intersections of a horizontal through this point with the minimum and maximum screw-stroke lines reveal a maximum usable screw displacement of about 500 cm^3, while Fig. 215 indicate a value of 800 cm^3.

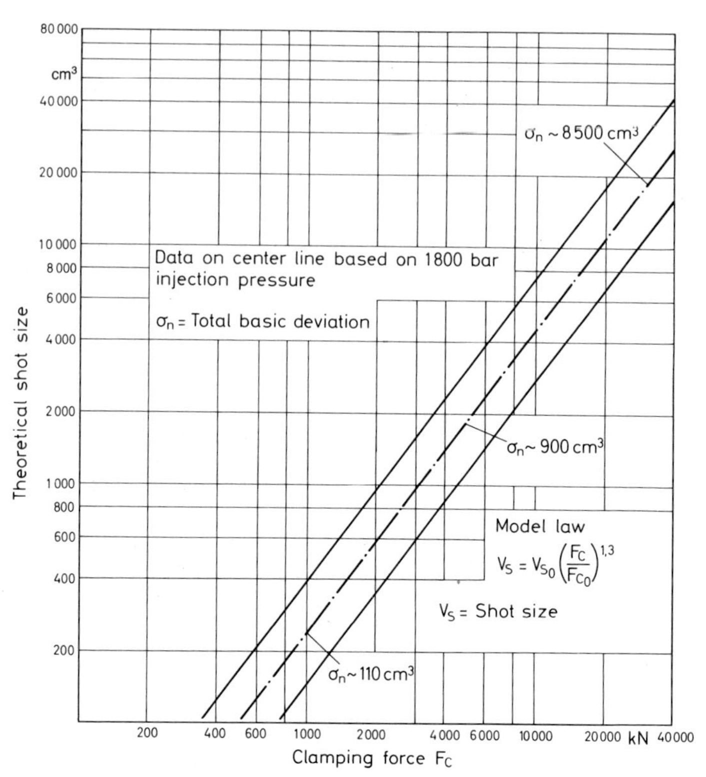

Fig. 215a Calculated shot capacity versus clamping force
Dashed line = averages. US machines.

The difference means that the listed machines provide for a feeding stroke of 4 x D. That much should not be utilized, though. Therefore the smallest usable injection capacity is 160 cm³ in accordance with Fig. 216.

The calculated shot capacity is, besides the clamping force, part of the specifications for an injection molding machine Thus a European listing of 500/180 or 500–180 means a shot capacity of 500 cm³ and 180 Mp (= 1800 kN) clamping force. These numbers may often be reversed. In the United States the clamping force is expressed in tons and the shot capacity, sometimes omitted, in ounces of polystyrene.

The option of combining injection and clamping units at will provides the processor with a modular system, with which he can adjust a machine to his needs in the most efficient way. Fig. 217 represents the ratio between working capacity (shot capacity) and clamping force versus the clamping force of a range of commercially available injection molding machines. The trend approximately follows the model law:

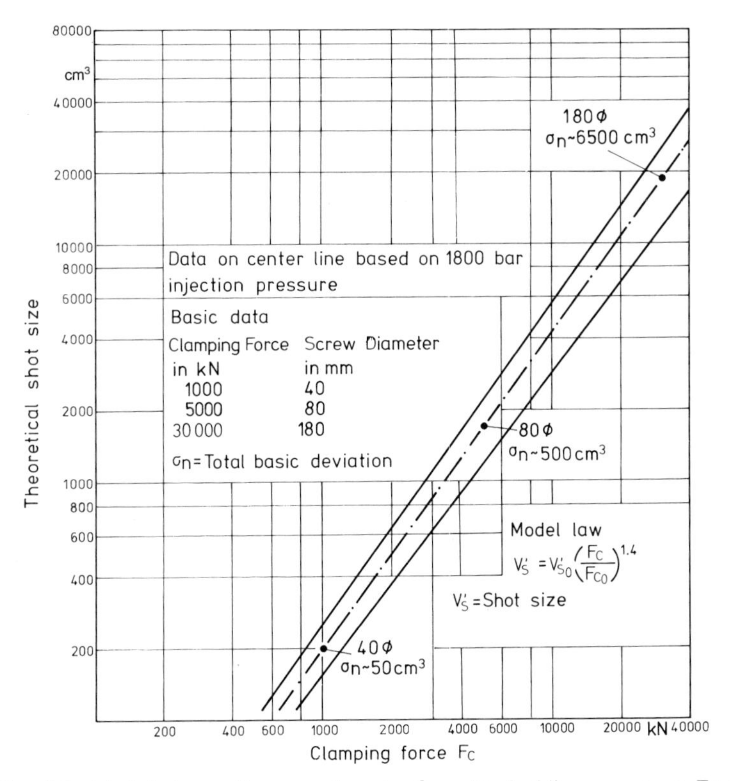

Fig. 215b Calculated shot capactiy versus clamping force. Dashed line = averages. European machines, V'_s = theoretical shot size based on 10^3 bar (10^2 MPa) injection pressure

$$a = a_0 \left(\frac{F_{cx}}{F_{cO}}\right)^{0.33} \qquad \textit{in Europe}$$

$$a = a_0 \left(\frac{F_{cx}}{F_{cO}}\right)^{0.27} \qquad \textit{in the USA}$$

Here a and a_0 are the ratios between working capacity (shot capacity) and clamping force and F_{cx} and F_{c0} the clamping forces. If one tries to formulate a standard, one obtains line (2) in Fig. 217 as an approximation.

The broad range of working (shot) capacity is striking. It can be compared with the spread of power input versus clamping force. Both are the result of the modular system of combining various injection and clamping units.

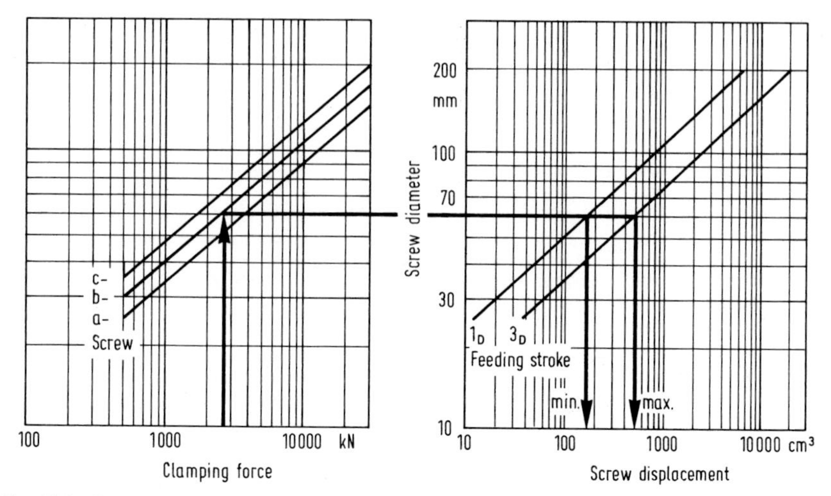

Fig. 216 Correlation between clamping force, optimum screw diameter, and maximum or minimum usable screw displacement (limits of feeding stroke: 1xD and 3xD)

The dashed lines 1, 2, and 3 in Fig. 217 indicate such cases, where three injection units of different capacities are available. The effective maximum shot weight can be computed as follows:

known: - Clamping force e.g. 3500 kN (350 Mp)
wanted: - Shot capacity at 100 MPa (cm^3)
- Shot capacity at 180 MPa (cm^3)
- Max. shot weight for PS at 180 MPa
calculation: - 3500 kN clamp force corresponds with a = 0.54 on the abscissa, hence 3500 x 0.54 = 1890
result: - maximum shot size at 100 MPa: 1890 cm^3
- maximum shot size at 180 MPa: 1890/1.8 = 1050 cm^3
- maximum shot capacity for PS (spec. weight at processing temperature): 1050 x 0.9 = 945 g
- usable effective shot capacity: 945 x 0.8 = 760 g

In this example the range of the maximum shot weights is about 945 ± 176 g. The overall range can be presented as follows:

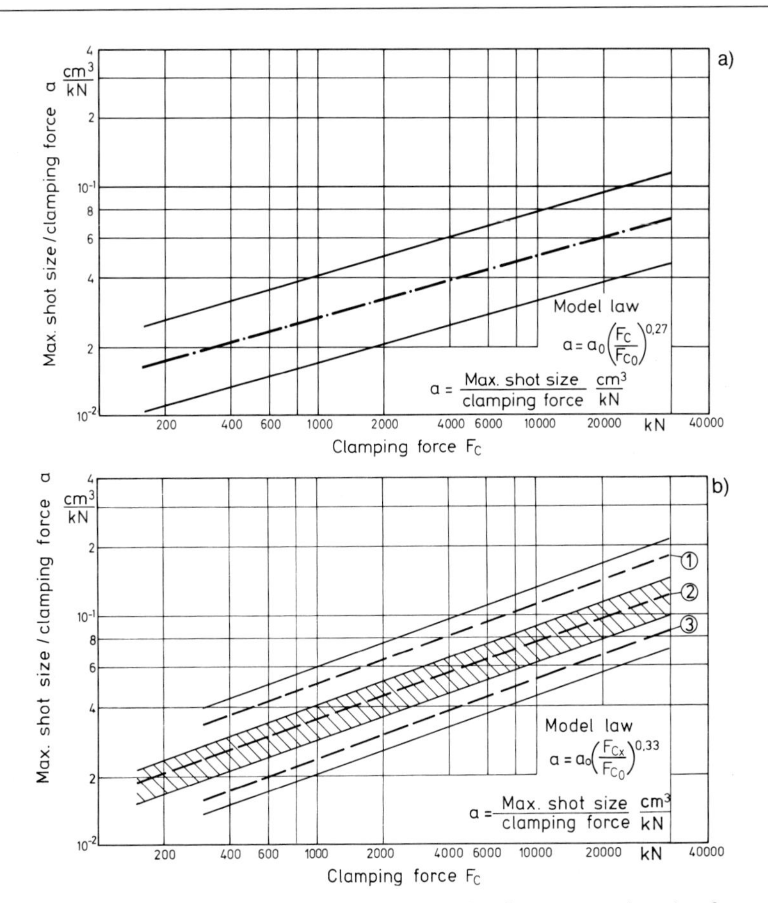

Fig. 217 Ratio of maximum shot capacity over clamping force versus clamping force. Dashed line = averages.
a) US machines, b) European machines. Lines 1, 2 and 3 are based on three different injection units for each clamping unit. Standard equipment = line 2 or cross hatched area respectively

Clamping force kN	max. shot capacity[1] g
1000	200 ± 110 (35)
5000	1600 ± 880 (300)
10000	4250 ± 2500 (750)
30000	18300 ± 10500 (3500)

[1)] Consider 3D rule according to Fig. 216

The data in parentheses reflect the deviations in the cross- hatched, so-called standard range. In special cases, curves 1 or 3 are representative.

Other significant performance data are injection pressure (Sect. 2.1) and injection rate. In the US, the injection rate (in^3/s) usually is the volume of the molding (in^3) divided by the injection time (s). According to European standards, the injection rate (g/s) is determined as the mass of the molding (g) divided by the injection time (s).

Injection Rate

The injection rate has received increasing attention in the design of injection molding machines following a demand for parts with thin walls to realize material savings, and since it is known that rapid injection results in smaller reactive forces from cavity pressure and lower residual stresses (less shrinkage) [41, 42, 43]. Fig. 218 shows the values of injection rates of today's commercially available machines. They should be viewed with caution, however, because they frequently are based on the theoretical maximum speed of the hydraulic piston or the advancing screw and the displaced volume or the weight of the displaced volume but without actual injection. Such data are not identical with the information of Fig. 218 but are higher only by 10 to 30% without considering the specific weight of the material.

There is very little uniformity in the injection rates of commercial machines. Machines of the lower quarter are not up to all requirements of injection molding but typical for equipment that is used exclusively for processing thermosets and elastomers. Data related to accumulator drives should be viewed with prudence because they also are not derived from actual injection but from the time needed for the screw stroke. If the injection pressure of an accumulator drive approaches the remaining pressure of the accumulator, the injection rate is reduced. While hydraulic pumps have an efficiency maximum at about 80% of the maximum pressure, accumulators always have a maximum at 66.7% of maximum pressure. This is illustrated with Fig. 129.

Available Injection Power

The available injection power for injecting the material from the barrel into the mold [xx] is calculated with the following equation:

$$E_i = \frac{\pi \times d^2 \times p_H \times s}{4 \times 10^2 \times 2 \times t} \ (kW)$$

with E_i = available injection power (kW), d = diameter of the hydraulic piston (m), p_H = hydraulic pressure during the time in which the piston moves from one-forth to three-forth of its entire stroke (MPa), s = the distance of travel (m), and t = the time (s) for the piston to move the distance s. This procedure of measuring in the middle range of the piston stroke is a European standard [68, 71]. Fig. 129 presents an example of the development of the available injection power versus the hydraulic pressure. There is

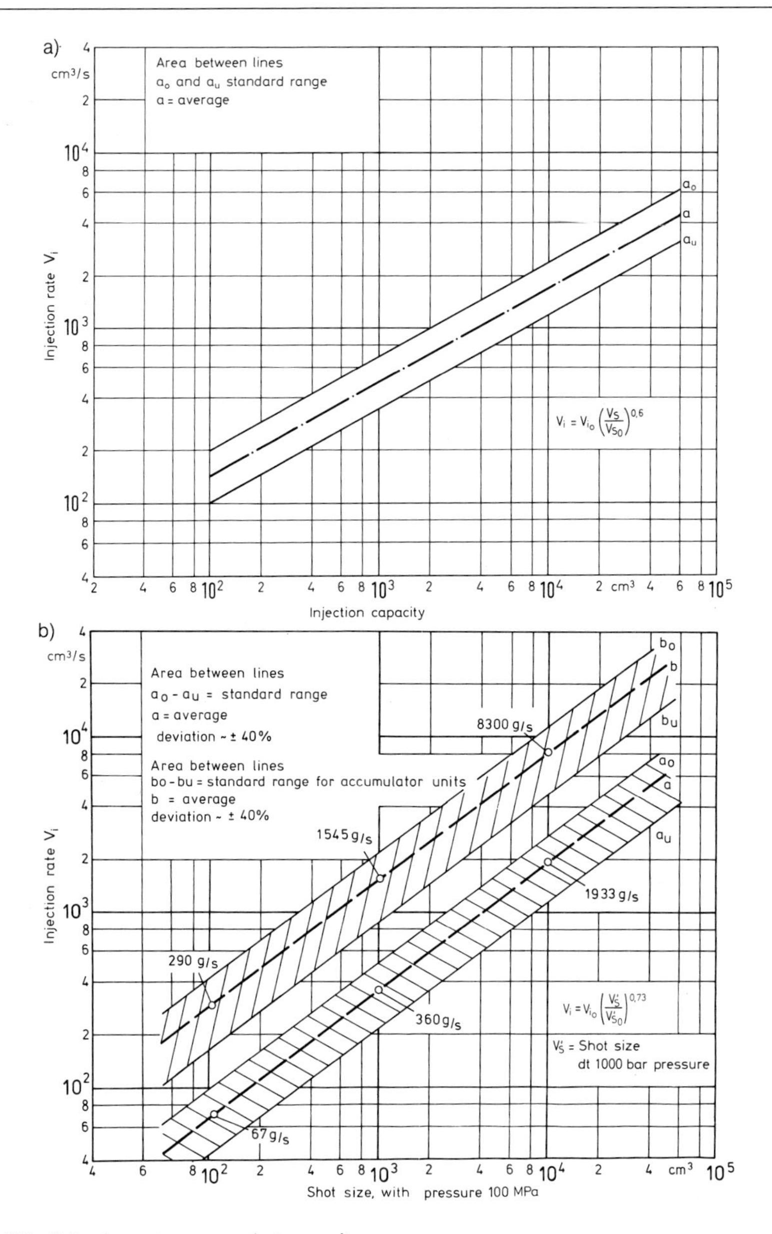

Fig. 218 Injection rate versus shot capacity
a) US machines, b) European machines, standard data for injection by pump (a) or accumulator (b) attention: Fig. 145b shows the injection rate in g/s. g/s = cm³/s when the specific weight is 1,0 g/cm³

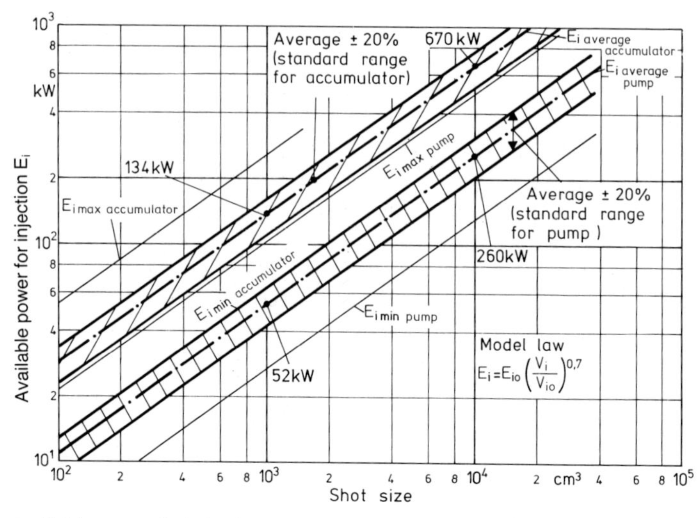

Fig. 219 Available power for injection (E_i) versus shot size. Information with reference to commercial availability and indication of averages (dashed lines) and standard range (cross hatched) each for pump and accumulator use.

an early power drop from the opening of the pressure relief valve, and the maximum power is at about 70 to 80% of the maximum pressure (Sect. 5.4).

The production of high-quality parts usually calls for a high injection speed, which should be reproducible cycle after cycle Such a steady state cannot be attained beyond the power drop. Variations in weight and dimensions would be the result.

Fig. 219, in conformity with Fig. 218 shows the available injection power dependent on the shot capacity. Its correlation with the listed power input is as follows:

$$E_i = n_H \times X_0 \times E_M \ (kW)$$

with E_i = pump output (kW), n_H = efficiency of the whole hydraulic system (0.7 to 0.9), X_0 = overload capacity of the electric motor, and E_M = electric motor input.

On an average, one can approximate $E_i \sim E_M/0.8$ as adequate injection power.

Plasticating Capacity

While the injection rate is a performance parameter during the injection stage, the recovery rate characterizes the plasticating capacity during the feeding stage. It is controlled by the geometry of the screw, its rotational speed, the back pressure, the design of the nonreturn valve, and the feeding travel.

The plasticating capacity, which is commonly expressed in lb/h in the United States and in g/s in Europe, is the mass of the molding divided by the feed time. This quantity can be easily found in practice with a stop watch and a scale. The measurement has to be taken during actual molding of polystyrene. The formerly common measuring during extrusion with an open nozzle and the screw in its foremost position is inadmissible.

Since determination and presentation of the plasticating capacity are still not uniform, information listed in company advertisements may be contradictory and often wrong. Fig. 220 has been conceived on the basis of representative data. It reflects the maximum specific output measured with screws of a design presented with Fig. 5 and related to a circumferential speed of 1 m/s or 0.1 m/s respectively.

With the area between the two lines, it provides almost the entire spectrum of the specific output of all universal screws. The back pressure in front of the screw tip is about 10 to 15 MPa. Depending on the processed material, a deviation of 10 to 20% is possible.

Amorphous plastics, such as PS, ABS, PC, PNNA, PC/PBT. and modified PPO mostly have a higher specific output than crystalline plastics, such as PE, PP, PA, and PBT. The common plasticating capacity of a screw with 100 mm diameter rotating with 20 rpm can be determined as follows:

$$v_c = \pi \times D \times N/60 \quad (m/s)$$

with vc = circumferential speed (m/s), D = diameter (m), and N = rpm.

$$v_c = \pi \times 0.1 \times 20/60 = 0.1 \quad (m/s)$$

With Fig. 220 the specific plasticating capacity is $1 \ \frac{g/s}{rpm}$

with D = 100 mm and $v_c = 0.1$.

Thus, the plasticating rate is:

$$Q_{pl20} = N \times M_{sp} \sim 20 \times 1 = 20 \ g/s$$

or with an rpm of 200 and therefore a specific plasticating capacity of 0.7:

$$Q_{pl200} = N \times M_{sp} \sim 200 \times 0.7 = 140 \ g/s$$

To be able to read also intermediate values, the diagram of Fig. 221 was developed from Fig. 220. It permits a direct reading of the plasticating capacity in dependence on the screw diameter and the circumferential speed.

When producing parts with heavy sections, circumferential speeds of 0.05 to 0.3 m/s are preferred [110]. Some publications indicate partly higher speeds [75]. High circumferential speeds, however, pose the danger of insufficient homogenization of the melt.

Much lower speeds are also chosen for the processing of thermosets and elastomers. For molding packaging articles, however, the recovery rate has to be as high as to work with speeds between 0.8 and 1.5 or even more.

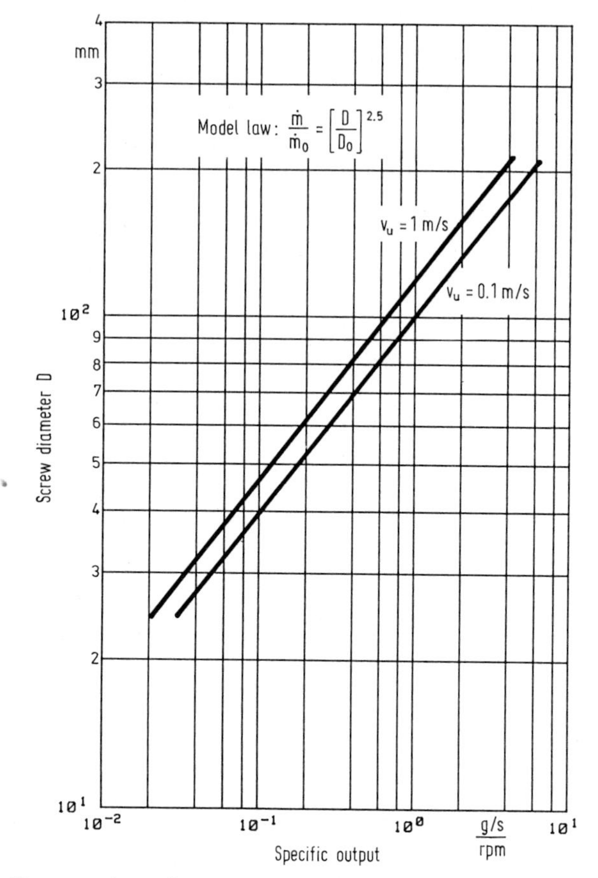

Fig. 220 Specific output depending on screw diameter (Universal screws according to Table 6), Parameter: circumferential speed

8.2.2 Performance Data of the Clamping Unit

The mold (or die) space is determined by the product of the horizontal and vertical distances between the tie bars (Fig. 222). This quantity permits comparing one machine with another. One should see to it that its magnitude does not affect the rigidity of the machine platens.

A large space between the tie bars calls for an appropriate thickness of the platens in accordance with physical laws. The deflection of the platens should not be more than 0.2 mm for 1 m distance between bars if the maximum clamping force acts upon a small area. Only such rigid platens can meet the demands of high-quality molding.

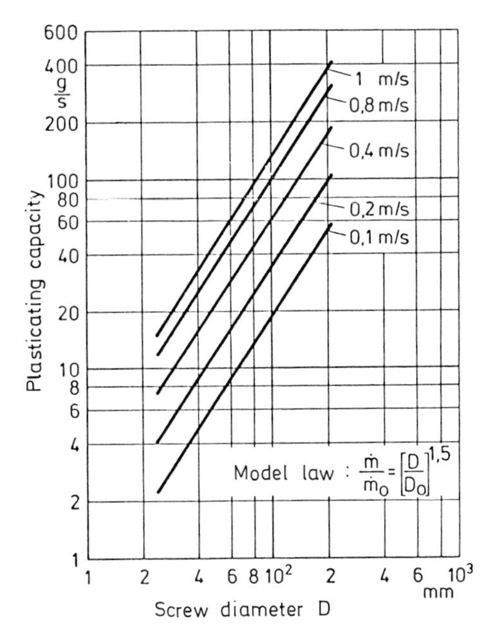

Fig. 221 Plasticating capacity versus screw diameter (Universal screws). Parameter: circumferential speed

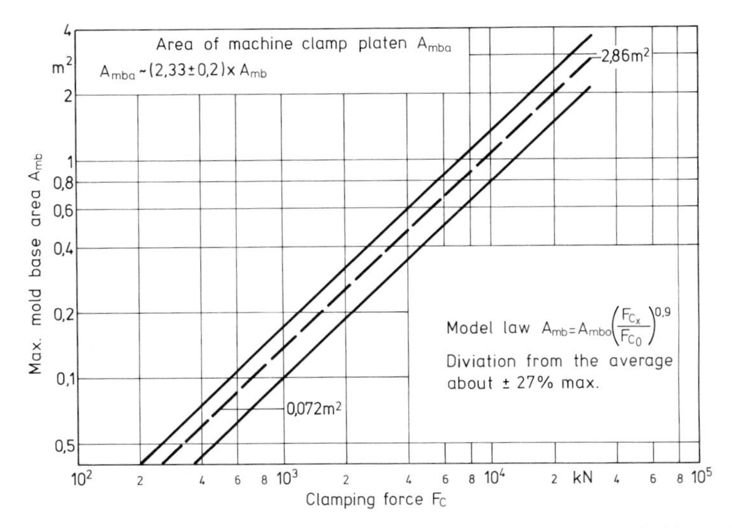

Fig. 222 Maximum mold base area versus clamping force (commercially available machines). Dashed line indicates standard equipment

In a comparison, that machine is better which has a larger usable platen area with the same rigidity of the platen. There must be a reasonable relationship between mold and platen size to avoid an adverse loading case of the machine platens.

Standards for the nozzle locating hole, platen bolting, and ejector hole pattern are picture in Figs. 223, 224, 225, and 226. They reflect the recommendations of the SPI Injection Machinery Division and the European Standard (Euromap), and permits an unrestricted exchange of molds and machines.

By now, the time has certainly arrived for manufacturers of quick-clamping systems to come to an agreement about standard dimensions, too, to allow for a mold exchange independent of the system.

With respect to the mold thickness that can be accommodated between the machine platens, one distinguishes between open or maximum daylight and closed daylight. Open daylight is the maximum distance between stationary and movable

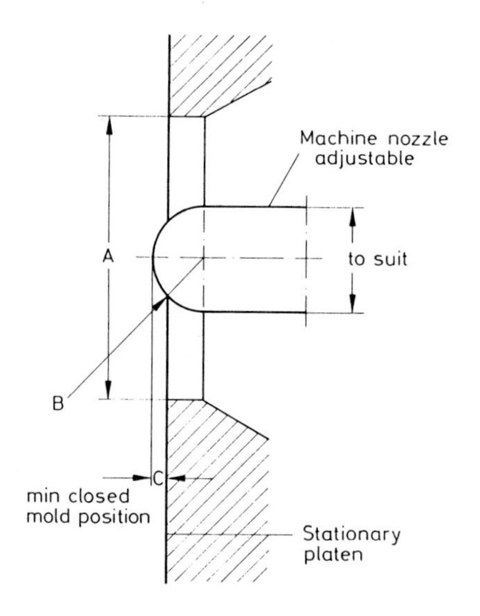

For machines up to and including 750 tons clamping force	For machines over 750 tons clamping force	Spher. radius*	Minimum closed mold position*
A $^{+.002}_{-.000}$		B $^{+.000}_{-.010}$	C
4	5	1/2 or 3/4	3/16

* B and C as common to both types of machine

Fig. 223a Machine nozzle and locating hole, SPI standard. Dimensions in inch

e_1	d_2 H 8[1]	d_3
> 00	80	140
> 224	100	160
> 280	125	200
> 355	125	250
> 450	160	315
> 560	160	400
> 710	200	500
> 900	250	630
> 1120	315	800

[1] If needed, the next larger dimension from the listed sequence 80 . . . 315 can be taken.

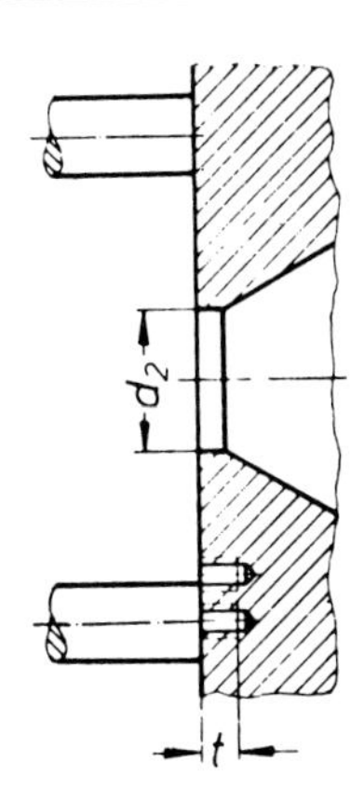

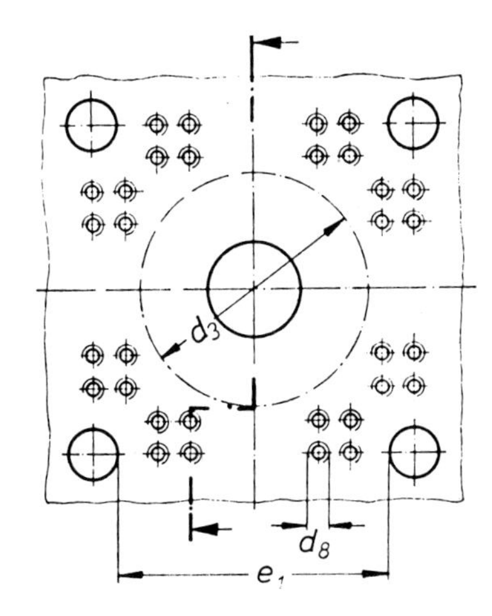

Threads or slots associated with their respective lines

Lines	Thread size d_8	t	T-slot dimensions (150 R 299) Toleranze range H 12
. . . > 140	M 12	30	14
140 . . . > 210	M 16	40	18
210 . . . > 350	M 20	45	22
350 . . .	M 24	50	28

Fig. 223b Nozzle locating hole in stationary platen, European standard

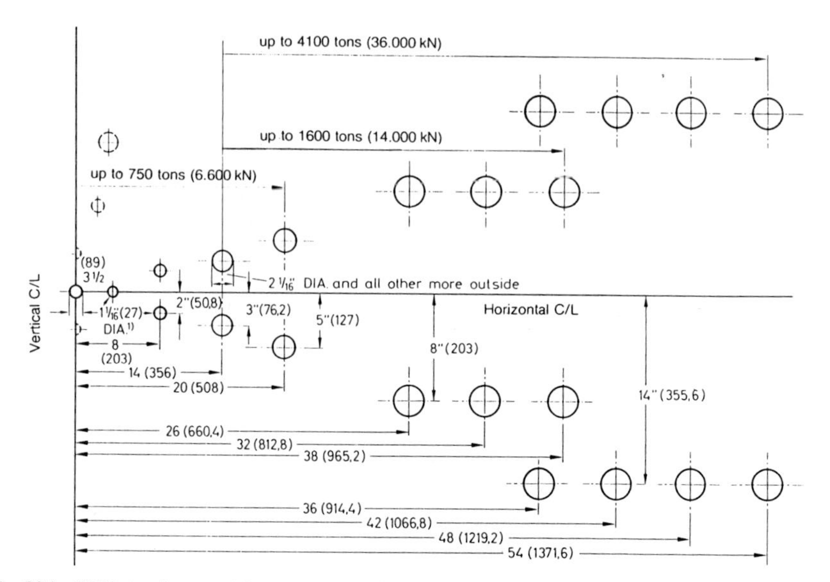

Fig. 224 SPI injection machinery standard for ejector hole pattern. Ejector pattern may be established along the horizontal or vertical centerline of the platen or a combination of both. Dimensions in in. and (mm).

1) Note: all holes $2^1/_{16}$" for machines with mor than 750 tons (6,600 kN) clamping force

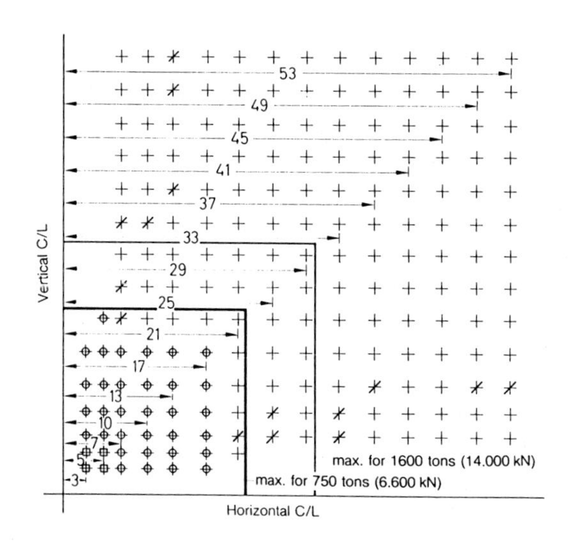

Fig. 225

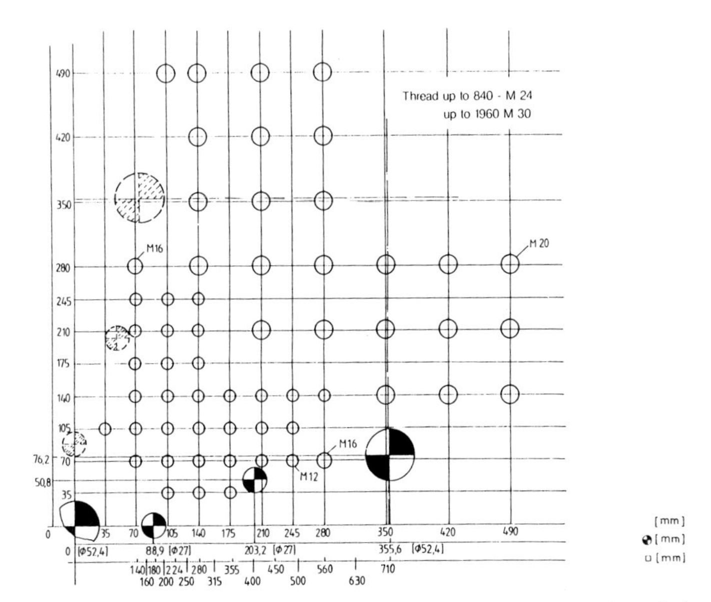

Fig. 226 European machinery standard (EUROMAP 2) for bolt and knockout hole pattern, one quadrant shown up to line 490.
Knockout hole pattern corresponding with SPI standard

Fig. 225 SPI injection machinery standard bolt hole pattern, one quadrant shown. Dimensions in inch.
For machine size over
350 tons omit 3 in
750 tons omit 3 and 5 in
2000 tons omit 3, 5, 7, 10, 13 and 17 in

✱ Holes omitted on movable platen
+ Holes for all other sizes

Standard Tap Holes

Machine size	Hole size	Thread	Minimum depth
up to 350 tons	$^5/_8$ in.	$^5/_8$ – 11	1 in.
350 to 750 tons	$^3/_4$ in.	$^3/_4$ – 10	$1^1/_4$ in.
750 to 1600 tons	1 in	1 – 8	$1^1/_2$
over 1600 tons	$1^1/_4$ in.	$1^1/_4$ – 7	2 in.

platens when the clamping mechanism is fully retracted and an eventual ejector box or spacers are not considered. Closed daylight is the distance between those platens when the clamping mechanism is fully extended. This dimension is identical with minimum mold thickness. The maximum mold thickness can be found by subtracting the minimum opening stroke required for releasing the part from the open daylight. Thus, this dimension primarily depends on the height of the part and the design of the mold. Mold thickness or closed daylight, respectively, versus clamping force of commercial machines are shown in Fig. 227. Since emphasis is placed on rigid, that is, thick mold halves, machines presented by the dashed line are mostly of some advantage.

The opening stroke is that distance from the stationary platen through which the movable platen is moved to release the part. The maximum stroke depends on the machine design and determines the maximum height of a part to be molded. Direct hydraulic systems usually have considerably longer opening strokes than mechanical systems. Listed is, most of the time, the difference between maximum and minimum distance of the platens. The size of toggle links is limited so that mold height and area of platens are confined. Machines with an opening stroke of less than the range pictured in Fig. 228 cannot meet all requirements. The opening strokes of machines for processing thermosets and elastomers generally are within the range of common machines. The gradient of the straight line is about $n = 0.66$ for American machines and $n = 0.5$ for European machines, which results in a model law of the form

$$S_{dx} = S_{d0} \left(\frac{F_{cx}}{F_{c0}} \right)^n$$

with S_{dx} the unknown stroke of a machine with a clamping force F_{cx} and S_{d0} and f_{c0} opening stroke and clamping force of a known machine.

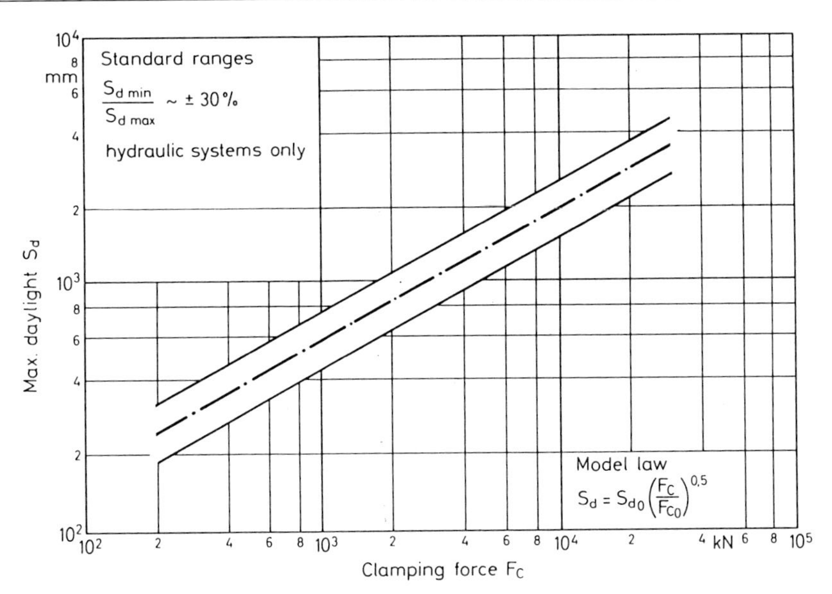

Fig. 227a Daylight versus clamping force. Dashed lines = averages. US machines, maximum daylight hydraulic of systems

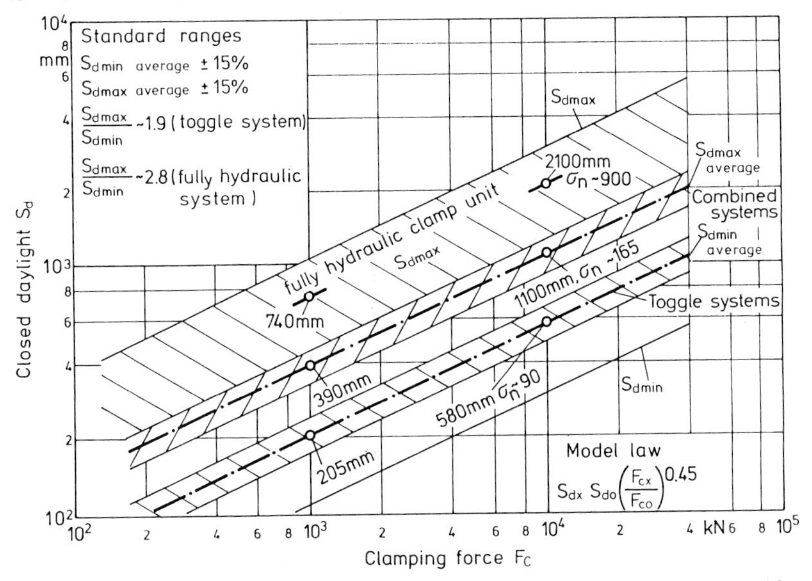

Fig. 227b Daylight versus clamping force. Dashed Lines = averages. European machines, three types of clamping systems separately shown; hydraulic - hydraulic-mechanical and mechanical (toggle) system.
Maximum daylight = Sd_{max}.
Minimum daylight = Sd_{min}.

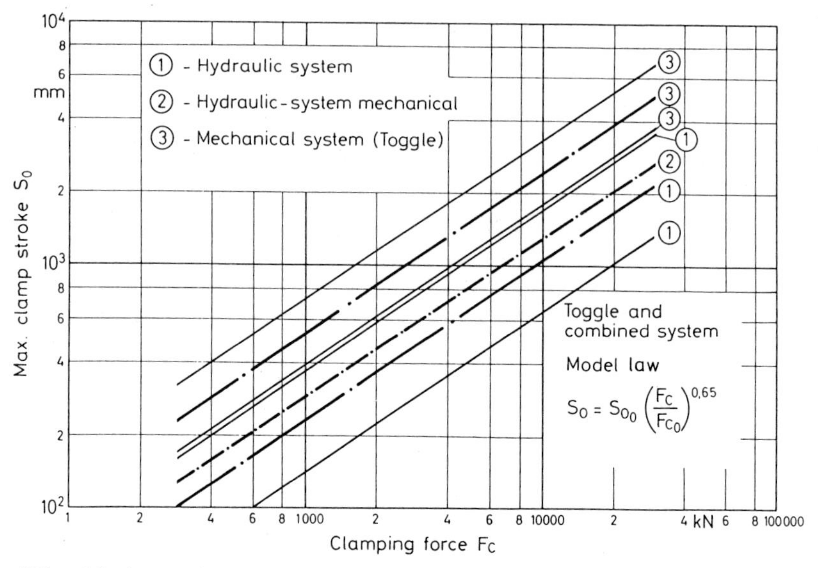

Fig. 228a Maximum clamp stroke versus clamping force for three systems (see Fig. 227). Dashed line = averages. US machines

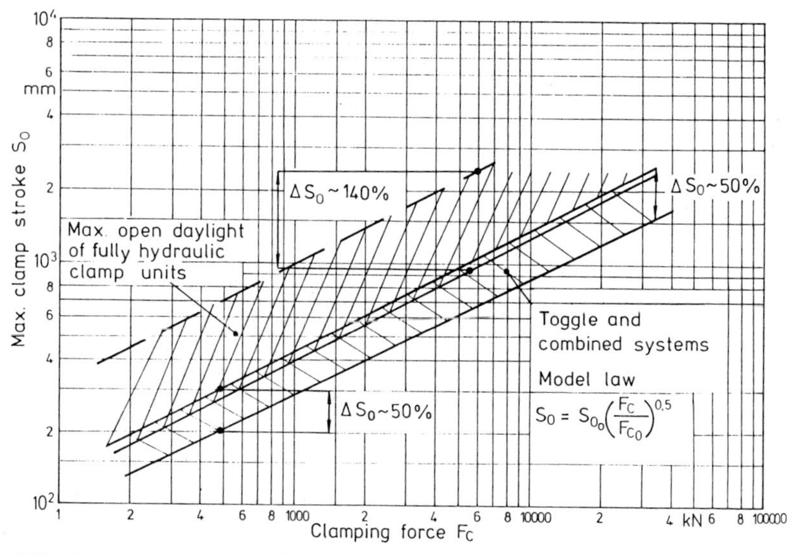

Fig. 228b Maximum clamp stroke versus clamping force for three systems (see Fig. 227) Dashed line = averages. European machines

8.2.3 General Performance Data

The name plate horse power is listed in specifications and provides basic information about the capacity of a machine with respect to speed of movements. It indicates the power input to the pumps. If an accumulator is used for actuating movements or especially for injection, specific data become applicable.

Fig. 229 presents an overview of the installed electric power input to the pump drive versus clamping force of commercial machines. The broad range comes as a surprise. There are obviously striking differences in efficiency. They can be explained to some degree by different drive systems such as accumulator or direct pump drive. Machines meeting modern standards have power inputs in the vicinity of the curves E_{P1} or E_{P2}. Machines with accumulator drive usually have a lower input, which does not adversely affect the efficiency within a single cycle. Efficiency may even be particularly high. These machines, however, do not permit a rapid sequence of fast cycles.

More meaningful information is provided by plotting power input versus working or shot capacity, because power input generally is adjusted to injection and total power input changes with the capacity of the injection unit if more than one injection unit can be combined with one clamping unit in a modular system. Fig. 230 confirms the expectation that plotting such a correlation makes the broad range disappear. The average data for fully hydraulic and toggle machines and those with accumulator drive are shown with dashed lines. Injection molding machines which are designed exclusively for processing thermosets and elastomers have about 50% less power input than comparable machines for thermoplastics.

Besides the power input as presented in Figs. 229 and 230, one has to consider that the heating capacity is included in the total power input, which, therefore, can be increased by 40 to 90% over the pump input data. An electromechanical screw drive has to provide the power listed in Section 3.3.3 separately. The installed heating capacity in machines for thermoplastics should be 0.4 to 0.5 kW for each gram per second (or 11.3 to 14.2 kW for each oz/s) of plasticating capacity to be able to provide a great portion of heat to the material by conduction if needed.

The dry-cycle time is a measurement of the basic velocity of the machine and comprises all stages of a cycle except plastication, mold filling, and cooling of the part. A particularly high basic velocity of a machine is called for in the production of small containers for packaging. Then a dry-cycle time between 0.8 and 1.4 sec. is common for machines with 500 to 2000 kN clamping force, in exceptional cases even lower [110]. Table 26 lists some data as guidelines.

Floor space, sometimes called foot print, requirement is the area calculated by multiplying length and width of a machine. Fig. 231 presents typical data for presently available commercial equipment, with the exception of some special designs. It is remarkable that floor space for a particular machine size (clamping force) may differ by as much as a factor of 1.8. The model law

$$A = A_0 \left(\frac{F_c}{F_{c0}}\right)^{0.78}$$

can be used to estimate the floor space requirement for a new machine.

A great deal of significance is often attached to the weight of an injection molding machine. The conclusion that the heaviest machine is the most 'valuable' one is correct only if all other, usually more important, features are in its favor. Heavy weight is not necessarily an indication of superior technology, nor is the opposite. An especially light-weight machine is likely to have disadvantages, mostly too thin machine platens.

Weight should not be saved with respect to

- thickness (rigidity) of machine platens,
- dimensions of toggles and links to ensure rigidity and long service life,
- design of bars and slides to avoid unacceptable deformations,
- support of movable platens. Weight can and should be reduced (within reason) with respect to
- all components the deformation of which does not affect processing,
- all movable components of the hydraulic system which are accelerated or braked down (improvement of reproducibility),
- all components of the injection unit which move during injec tion (improvement of reproducibility).

The logical consequence of such a value-oriented concept is neither a very light-weight nor an extremely heavy machine. A light- weight machine – its weight can vary, after all, by a factor of 1.6 for the same capacity – may have a considerable advantage, if limited loading capacity of the floor space is decisive for the purchase.

Aside from special types, Fig. 232 covers about all commercially available machines. The model law

$$W_M = W_{M0} \left(\frac{F_c}{F_{c0}}\right)^n$$

with $n = 1.24$ for $F_c > 1000$ kN and $n = 0.73$ for $F_c < 1000$ kN, can be applied to estimate the weight of a machine if its clamping force F_c is known and it can be compared with the weight W_{M0} and the clamping force F_{c0} of another machine.

All data compiled with the figures in Section 8.2 can be used in connection with Table 25 to determine the respective performance data for a machine of a low, medium, or high standard.

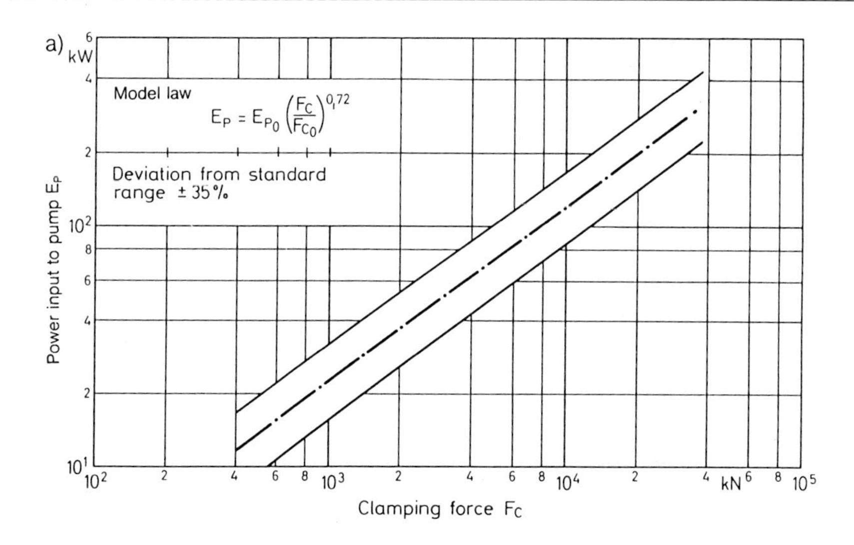

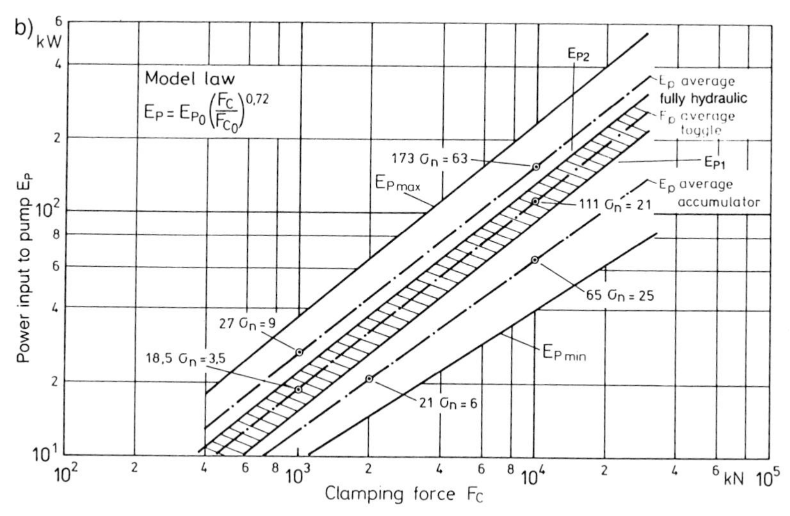

Fig. 229 Power input to pump versus clamping force. Dashed lines = averages
a) US machines
b) European machines. E_p for machines with fully hydraulic clamp unit and pump injection, for machines with toggle clamp unit or combined systems and direct pump injection and for machines with different clamp systems and accumulator injection system

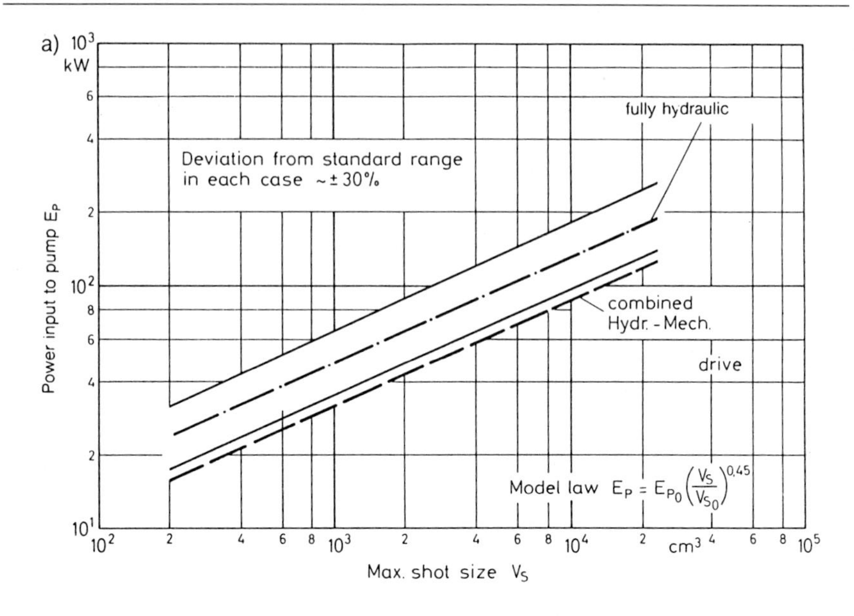

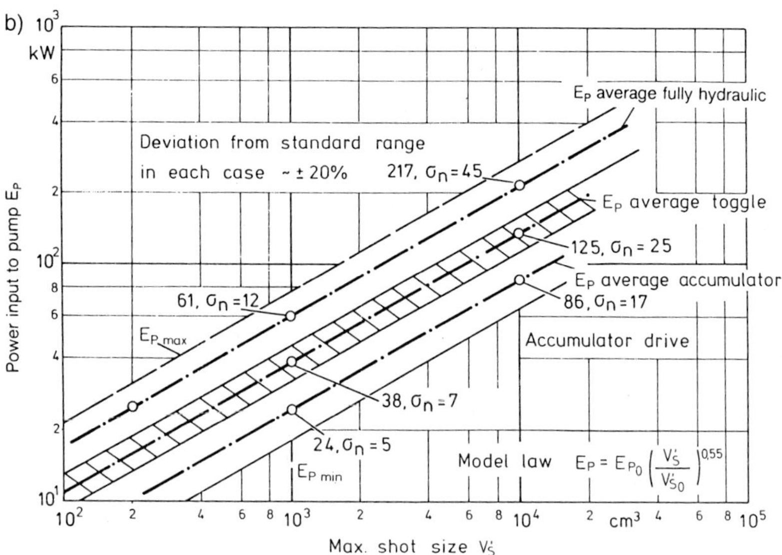

Fig. 230 Power input to pump versus maximum shot size. Dashed line = averages.

a) US machines, hydraulic clamp.

b) European machines, standard ranges for three typical clamping systems, fully hydraulic, combined and toggle. Attention: V_s = theoretical shot size based on 10^3 bar (10^2 MPa) injection pressure

Table 26 Dry Cycle Times of Injection Molding Machines

Clamping force (kN)	Dry cycle times for processing of	
	Thermopolastics (s)	Thermosets and elastomers (s)
25	0.8-1.8	–
50	0.9-1.5	up to 3
100	1.1-2	up to 6
200	1.7-4	up to 12
500	2 -7	up to 25
1000	5.5-10	–
≥ 3000	15 -25	–

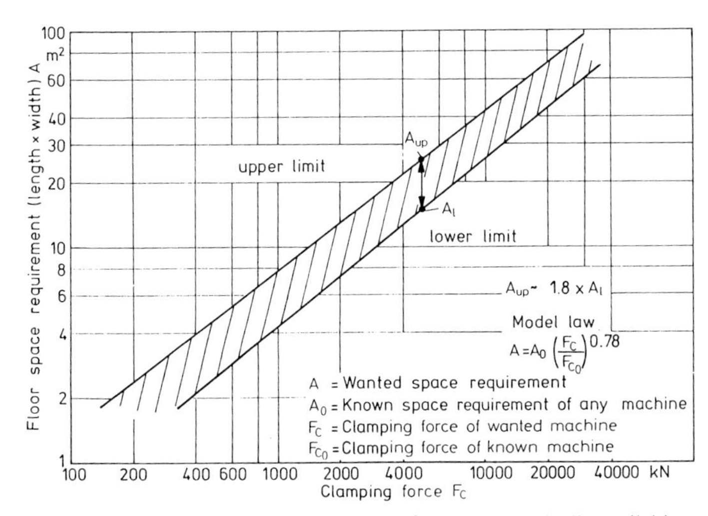

Fig. 231 Floor space requirement versus clamping force (commerically available machines)

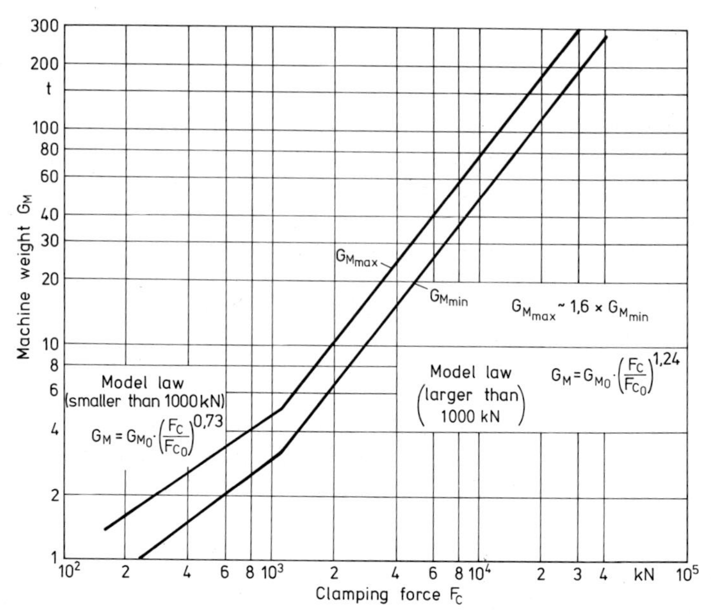

Fig. 232 Machine weight versus clamping force (commerically available machines)

9 Safety

In the United States the Occupational Safety and Health Act of 1970 (OSHA) has the declared Congressional purpose to assure as far as possible every working man and woman in the Nation safe and healthful working conditions and to preserve our human resources.

Each employer under the act has the duty to furnish his/her employees a place of employment free from recognized hazards causing or likely to cause death or serious physical harm. The employer has the specific duty of complying with safety and health standards issued under this act. Each employee has the duty to comply with these safety and health standards and all rules, regulations, and orders issued which are applicable to his/her own activities and conduct.

Since these standards deal with almost all manufacturing operations, the Society of the Plastics Industry, recognizing the inherent hazards of operating injection molding machines, has developed specific standards. Published by the American National Standard Institute (ANSI) in 1976, these standards provide for basic safety features such as safety gates; hydraulic, electrical, and mechanical interlocks; rear guards; etc. New machines meet or exceed these standards.

Some machine manufacturers have adopted a no-motion concept which blocks every movement on the machine as long as the safety gate is open, thus providing the greatest safety to the operator.

Harmful noise is discussed in the appropriate section.

As far as other countries are concerned, it is strongly suggested to familiarize oneself with local safety laws and regulations.

10 Plastic Materials for Injection Molding

Injection molding machines are well suited for processing plastics and have a very broad range of usefulness. All thermoplastic materials can be processed if they have a processing range between melting point (melting range) and the beginning of decomposition. Most thermosets and elastomers can also be fed into screws, plasticated, and processed on injection molding machines.

The processing of ceramic/thermoplastic compounds or those made with powdered metals in the production of technical parts is materializing on an increasing scale [454]. The injection molding machine is utilized to produce preformed items of ceramic or metal powder mixed with some amount of plastic, which are afterwards rendered to a finished article by firing.

The fluidity of the material, with which the feed hopper is filled, is a favorable precondition for an easy processability. It is not, however, the only precondition. Sticky (polyester with long fibers) or ribbon-shaped materials (rubber) can be processed with plunger or screw units.

Table 27 provides a brief summary of common plastic materials, processing parameters, and their shrinkage data. It also lists the abbreviations for these materials as suggested by ASTM D 1600-92 and used in this book.

Drying: Some thermoplastics may contain volatile ingredients such as moisture or monomers. Moisture can result in visible bubbles or streaking or in degradation during processing. Monomers may exude later during use and will jeopardize applications in contact with foodstuffs. To remove moisture, materials are dried either in their solid state or as melt (Fig. 233). Nonhygroscopic and less sensitive plastics can be dried with hot-air dryers, which usually are of simple design and correspondingly inexpensive [445]. Fresh air is heated, frequently mixed with a certain percentage of recycled hot air, and absorbs and carries away surface moisture from the material, which is kept in either a separate oven or the hopper dryer. It is rather easy to maintain a constant temperature of the air and, with it, of the material.

This does not ensure, however, a constant residual moisture content in the pellets, because the efficiency of the dryer depends on the humidity of the air taken in. Therefore, the moisture content of the material literally depends on the weather. Such dryers usually perform adequately in relatively dry moderate or cold climates.

Reliable drying, even under severe climatic conditions and for sensitive hygroscopic materials is achieved with a dehumidifying dryer [445]. In contrast to the aforementioned equipment, a heat carrier (usually air) is dried by means of a chemical desiccant to a low dew point level and heated to a controlled temperature. It enters the hopper dryer at the bottom, is distributed through a perforated conical device into the material and drives out and carries away moisture. While at least one desiccant bed is in use for drying air, another one is regenerated by removing the absorbed moisture with hot air.

Such dryers operate especially economical if they serve several machines at the same time. The drying occurs with a heat-carrier temperature between 70 and 130°C, corresponding to partial pressures of water vapor of 0.3 to 2 bar [81]. Only occasionally an inert gas is employed instead of air.

The drying of melt in vented barrels had gained importance between 1976 and 1985 and has been successfully done with ABS, CA, CAB, PA 6, PA 66, PBT, PC, PMMA, PPO, and SAN [99 to 109]. With common processing temperature of 200 to 340°C partial pressures of 15 to 115 bar are generated. These pressures are sufficiently high to remove moisture without applying a vacuum, which would be insignificant as an additional effect. This method is on its way out because of inherent technical difficulties and general availability of efficient dryers.

If one disregards the kind of plastic and only considers the effect of processing on the quality level of moldings, then the injection and the holding-pressure stage have the greatest influence besides the temperatures of melt, oil, and mold. The correlation of quality features to the pressure profile in the mold is presented with Fig. 12. From this, one can deduce how important it is to ensure or improve the quality level of injection-molded parts by a quantitative of the optimized pressure control and reproducibility. In the last years and after detailed process analyses, one succeeded in optimizing the holding-pressure stage with the help of the pressure-volume-temperature (P-V-T) diagram [286, 289, 292 to 295, 317, 322, 326, 352, 469, 470, 472, 475, 481, 482].

Predicting the shrinkage, carries great weight if high demands on dimensional accuracy have to be met. Shrinkage is anisotropic. Reason for this is e.g. molecular orientation, which results from deformation of molecular chains during the feeding stage and the time provided for relaxation (time from the onset of injection until freezing). The dimensions of a molding are in linear correlation with its specific volume (Fig. 234).

During cooling, residual stresses are generated. Compressive stresses are mostly in the outer layers where the melt freezes first, while tensile stresses act in the interior. These stresses are cooling stresses [41, 43 to 55]. They are superimposed by other stresses, which have different causes. Altogether, they can result in critical regions. The processing of thermosets is distinguished by its mold temperature besides its special machinery. A thermoset must not, by all means, cure in the barrel but in the mold. Therefore the mold is always heated to 150 to 200°C primarily by electric heating elements.

The thermoset injected into the cavity is very fluid with a large ratio of flow length to wall thickness if no premature cross linking occurs. This is the reason for a strong tendency to flashing. It calls for good sealing edges around the mold and closely fitted ejector pins. The dwell time in the mold is mostly longer than that for thermoplastics because complete curing has to be achieved. The available injection pressure should be higher by about 30 MPa to have the capability of injecting already slightly precured material from the nozzle.

Processing elastomers has a great similarity with the molding of thermosets. Barrel and mold temperature are comparable, and so are machine capacity, flow properties,

Table 27 Processing Temperatures, Mold Temperatures, and Shrinkage of Most Common Plastics Used in Injection Molding

Material	Symbol	Density [g/cm³]	Glass fiber content [%]	Average specific heat [kJ/(kg × K)]	Processing temperature [°C]	Mold temperature [°C]	Shrinkage* [%]
Polystrene	PS	1.05		1.3	180 – 280	10 – 40	0.3 – 0.6
Styrene-butadiene	SB	1.05		1.21	170 – 260	5 – 75	0.5 – 0.6
Styrene-acrylonitrile	SAN	1.08		1.3	180 – 270	50 – 80	0.5 – 0.7
Acrylonitrile-butadiene-styrene	ABS	1.06		1.4	210 – 275	50 – 90	0.4 – 0.7
Acrylonitrile-styrene-acrylate	ASA	1.07		1.3	230 – 260	40 – 90	0.4 – 0.6
Low-density polyethylene	LDPE	0.954		2.0 – 2.1	160 – 260	50 – 70	1.5 – 5.0
High-density polyethylene	HDPE	0.92		2.3 – 2.5	260 – 300	30 – 70	1.5 – 3.0
Polypropylene	PP	0.917		0.84 – 2.5	250 – 270	50 – 75	1.0 – 2.5
Polypropylene-GR	PP-GR	1.15	30	1.1 – 1.35	260 – 280	50 – 80	0.5 – 1.2
Polyisobutylene	PIB	0.93		–	150 – 200	50 – 80	–
Poly(4-methylpentene-1)	PMP	0.83		–	280 – 310	70	1.5 – 3.0
Poly(vinyl chloride)	PVC soft	1.38		0.85	170 – 200	15 – 20	< 0.5
Poly(vinyl chloride)	PVC rigid	1.38		0.83 – 0.92	180 – 210	30 – 50	~ 0.5
Poly(vinylidene flouride)	PVDF	1.2			250 – 270	90 – 100	3–6
Polytetraflouroethylene	PTFE	2.12 – 2.17		0.12	320 – 360	200 – 230	3.5 – 6.0
Poly(methyl methacrylate)	PMMA	1.18		1.46	210 – 240	50 – 70	0.1 – 0.8
Polyoxymethylene, polyacetal	POM	1.42		1.47 – 1.5	200 – 210	> 90	1.9 – 2.3
Poly(phenylene oxide)	PPO	1.06		1.45	250 – 300	80 – 100	0.5 – 0.7
Poly(phenylene oxide)-GR	PPO-GR	1.27	30	1.3	280 – 300	80 – 100	> 0.7
Cellulose acetate	CA	1.27 – 1.3		1.3 – 1.7	180 – 230	50 – 80	0.5
Cellulose acetate butyrate	CAB	1.17 – 1.22		1.3 – 1.7	180 – 230	50 – 80	0.5
Cellulose propionate	CP	1.19 – 1.23		1.7	180 – 230	50 – 80	0.5

Polycarbonate	PC	1.2		1.3	280 - 320	80 - 100	0.8
Polycarbonate-GR	PC-GR	1.42	10 - 30	1.1	300 - 330	100 - 120	0.15 - 0.55
Poly(ethylene terephthalate)	PET	1.37			260 - 290	140	1.2 - 2.0
Poly(ethylene terephthalate)-GR	PET-GR	1.5 - 1.53	20 - 30		260 - 290	140	1.2 - 2.0
Poly(butylene terephthalate)	PBT	1.3			240 - 260	60 - 80	1.5 - 2.5
Poly(butylene terephthalate)-GR	PBT-GR	1.52 - 1.57	30 - 50		250 - 270	60 - 80	0.3 - 1.2
Polyetheretherketone	PEEK	1.32			350 - 390	120 - 150	1.1
Polyetheretherketone-GR	PEEK-GR	1.49	30		350 - 400	120 - 150	0.2 - 1
Polyamide 6 (nylon-6)	PA 6	1.14		1.8	240 - 260	70 - 120	0.5 - 2.2
Polyamide 6-GR	PA 6-GR	1.36 - 1.65	30 - 50	1.26 - 1.7	270 - 290	70 - 120	0.3 - 1
Polyamide 66 (nylon-66)	PA 66	1.15		1.7	260 - 290	70 - 120	0.5 - 2.5
Polyamide 66-GR	PA 66-GR	1.20 - 1.65	30 - 35	1.4	280 - 310	70 - 120	0.5 - 1.5
Polyamide 11	PA 11	1.03 - 1.05		2.4	210 - 250	40 - 80	0.5 - 1.5
Polyamide 12	PA 12	1.01 - 1.04		1.2	210 - 250	40 - 80	0.5 - 1.5
Polyamide-imide	PAI	1.4		-	330 - 380	230	-
Poly(phenylene sulfide)	PPS	1.64	40		370	> 150	0.2
Poly(etherimide)	PEI	1.27		-	340 - 425	65 - 175	0.4 - 0.7
Poly(ether sulfone)	PES	1.6	30	-	360 - 390	140 - 190	0.2 - 0.5
Polyether ketone	PEK	1.3		-	360 - 420	120 - 160	1.1
Polysulfone	PSU	1.24		-	330 - 380	110 - 180	0.7
Polyurethane (thermopl.)	PUR	1.2		1.85	195 - 230	20 - 40	0.9
Phenol-formaldehyde resin	PF	1.4		1.3	60 - 80	170 - 190	1.2
Melamine-formaldehyde resin	MF	1.5		1.3	70 - 80	150 - 165	1.2 - 2
Melamine/phenol-formaldehyde resin	MPF	1.6		1.1	60 - 80	160 - 180	0.8 - 1.8
Unsaturated polyester	UP	2.0 - 2.1		0.9	40 - 60	150 - 170	0.5 - 0.8
Epoxy, epoxide	EP	1.9	30 - 80	1.7 - 1.9	ca. 70	160 - 170	0.2

* Note difference in shrinkage in flow direction and transverse, processing influence.

and curing. Demolding with mechanical ejection does not work generally. It has to be accomplished manually or by compressed air. The time needed for plastication is usually much shorter than that for curing. Therefore, several clamping units are often combined with one injection unit.

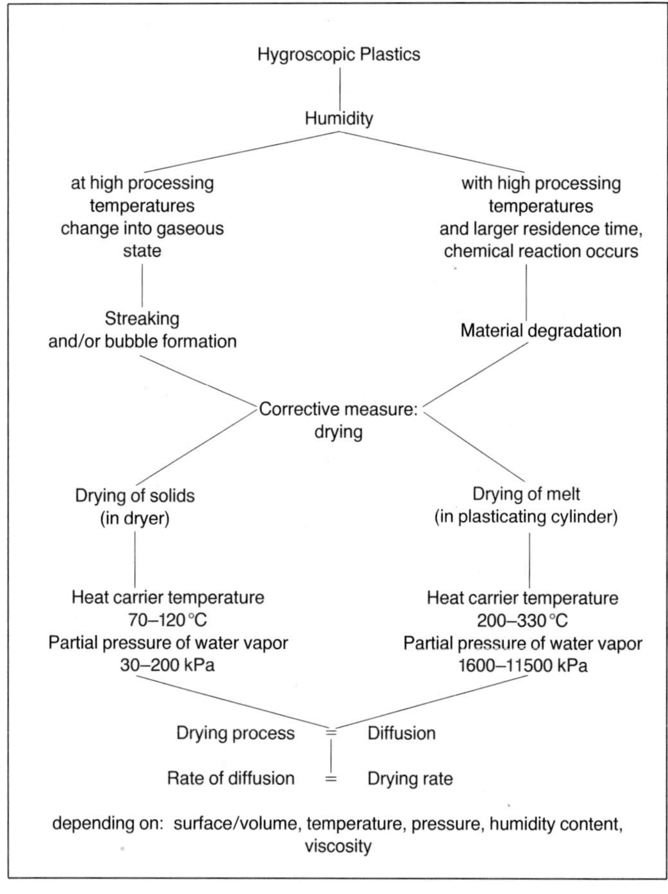

Fig. 233 Problem circle for hygroscopic plastics [81]

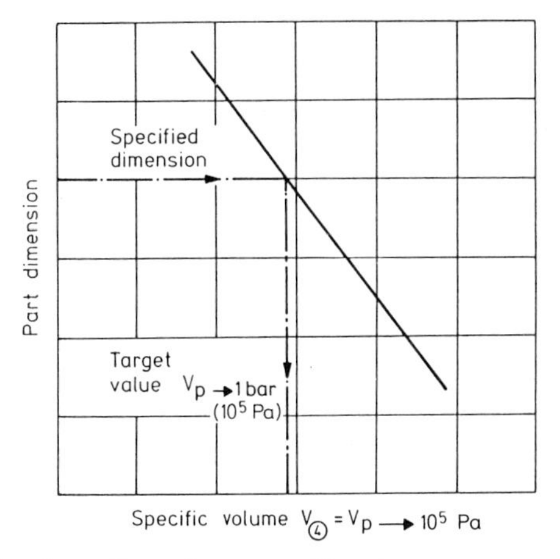

Fig. 234 Relationship of part dimension to specific volume

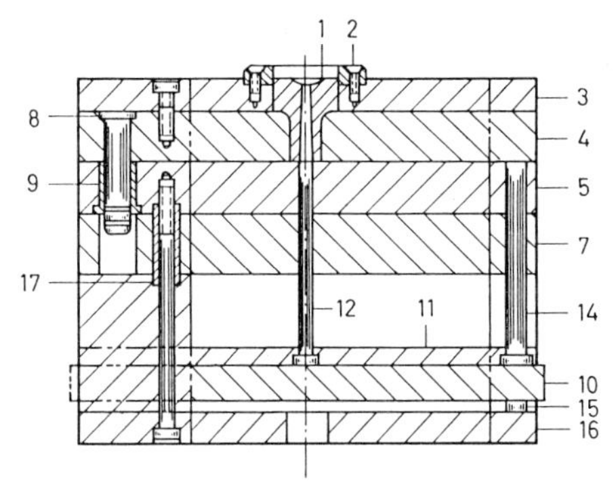

Fig. 235 Cross section through a two-plate injection mold (based on a D-M-E standard mold base

1: Sprue bushing. 2: Locating ring. 3: Clamping plate. 4: Cavity retainer plate (A-plate). 5: Cavity retainer plate (B-plate). 6: Cavity insert. 7: Support plate. 8: Leader pin. 9: Shoulder bushing. 10: Ejector plate. 11: Ejector retainer plate. 12: Sprue puller pin. 13: Ejector pin. 14: Return pin. 15: Stop pin. 16: Ejector housing. 17: Tubular dowel

11 Accessories

11.1 Injection Molds

An injection molding machine, by itself, does not constitute a complete operational installation. An injection mold is needed to provide shape and cooling (or curing or vulcanization). It comprises the forming cavity and all the necessary equipment for cooling (or heating) and ejecting the part. In the simplest and most frequent case, the mold consists of two halves, one of which is mounted to the stationary machine platen, the other one to the movable platen and moves with it. These two basic components are part of every mold. Parts and pertinent cavities must be designed in such a way that demolding is possible after the mold halves are separated in the plane of the parting line and the ejectors activated. Fig. 235 shows the principle of an injection mold assembly.

During mold opening, the molding usually remains in the movable mold half. At the same time or shortly thereafter, the ejector is actuated. The ejector plate of the mold is moved against the ejector bolt of the machine or actuated by outside means. This activates the ejector or stripper mechanism, which pushes the molding out of the cavity or off the core. Closing the mold returns the ejector mechanism to its original position.

A leakproof connection between nozzle and sprue bushing, even under high pressure, is accomplished by the machine pressing the injection unit against the stationary platen (Sect. 3.2, Table 5). The reactive pressure in the mold during injection and holding pressure is between 15 and 80 MPa (higher in special cases). This pressure depends on melt and mold temperatures, and the cross sections of the molding. Pressures generated in the cavity can be reliably estimated today by simulation programs (Cadmould, Moldflow et al.) [423 to 426, 431 to 434, 481]. One can roughly assume that about one-third of the maximum injection and holding pressure are effective in a cavity with a pinpoint gate and about two-thirds with a sprue gate. Molds have to withstand this forces as well as the clamping force for a long period of time. There are reliable computer programs for mold design [313, 423, 431, 434, 481 et al.].

The gate provides the connection between nozzle or runner and cavity. There is a multitude of shapes and sizes of gates. The choice of the gate is determined by the demands of processing technique and economics. From a processing point of view, direct gating with sprue or restricted gate is most beneficial. Table 28 presents a summary of gates and their application.

There are single- and multi-cavity molds. The decision in favor of a multi-cavity mold depends on the part weight and the capacity of the machine, particularly its plasticating capacity, usable shot capacity, and clamping force. Costs of the mold and quantity of parts are additional economic factors. It is known that molds with 148 cavities have been built.

Complicated parts frequently call for special provisions in the mold to permit demolding of undercuts, threads, or separation of part and gate during ejection. Depending on the demolding mechanism, one can differentiate among

- standard molds,
- cam-acting molds (for undercuts, cores),
- split-cavity molds (for long cylindrical parts),
- unscrewing molds (for threads),
- three-plate molds (for in-mold separation of runner from part),
- stack molds,
- special molds (for multicolor molding).

Molds for multicolor molding contain several cavities, transposable against one another and providing space after the first injection for subsequent injection of different colors. Such molds are employed to a large extent for the production of typewriter keys, other parts with molded-in letters or symbols, automotive taillights, toys et al (Fig. 172).

Fig. 236 presents an injection mold designed for a rapid exchange of standardized cavity retainer plates [421, 432]. It is employed for molding test specimens for the verification of mutually agreed comparable properties. The exchange with automatic coupling of cooling lines takes about 10 to 15 sec. It is done manually here [429].

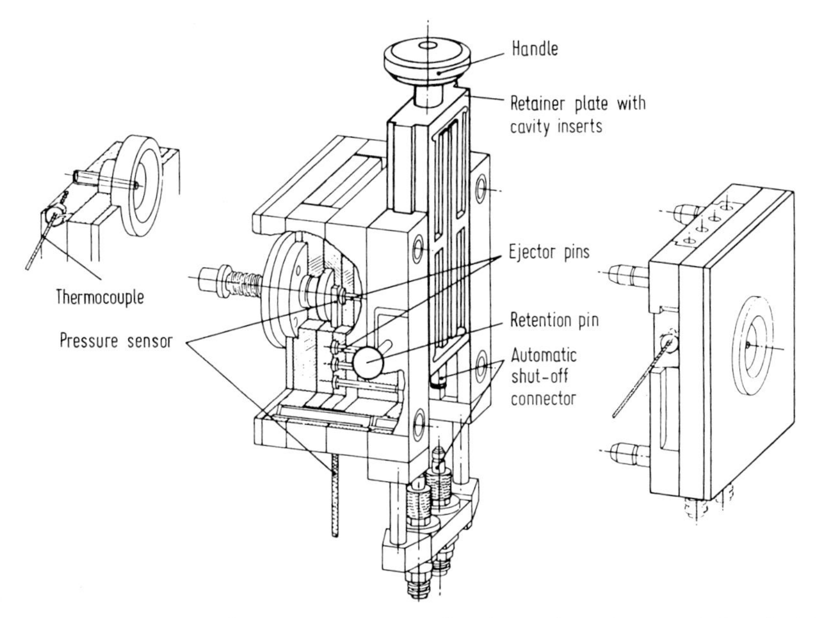

Fig. 236 Injection mold for manual exchange of cavity retainer plate especially suited for molding test specimens (So-called Campus mold) (System Bayer)

Table 28 Common Gating Systems and Their Application

Gate type	Priciple	Gate removal	Application
Sprue type	Direct central feed	Part must be degated manually	Single-cavity molds, parts with heavy sections, demand for good appearence, requires post-operation
Pinpoint gate	Direct central or edge feed	Part is torn off	Most common gate for single- and multicavity molds, generally no post-operation. With runner systems: Separation during mold opening and removal of runner in the plane of an additional parting line. Three-plate mold
Fan gate	Direct feed through fanned out thin channel	Manual separation	Need for symmetrical or linear filling, especially to prevent warpage
Tunnel gate (submarine gate)	Runner with pinpoint gates machined into cavity steel, without additional parting line	Gate is shorn off during mold opening	Multicavity mold with automatic gate removal
Hot tip, hot manifold	Hot runner or probe filled with melt, temperature-controlled feed through sprue or pinpoint gate	Separation at the melt-solid interface, sprue remains with part	Multiple gating of parts with long flow distances and multicavity molds (stack molds)
Insulated runner	Runner with large cross section, melt in its center, feed generally with pinpoint gate	Separation during mold opening	Multiple gating of large parts and multicavity molds

Figs. 237 and 238 demonstrate methods of rapid mold exchange or automated mold setup as examples. If all mold plates such as top clamping plates and ejector housings were standardized, this would be a great step forward in rationalizing the injection molding technique and its automatization. One can estimate that every setup of a mold could be faster by 50 to 90%. A first step into the right direction is the adapter plate, which has been designed by a supplier of mold standards to fit all his mold bases.

Injection molds for production purposes call for high-quality steels, which are mostly hardened or surface-treated [118, 426] to be able to withstand high loads. For internal pressure of 90 MPa a yield strength of 1300 MPa is required today.

Fig. 237 Mold with couplings for heat control and energy supply exchangeable by machine-connected hoist or overhead crane (System Netstal)

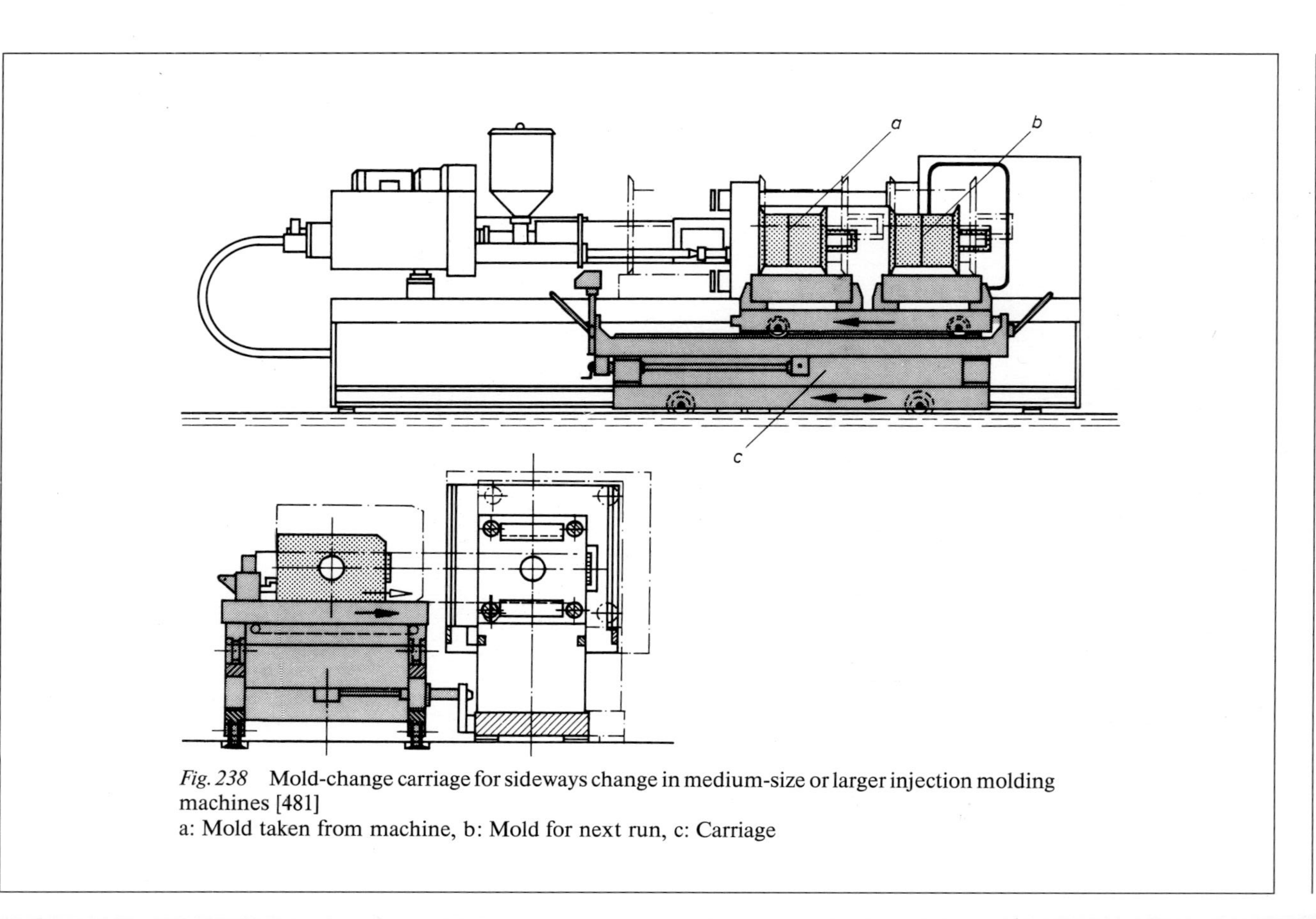

Fig. 238 Mold-change carriage for sideways change in medium-size or larger injection molding machines [481]
a: Mold taken from machine, b: Mold for next run, c: Carriage

11.2 Auxiliary Equipment for Automation

The injection molding machine must be considered the central unit of a modern, largely automated production line (Fig. 239). Equipment for drying, material supply, ventilation, coloring, loading, demolding, stacking, and post treatment are fully integrated in the process. It were beyond the scope of this book to report about these peripherals in their entirety. Some of them have been discussed in other sections.

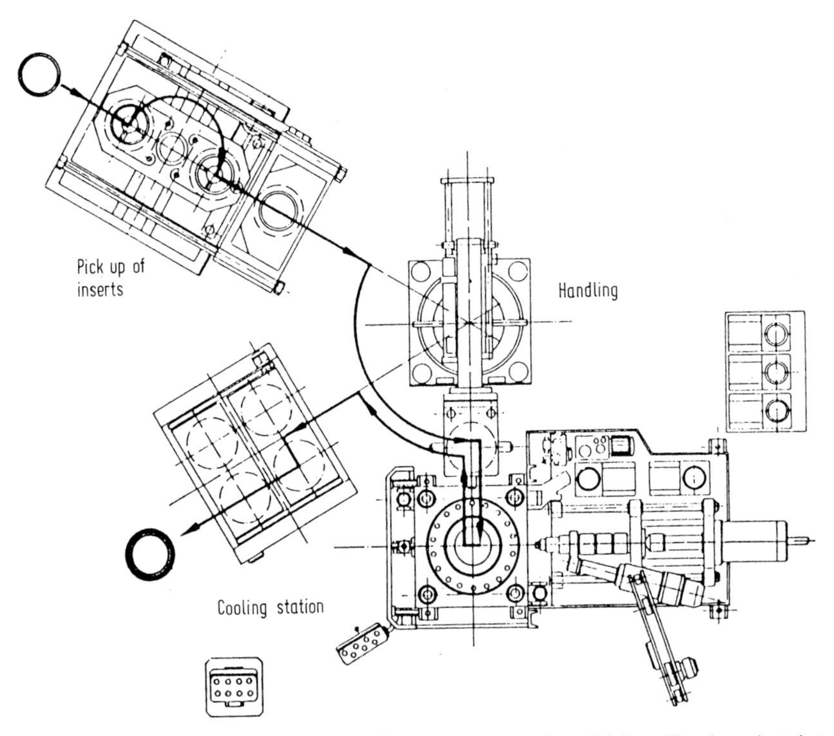

Fig. 239 Lay-out of a molding operation for rubber processing with handling by robot, intermediate storage of moldings and inserting function, arbitrarily programmable control

11.3 Heat Exchangers

Equipment for temperature control in the machine itself has been discussed in the appropriate sections. Especially important is the temperature control of the mold, because it exerts a significant influence on surface quality, dimensional accuracy, and overall cycle time [443]. Required mold temperatures can be taken from Table 27. Table 29 summarizes the requirements for heat-exchange units. Heat exchangers should never be selected by price, because inexpensive devices meet the demands on heat-exchange capacity only poorly. Significant for a good heat exchange is an adequate pump efficiency, since the necessary flow rate has to be maintained even with long and narrow cooling channels. For connections, unnecessarily long hoses should be avoided. All hoses should also have a large cross section.

On the clamping side, injection molding machines have to be equipped with a visual or measuring control for the coolant circuit. For precision molding, monitoring the flow rate is mandatory. The simplest solution is the use of a rotameter.

Table 29 Minimum Requirements on Heat Exchangers for High-Qualitiy Molding [441, 442]

Feature	Dimen-mension	Standard-injection molding Heat transfer medium		High quality injection molding Heat transfer medium	
		Water	Oil	Water	Oil
Heating capacity	kW	6	6	6	6
Cooling capacity	kW	11 kW at 333 K	6 kW at 343 K	11 kW at 333 K	6 kW at 343 K
Specific heating capacity	$kW \cdot m^{-2}$	30	25	30	25
Temperatur-control range, max.	°C	100	200	100	200
Temperature constancy in supply line	K	± 3	± 3	± 1	± 1
Pump output at 15 L/min.	kPa	200 – 400	~ 400	~ 700	~ 1000
Volume of coolant	L	7 – 14	7 – 14	10 – 20	10 – 20
Noise	dB(A)	≤ 75	≤ 75	≤ 75	≤ 75

Control of the mold temperature can be carried out as follows:

a) Temperature control of the coolant supply. Temperature variations of the mold are not considered. The temperature of the mold may deviate and usually differs from that of the coolant.
b) Control of the coolant temperature with sensors in the mold. Mold temperature remains constant. Cascade control to shorten thc startup time is difficult with respect to the correct identification of parameters. It should be selected only after a thorough evaluation of its usefulness.
c) Self-adapting control. Best solution so far. The entire dynamic adaptation of all control parameters is done automatically and results in good performance even with variations in the production process.

References

[1] ESSO Annual Report 1988

[2] Kunststoff Magazin Prodoc (1990) 6, p. 14

[3] *Pichota, H.:* Kunststoffe - German Plastics 79 (1989) p.893

[4] *Walter, G.:* Die heutigen Anforderungen der Automobilindustrie an die Kunststoffhersteller. 2nd SKZ-Symposium, Nov. 22, 1988, Würzburg/Germany

[5] *Haldenwanger, H.G., Schaper, S.:* Öko-/Energiebilanzierung im PKW-Bau mit verschiedenen Werkstoffen. In Kunststoffe im Automobilbau, VDI, Düsseldorf, 1994

[6] *Wahl, J.:* Swiss Plastics 11 (1989) 7/8, p.23

[7] *Eberhard, L.:* Swiss Plastics 12 (1990) 6, p.19

[8] *Jäger, J.:* Die Kunststoffverarbeitung in den 90-er Jahren. Hanser, Munich, New York, 1989

[9] Kunststoffrecycling - Verwerterbetriebe von Kunststoffabfällen. Hanser, Munich, New York, 1994

[10] Verwertung - Recycling - von GFK am Beispiel von SMC/BMC. Arbeitsgemeinschaft verstärkter Kunststoffe, Frankfurt/Germany

[11] Outlook Remains Buoyant in EC. European Plastics News (1993) 7/8

[12] *Orth, P.:* Recyclinggerechte Konstruktion im Automobilbereich mit technischen Thermoplasten. Print. information Bayer A.G., Leverkusen, 1993

[13] *Rothe, J.:* Kunststoffe - German Plastics 82 (1992) 12, p.1212

[14] *Schreiner, H., Heyden, H.:* Kunststoffe - German Plastics 82 (1992) 10, p.887

[15] Print. information Cincinnati Milachron, Batavia, OH - USA

[16] *Thoma, H., Stillhard, B.:* Kunststoffe - German Plastics 82 (1992) 10, p.891

[17] *Juster, H., Winninger, H.:* Kunststoffe - German Plastics 79 (1989) 10, p.1139

[18] *Stillhard, B., Weber, O.:* Kunststoffberater (1993) 12, p.24

[19] *Rubin, I. I.:* Injection Molding - Theory and Practice. Wiley, New York, 1972

[20] *Laeis, M. E.:* Der Spritzguß thermoplastischer Massen. Hanser, Munich, New York, 1959

[21] *Glenz, W.:* Kunststoffe - ein Werkstoff macht Karriere. Hanser, Munich, New York, 1985*

[22] *Woebken, W.:* Durchbruch in der Spritzgießtechnik vor 34 Jahren. VDI, Düsseldorf, 1990*

[23] *Woebken, W.:* Kunststoffe - German Plastics 80 (1990) 1, p.5

[24] *Schenkel, G.:* Intern. Polymer Processing 3 (1988) p.3

[25] DE-PS 485362

[26] DE-PS 858310

[27] *Leng, R.:* Partnerschaftliche Kooperation - ein möglicher Weg in die Zukunft. VDI, Düsseldorf, 1992

[28] *Catic, I.:* Kunststoffe - German Plastics 65 (1975) p.122

[29] *Sonntag, R.:* Entwicklung der Spritzgießtechnik. In [21]

[30] *McKelvey, J. M.:* Polymer Processing. Reinhold, New York, 1962

[31] *Michaeli, W.:* Einführung in die Kunststoffverarbeitung. Hanser, Munich, New York, 1992

[32] *Stitz, S.:* Diss. RWTH Aachen, 1973

[33] *Bodini, G., Pessani, F. C.:* Moulding Machines and Moulds for Plastics Processing. Negri Bossi, Milano, 1987

[34] *Jung, P., Patschke, H.:* Spritzgießen von Thermoplasten. VEB Deutscher Verlag für Grundstoffindustrie, Leipzig, 1988

[35] *Sarholz, R., Beese, U., Hengesbach, H. A., Wübken G.:* Spritzgießen: Verfahrensablauf, Verfahrensparameter, Prozeßführung. Hanser, Munich, New York, 1979
[36] *Wintergerst, S.:* Kunststoffe - German Plastics 63 (1973) p. 636
[37] *Bauer, W.:* Kunststoffe - German Plastics 53 (1963) p. 210
[38] *Wiegand, H., Vetter, H.:* Kunststoffe - German Plastics 56 (1966) p. 761; 57 (1967) p. 276
[39] *Menges, G., Leibfried, D.:* Plastverarbeiter 21 (1970) p. 951
[40] *Leibfried, D.:* Diss. RWTH Aachen, 1971
[41] *Wübken, G.:* Diss. RWTH Aachen, 1974
[42] *Wübken, G.:* Kunststoffe - Plastics 24 (1977) p. 37
[43] *Geyer, H.:* Kunststoffe - German Plastics 65 (1975) p. 7
[44] Gleichmäßigkeit und Richtungsabhängigkeit der mechanischen Eigenschaften von Cellidor Spritzgußteilen. Print. information 154/77, Bayer AG, Division KL
[45] *Knappe, W.:* Kunststoffe - German Plastics 51 (1961) p. 562
[46] *Grosskurth, K. P.:* Kautschuk, Gummi, Kunststoffe 29 (1976) p. 392
[47] *Wübken, G.:* Plastverarbeiter 26 (1975) p. 392
[48] Abschätzung der Größe innerer Spannungen in Makrolon Formteilen. Print. information 104/77, Bayer AG, Division KL
[49] *Racke H., Fett, T.:* Materialprüfung 13 (1972) 2, p. 37
[50] *Fett, T.:* Plastverarbeiter 24 (1973) p. 665
[51] *Zachmann, H. G.:* Fortschr. Hochpolym. Forsch. 3 (1974) p. 581
[52] *Menges, G., Wübken, G., Horn, H.:* Colloid Polymer Sci. 252 (1974) p. 267
[53] *Heckmann, W.:* Colloid Polymer Sci. 252 (1974) p. 826
[54] *Hindle, C. S.:* Reprint of paper, SPE-Antec, May 1981
[55] *Schauf, D.:* Paper at the International Conference, Dresden/Germany, April 1978
[56] *Michaeli, W., Breuer, P., Hohenauer, K., von Oepen, R., Philipp, M., Pötsch, G., Recker, H., Robers, Th., Vaculik, R.:* Kunststoffe - German Plastics 82 (1992) 12, p. 1167
[57] *Wunck, Ch.:* Diss. RWTH Aachen, 1991
[58] *Johannaber, F.:* Kunststoffe - German Plastics 63 (1973) p. 490
[59] *Just, B.:* Bosch, Technical Report 7 (1981) 3, p. 104
[60] *Paffrath, H. W., Rupprecht, L.:* Plastverarbeiter (1982) 7, p. 416
[61] *Johannaber, F.:* Kunststoffe - German Plastics 74 (1984) 1, p. 2
[62] *Lampl, A., Gissing, K., Painsith, H.:* Plastverarbeiter 34 (1083) p. 427
[63] *Gissing, K., Lampl, A.:* Plastverarbeiter 34 (1983) 5, p. 427
[64] *Anders, S., Salewski, K., Steinbüchl, R., Rupprecht, L.:* Kunststoffe - German Plastics 81 (1991) p. 576
[65] *Bielfeldt, F. B., Herbst, R.:* Kunststoffe - German Plastics 63 (1973) p. 576
[66] *Renger, M.:* Ertragssteigerung durch Prozeßoptimierung. Print. information Battenfeld, Meinerzhagen 1991
[67] *Burr, A.:* Qualität sichern im Spritzgießbetrieb. VDI, Düsseldorf, 1993
[68] *Meridies, R.:* Leistungsdaten und Abnahmebedingungen von Spritzgießmaschinen. Paper VDI-Bildungswerk, 1973
[69] *Jäger, E. M.:* Ind. Anz. 97 (1975) p. 107
[70] *Elbe, W.:* Diss. RWTH Aachen, 1973
[71] *Meridies, R.:* Plastverarbeiter 21 (1970) p. 11
[72] *Fischer, P.:* Diss. RWTH Aachen, 1976
[73] *Potente, H.:* Kunststoffe - German Plastics 71 (1981) p. 374; *Potente, H.:* Auslegung von Schneckenmaschinen-Baureihen. Modellgesetze und ihre Anwendung. Hanser, Munich, New York, 1981*

[74] *Menges, G.:* Plastifizier- und Mischverhalten von Schnecken-Spritzgießmaschinen. Research report IKV Aachen, 1978*
[75] *Bürkle, E.:* Diss. RWTH Aachen, 1989*
[76] *Beck, H.:* Spritzgießen. Hanser, Munich, New York, 1963
[77] Injection Moulding Machine-Screws for Bayer Thermoplastics. Print. information, ATI 173e, Leverkusen 1988
[78] *Bossecker, S.:* IKV Aachen, 1988
[79] *Steinbüchl, R.:* Bestimmung der Verweilzeit beim Spritzgießen technischer Thermoplaste, ATI No. 634 Bayer AG.
[80] *Langecker, G.:* Verbesserte Qualität bei höherer Plastifizierleistung - ein Widerspruch? Print. information Battenfeld, 10/87
[81] *Rheinfeld, D.:* Plastverarbeiter 29 (1978) p. 413
[82] *Renger, M.:* Kunststoffe - German Plastics 79 (1989) 11, p. 1113
[83] *Schulte H.:* Diss. University-Gesamthochschule Paderborn/Germany, 1990*
[84] Research project GSHS-Unversity Paderborn
[85] *Potente, H., Schulte, M., Hansen, M.:* Plastverarbeiter 40 (1989) 12, p. 30
[86] Print information by IKP, Paderborn, Bayer AG, Leverkusen, and Mannesmann Demag Kunststofftechnik
[87] Korrelation zwischen Schneckendurchmesser, Dosiervolumen, Dichte und Schußgewicht. Print. information Bayer AG, 11/1989
[88] *Johannaber, F.:* Kunststoffe - German Plastics 79 (1989) 1, p. 25
[89] *Johannaber, F.:* Kunststoffe - German Plastics 79 (1989) 1, p. 5
[90] *Limper, A.:* Der Extruder im Extrusionsprozeß. VDI, Düsseldorf, 1989
[91] *Verbraak, C.P.J.M., Melfer, H.E.H.:* Polym. Eng. Sci. 29 (1989) 7
[92] *Wortberg, J.:* Der Extruder im Extrusionsprozeß. VDI, Düsseldorf, 1989
[93] *N.N.:* Schnecken aller Art. Kunststoffe 82 (1992) 10
[94] *Hähnsen, H., Johannaber, F., Orth, P.:* Recycling im Spritzgießbetrieb. VDI, Düsseldorf, 1993
[95] Relationship between Screw Diameter, Metered Volume, Density and Shot Weight. Print. information No. 765e, Bayer AG, Leverkusen
[96] Print. information Bayer AG, Leverkusen
[97] *Sokolow, N.N.:* Mod. Plast. Int. 9 (1979) 3, p. 54
[98] *Friedrich, E.:* Kunststoffe 47 (1957) p. 218
[99] *Loske, V., Rothe, J.:* Kunststoffe - German Plastics 68 (1978) p. 335
[100] *Rheinfeld, D.:* Plastverarbeiter 29 (1978) p. 413
[101] *Graf, H., Mayer, F., Lampl, A.:* Kunststoffe - German Plastics 71 (1981) p. 466
[102] *Whealan, A., Craft, J.L.:* Developments in Injection Molding. Vol. 1 and 2. Applied Science Publishers, London 1978, 1981
[103] *Ronzini, I., Casale, A., de Marosi, G.:* SPE J. 27 (1971) p. 74
[104] *Nunn, R.E.:* Plast. Eng. 36 (1980) 2, p. 35
[105] *de Capite, R., Gudermuth, C.S.:* Plast. Eng. 32 (1976) p. 37
[106] *Rosato, O.V.:* Plast. World 4 (1977) p. 53
[107] *Ladwig, H.:* Kunstst. Plast. 25 (1978) p. 66
[108] *Backhoff, W., van Hooren, R., Johannaber, F.:* Kunststoffe - German Plastics 67 (1977) p. 307
[109] *van Hooren, R., Kaminski, A.J.:* Plastverarbeiter 8 (1980) p. 441
[110] Processing Data for Injection Moulder. Print. information Bayer AG, Leverkusen, 2/1982

[111] *Bovensman, W.:* Plastifizieren und Homogenisieren von Duromeren mit Schnecken-Plastifiziergeräten. Series Ingenieurwissen. VDI, Düsseldorf, 1973
[112] *Schönthaler, W. et al.:* Plastverarbeiter 29 (1978) 1, p. 243
[113] *Bode, M.:* Spritzgießen von Elastomeren. VDI Düsseldorf, 1978, p. 39
[114] *Merkt, L.:* Die Verarbeitung von Kautschuk auf Schneckenspritzgießmaschinen. Print. information Arburg, Lossburg
[115] *Merkt, L.:* Paper reprint, SKZ, Würzburg, 21./22.5.1987
[116] *Limper, A., Barth, P., Grajewski, F.:* Technologie der Kautschukverarbeitung. Hanser, Munich, New York, 1989
[117] *Benfer, P., Fischbach, G., Schneider, W., Weyer, G.:* Paper reprint, colloquium IKV Aachen, 1986
[118] *Menning, G.:* Verschleißschutz in der Kunststoffverarbeitung. Hanser, Munich, New York, 1990*+
[119] *Meridies, R., Bassner, F.:* Plastverarbeiter 21 (1970) p. 617
[120] *Mahler, W.-D.:* Diss. TH Darmstadt, 1975
[121] *Brinkschroder, J., Johannaber, F.:* Print. information Bayer AG, Leverkusen, 1978
[122] *Lülsdorf, P.:* Verschleißprobleme mit Zylinder und Schnecke beim Extrudieren. VDI instruction, August 1975
[123] *Lenze, G.:* Plastverarbeiter 25 (1974) p. 209 and 347
[124] *Volz, P.:* Kunststoffe - German Plastics 70 (1979) p. 738
[125] *Menning, G., Volz, P.:* Kunststoffe - German Plastics 70 (1980) p. 385
[126] *Volz, P.:* Diss. TH Darmstadt, 1981
[127] *Volz, P.:* Kunststoffe - German Plastics 72 (1982) p. 337
[128] *Volz, P.:* Kunststoffe - German Plastics 72 (1982) p. 71
[129] *Kaminski, A.:* Spritzgießen. VDI, Düsseldorf, 1983 and Print. information Bayer AG, Leverkusen, 1984
[130] Wege der Verschleißminderung in der Kunststoffverarbeitung. VDI Convention 4/1985
[131] *Kaminski, A.:* Kunststoffe - German Plastics 66 (1976) p. 208
[132] *Fleischer, G.:* Verschleiß und Zuverlässigkeit. VEB Technik, Berlin, 1982
[133] *Kaminski, A.:* Verschleißschutz beim Spritzgießen. Print. information, No. 458, Bayer AG, Leverkusen, 1984
[134] *Johannaber, F.:* Kunststoffe - German Plastics 75 (1985) 9, p. 560
[135] *Speuser, G.:* Konstruktion einer Spritzgießschnecke für die Pulverkautschukverarbeitung. IKV Aachen, 1985
[136] *Menges, G., Elbe, W.:* Plastverarbeiter 24 (1973) p. 137
[137] *Menges, G., Elbe, W.:* Plastverarbeiter 23 (1972) p. 312
[138] *Lampl, A., Lindorfer, B.:* Kunststoffe - German Plastics 79 (1989) 11, p. 1097
[139] *Renger, M.:* Kunststoffe - German Plastics 79 (1989) 11, p. 1113
[140] DE-PS 3676334
[141] Automatischer Zylinderwechsel, Battenfeld, Meinerzhagen
[142] *Jäger, F.:* Rationalisierung im Spritzgußbetrieb. Netstal News No. 20 (1986)
[143] *Hotz, A.:* Kunstst. Ber. 21 (1976) p. 194
[144] *Schröder, K.:* Qualitätsoptimierende Verarbeitungssysteme bei der Verarbeitung von Duroplasten. Battenfeld Convention 9/1991
[145] *Killinger, H.:* Kunststoffe - German Plastics (1989) 2, p. 137
[146] *Gissing, K.:* Spritzgießen und Spritzprägen. Plastics Colloquium, Leoben/Austria, Nov. 1982
[147] Duroplast Technologie der 90er Jahre. Battenfeld, Meinerzhagen, convention Sept. 1990

[148] *Langecker, G.R.:* Kunststoffe - German Plastics 74, (1984) 5, p.258
[149] *Wübken, G.:* Kunststoffe - German Plastics 71 (1981) 12, p.258
[150] *Henschely, F.:* Auslegung und Erprobung von Verschlußdüsen für das Spritzgießen. Graduation paper, FH Osnabrück 4/92
[151] *Brünger, D.:* Kunststoffe - German Plastics 82 (1992) p.971
[152] Print. information, Kona Corp. Gloucester, MA/USA
[153] *Swenson, P., Ladwig, H.:* Kunststoffe - German Plastics 82 (1992) 12, p.1157
[154] *Speckenheuer, G.P., Stracke, A.:* Kunststoffe - German Plastics 83 (1993) p.171
[155] *Hengesbach, A.H., Schramm, K., Wübken, G., Sarholz, R.:* Plastverarbeiter 27 (1976) p.583
[156] *Lidl, R., Haberer, H.:* Krauss Maffei Kunststoffmaschinenjournal 5 (1978)
[157] *Trepte, H.:* Plaste und Kautschuk 39 (1992) 4, p.131
[158] *Jürgens, W.:* Diss. RWTH Aachen, 1969
[159] *Gissing, K.:* Diss. Montanuniversität Leoben/Aus.
[160] Colloquium IKV Aachen, reprint 1992
[161] *Auffenberg, D.:* Diss. RWTH Aachen, 1975
[162] *Schwarz, O., Ebeling, F.-W., Lüpke, G., Schelter, W.:* Kunststoffverarbeitung. Vogel, Würzburg/Germany 1985+
[163] *Mink, W.:* Grundzüge der Spritzgießtechnik. Zecher & Hüthig, Speyer/Germany, 1971+
[164] Print. information Mannesmann Demag Kunststofftechnik, 1972
[165] *Keller, H.R.:* Plastverarbeiter 18 (1967) p.447
[166] *Naetsch, H., Nikolaus, W.:* Plastverarbeiter 28 (1977) p.169
[167] Husky Newsletter Dec. 1981
[168] Print. information Werner & Pfleiderer, 1977
[169] *Naetsch, H., Nikolaus, W.:* Plastverarbeiter 33 (1982) p.201
[170] *Sauerbruch, E.:* Kunststoffe - German Plastics 61 (1971) p.471
[171] Netstal News 23
[172] *Urbanek, O., Leonhartsberger, H., Steinbichler, G.:* Kunststoffe - German Plastics 81 (1991) 12, p.1081
[173] *Urbanek, O.:* Spritzgießen von Präzisionsteilen auf der holmlosen Spritzgießmaschine. Paper, Vereinigung Österr. Kunststoffverarbeiter 11/1991
[174] Plastverarbeiter 43 (1992) 2, p.138
[175] Print. information, Engl, Schwertberg/Aus
[176] *Yamazaki, Y.:* Japan Plastics (1985) 3/4, p.34
[177] *Ulmer, D.:* Handbuch der Hydraulik. Vickers GmbH, Bad Homburg v.d.H./Germany
[178] Hydraulik in Theorie und Praxis. Bosch, 1990
[179] *Hengesbach, H.A., Schramm, K., Wobben, D., Sarholz, R.:* Paper at 8. Kunststofftechnisches Colloquium, IKV Aachen, March 24-26, 1976
[180] *Hengesbach, H.A.:* Gesichtspunkte der hydraulischen Leistungssteuerung. Paper, IKV Aachen, Mat 31, 1975
[181] *Rindt, H.:* Plastverarbeiter 28 (1977) p.15
[182] *Morell, H.-J.:* Ölhydraulik und Pneumatik 20 (1976) 7, p.466
[183] *Copetti, Th.:* Plastverarbeiter 31 (1980) p.655
[184] *Lange, H.-J.:* Kunststoffe - German Plastics 74 (1984) p.130
[185] *Krines, H.G.:* Kunststoffe - German Plastics 71 (1981) p.558
[186] Print. information Mannesmann Demag Kunststofftechnik 1981
[187] *v. d. Meulen, W.:* Kunstst. Plast. 27 (1980) p.9
[188] Demag Nachrichten, 1/1978

[189] *Blüml, H.:* VDI Berichte No. 228 (1975) p. 125
[190] Spritzgießen. VDI-Ges. Kunststofftechnik, VDI, Düsseldorf 1983
[191] *Schreiner, H.:* Mit Digitalhydraulik zu höherer Produktqualität. Paper, Fluid Convention, München, 1988
[192] *Fischbach, G.:* Kunststoffe - German Plastics 79 (1989) 11, p. 15
[193] *Schreiner, H.:* Schweizer Maschinenmarkt (1989) 5, p. 18
[194] *Rothe, J.:* Pumpen, Motoren und Zylinder für die Kunststoffverarbeitung. Paper, convention at Rexroth, Sept. 1991
[195] *Dantlgruber, E.:* Neue Pumpensteuerungen für Spritzgießmaschinen. Paper, convention at Rexroth, Sept. 1991
[196] *Menges, G., Bourdon, F.:* Energiebilanz an Kunststoffspritzgießmaschinen, MAV Oct. 1986
[197] *Zehner, F., Dischnik, M.:* Einfluß des Dithers auf das Durchflußgesetz hydraulischer Brückenschaltungen. Ölhydraulik + Pneumatik 39 (1982) p. 457/461
[198] *Brasch, H.:* Moderne Pressenhydraulik. Ölhydraulik + Pneumatik 39 (1989) No 4
[199] *Janke, W., Verstegen, C.:* Kunststoffberater 12/1983
[200] *Götz, W.:* Elektrohydraulische Proportional- und Regeltechnik in Theorie und Praxis. Print. information Bosch, 1989
[201] *Götz, W.:* Hydraulik in Theorie und Praxis. Print. information Bosch, 1983
[202] *Götz, W.:* see [200]
[203] *Rothe, J.:* Mannesmann Demag Report 4 (1981) p. 4
[204] *Rothe, J.:* Rationalisieren im Spritzgießbetrieb. VDI, Düsseldorf, 1981
[205] *Schreiner, H.:* Energiesparende Konstruktion bei Spritzgießmaschinen. Paper, 27. Colloquium SKZ, Würzburg, 1979
[206] *Rothe, J.:* Spritzgießen, VDI, Düsseldorf, 1983
[207] *Nunn, R.E., Ackermann, K.:* Reprint of paper, SPE-ANTEC 7 (1981) 5, p. 786
[208] *Rheinfeld, D.:* Krauss-Maffei Journal 12 (1981) p. 3
[209] *Johannaber, F.:* Kunststoffe - German Plastics 71 (1981) p. 44 and p. 702
[210] *Johnson, T.:* Print. information, Husky Injection Molding Systems Inc. Bolton, Ont.
[211] Mod. Plast. 9 (1979) 5, p. 8
[212] *Johansen, O., Haudrum, J., Knudsen, J.:* Plastverarbeiter 42 (1991) 6, p. 70
[213] *Lange, H.-J.:* Kunstst. Ber. 21 (1976) p. 260
[214] *Krines, H.G.:* Kunststoffe - German Plastics 71 (1981) p. 558
[215] *Rheinfeld, D.:* Krauss Maffei Journal 8 (1979)
[216] Qualitätsoptimierte Spritzgießtechnik. Reprint from convention SKZ, Würzburg, Nov. 1991
[217] *Späth, W., Walter, H.:* Kunststoffe - German Plastics 66 (1976) p. 59
[218] *Ulmer, O.:* Handbuch der Hydraulik. Sperry Rand Vickers, Bad Homburg v.d.H., 1971
[219] *Lafreniere, E.:* Reprint of paper, SPE-Antec 27 (1981) 5, p. 812
[220] *Rindt, H.:* Plastverarbeiter 28 (1977) p. 15
[221] *Morell, H.-J.:* Ölhydraulik und Pneumatic 20 (1976) 7, p. 466
[222] *Johannaber, F.:* Kunststoffmaschinenführer, 3rd edition. Hanser, Munich, New York, 1992
[223] *Meridies, R.:* Plastverarbeiter 27 (1976) p. 291
[224] Demag Nachrichten 1/78
[225] *Blüml, H.:* VDI-Ber. 228 (1975) p. 125
[226] *Sonntag, R., Rothe, J., Elbe, W.:* Kunststoffe - German Plastics 68 (1978) p. 130
[227] *Hemmi, P.:* Kunststoffe - German Plastics 63 (1973) p. 198
[228] *Sitz, S., Hengsbach, H.A.:* Plastverarbeiter 25 (1974) p. 201

[229] *Stitz, S., Hengesbach, H. A., Pütz, D.:* Kunststoffe - German Plastics 63 (1973) p. 777
[230] *Menges, G., Recker, H.:* Automatisierung in der Kunststoffverarbeitung. Hanser, Munich, New York, 1986
[231] *Daca, W., Kögler, M.:* Kunststoffe - German Plastics 81 (1991) p. 1092
[232] *Recker, H., Berndtsen, N., Sarholz, R., Dormeier, S., Hengesbach, H. A., Hellmeyer, H. O., Hensel, H.:* Plastverarbeiter 28 (1977) p. 1 and 81
[233] *Lampl, A., Lindorfer, B.:* Kunststoffe - German Plastics 79 (1989) 11, p. 1097
[234] *Menges, G., Sarholz, R., Schmidt, L., Thienel, P.:* Plastverarbeiter 29 (1978) p. 295, 369 and 425
[235] *Rheinfeld, D.:* Kunstst. Ber. 20 (1975) p. 597
[236] *Stillhard, B.:* Kunststoffe - German Plastics 66 (1976) p. 710
[237] *Stitz, S.:* Kunstst. Pla. 25 (1978) p. 33
[238] *Stillhard, B., Hengesbach, H. A.:* Kunststoffe - German Plastics 65 (1975) p. 406
[239] *Copetti, T.:* Kunstst. Ber. 21 (1976) p. 667
[240] *Hardt, B., Johannaber, F.:* Plastverarbeiter 29 (1978) p. 475
[241] *Menges, G., Stitz, S., Vargel, J.:* Kunststoffe - German Plastics 61 (1971) p. 74
[242] *Stitz, S., Hengesbach, H. A.:* Plastverarbeiter 25 (1974) p. 201
[243] *Thienel, P.:* Diss. RWTH Aachen, 1977
[244] *Rheinfeld, D., Jensen, R.:* Kunststoffe - German Plastics 65 (1975) p. 448
[245] *Offergeld, H., Haupt, M., Reichstein, H., Brinkmann, T.:* Kunststoffe - German Plastics 82 (1992) 8, p. 636
[246] *Henschen, S., Brinkmann, T.:* Kunststoffe - German Plastics 82 (1992) 12, p. 1274
[247] *Steinbichler, G.:* Kunststoffe - German Plastics 82 (1992) p. 902
[248] *v. d. Meulen, W., Allerdisse, W.:* Kunststoffe - German Plastics 67 (1977) p. 189
[249] *Allerdisse, W.:* Kunstst. Ber. 23 (1978) p. 475
[250] *Copetti, T.:* Kunststoffe - German Plastics 73 (1983) p. 170
[251] *v. d. Meulen, W.:* Automatisierung in der Kunststoffverarbeitung. Hanser, Munich, New York, 1985
[252] *v. d. Meulen, W.:* Kunststoffe / Plastics 27 (1980) 9, p. 9
[253] *Neuhäusl, E. et al.:* Plaste und Kautschuk 39 (1992) 2, p. 49
[254] *Laimer, F.:* Spritzgießen. VDI, Düsseldorf, 1983
[255] Das Mikroprozessor-System MPC 80. Print. information GKN-Windsor GmbH, Maintal / Germany, Sept. 1979
[256] K-Plastik & Kautschuk Zeitung, No. 4 (1991) p. 9
[257] K-Plastik & Kautschuk Zeitung, No. 4 (1991) p. 19
[258] *Lidl, R.:* Kunststoffe - German Plastics 71 (1981) p. 342
[259] Mod. Plast. Int. (1979) 12, p. 8
[260] *Davis, M. A.:* Plastic. Eng. 33 (1977) 4, p. 26
[261] *Menges, G., Bongardt, W.:* Kunstst. Plast. 27 (1980) 2, p. 10
[262] Mod. Plast. Int. 11 (1981) 10, p. 14
[263] *Bongardt, W.:* Diss. RWTH Aachen, 1982
[264] Reprint of paper, colloquium IKV Aachen, 1984
[265] *Matzke, A.:* Diss. RWTH Aachen, 1985
[266] *Wippenbeck, P.:* Kunststoffberater 33 (1988) 11, p. 30
[267] DE 3927995 A1, Aug. 24, 1989
[268] *Michaeli, W., Lauterbach, M.:* Kunststoffe - German Plastics 79 (1989) 9, p. 852
[269] *Böhm, Th., Lambert, C.:* Kunststoffe - German Plastics 79 (1989) 10, p. 965
[270] *Köhler, S.:* Unilog TC 40. Echtzeit-Prozeßregelung von Kunststoffverarbeitungsmaschinen. Paper at Battenfeld Symposium, 1991

[271] *Huber, A.:* Möglichkeiten der Qualitätssteigerung durch Transputertechnik. Paper at Battenfeld Symposium, 1991

[272] *Köhler, S.:* Plastverarbeiter 42 (1991) 11, p.138

[273] *Dahmen, H.J.:* Graduation paper, IKV Aachen, 1988

[274] *Marschall, U.:* K-Plastic & Kautschuk Zeitung, Febr. 1992 p.9

[275] Patent specification DE 3524310 C1, July 6, 1985

[276] *Paulson, D.C.:* Plast. Design Process. (1968) 10

[277] *Langecker, G.:* Wann kommt die selbstoptimierende Spritzgießmaschine? Print. information Battenfeld, Meinerzhagen/Germany, 10/1987

[278] *Bauer, R.:* Kunststoffe - German Plastics 79 (1989) 11, p.1135

[279] *Lampl. A.:* Plastverarbeiter 38 (1987) 5, p.14

[280] *Bockenheimer, A.:* Welche Qualitätsverbesserung bietet die PVT-Nachdruckoptimierung? Print. information Battenfeld, Meinerzhagen/Germany, 10/1989

[281] *Thurau, O.:* Kunststoffe - German Plastics 79 (1989) 10, p.943

[282] *Michaeli, W., Grundmann, M., Pichler, F.:* Kunststoffe - German Plastics 81 (1991) 11, p.1023

[283] *Steinbichler, G.:* Swiss Plastics 8 (1986) 9, p.47

[284] *Schaffner, W., Oertle, P., Rusterholz, R.:* Kunststoffe - German Plastics 64 (1974) p.34

[285] *Hellmeyer, H.O., Lixfeld, H.-D., Menges, G.:* Kunststoffe - German Plastics 67 (1977) p.184

[286] *Menges, G., Müller, H.:* Geisterschichten in der Kunststoffverarbeitung. Reprint, colloquium IKV Aachen, 1986*

[287] *Geisbüsch, P.:* Plastverarbeiter 42 (1991) 11, p.126

[288] *Reiling, A.:* Kunststoffe - German Plastics 79 (1989) 1, p.29

[289] *von Eysmont, B.:* Diss. RWTH Aachen, 1989

[290] *Stitz, S.:* Reprint of paper, SPE-Antec, May 1981

[291] Der konsequente Weg zur Qualtitätssicherung im Spritzgießbetrieb. Print. information Engel, Schwertberg/Austria, 1989

[292] CIM unser Konzept für die Zukunft. Print. information Mannesmann Demag Kunststofftechnik, 1989

[293] *Gliese, F.R.:* Diss. RWTH Aachen, 1987

[294] *Bongart, W.:* Kunststoffberater (1982) 10, p.12

[295] *Kornmayer, H.:* Plastverarbeiter (1990) 9, p.178

[296] *Jäger, F.M.:* Netstal News, No.20, Febr. 1986

[297] *Krist, Th.:* Hydraulik, Pneumatik, Fluidik/Pneulogik. Hoppenstedt Technik Tabellen Verlag, Darmstadt, 1991

[298] *Vargel, J.:* Diss. RWTH Aachen, 1974

[299] *Berger, J., Eder, S., Winkler, H.-P.:* Kunststoff - German Plastics 83 (1993) 3, p.176

[300] *Schäfer, O.:* Grundlagen der selbsttägigen Regelung. Technischer Verlag, Gräfelfing/Germany, 1970

[301] *Recker, H.:* Automatisierung in der Kunststoffverarbeitung. Hanser, Munich, New York, 1986

[302] DIN 19226, May, 1968

[303] *Görmar, H., Pütz, H.:* Kunststoffe - German Plastics 69 (1979) p.588

[304] *Görmar, H., Pütz, H.:* Messen an Extrusionsanlagen. VDI, Düsseldorf, 1978

[305] *Wortberg, J.:* Überwachung des Schmelzezustandes bei der Verarbeitung von Plastomeren. Paper, IKV Aachen, 1978

[306] Print. information Dynisco Gerate GmbH, Heilbronn

[307] *Wittemeier, R.:* Research project, University GSHS, Paderborn/Germany, 1990

[308] Print. information Schlaepfer AG, Winterthur/Switz.
[309] Print. information Kistler Instrumente AG, Winterthur/Switz.
[310] Print. information Hartmann & Braun, Frankfurt
[311] Print. information Hottinger Baldwin Meßtechnik GmbH, Darmstadt/Germany
[312] *Menges, G., Recker, H.:* AIF-Forschungsbericht No. 5020. Köln, 1982
[313] CADMOULD-3D, program system, IKV Aachen
[314] MOLDFLOW, program system, Moldflow Ltd.
[315] I-DEAS, program system, SDRC, Frankfurt/Germany
[316] *Sarholz, R., Beese, U.:* Prozeßführung beim Spritzgießen. Research project IKV Aachen, 1977
[317] *Stitz, S.:* Kunststoffe - German Plastics 63 (1973) 11, p. 777
[318] *Rörick, W.:* Diss. RWTH Aachen, 1979
[319] *Sarholz, R.:* Diss. RWTH Aachen, 1980
[320] *Hellmeyer, H. O.:* Diss. RWTH Aachen, 1977
[321] *Wiegand, H.:* Prozeßautomatisierung beim Extrudieren und Spritzgießen von Kunststoffen. Hanser, Munich, New York, 1979
[322] *Schwab, E.:* Automatisierung in der Kunststoffverarbeitung. Hanser, Munich, New York, 1986
[323] *Isermann, R.:* Digitale Regelsysteme, 2nd edition, vol. I. Springer, Berlin, Heidelberg, New York, London, Paris, Tokyo, 1988
[324] *Gordon, M.J.:* Total Quality Process Control for Injection Molding. Hanser, Munich, New York, 1992
[325] *Recker, H., Spix, L.:* Kunststoffmaschinenführer, 3rd ed. Hanser, Munich, New York, 1992
[326] *Buschhaus, F.:* Diss. RWTH Aachen, 1982
[327] *Lauterbach, M.:* Diss. RWTH Aachen, 1989*
[328] Transputers. European Plastics News (1991) 11, p. 22
[329] Messen, Steuern, Regeln. Documentation IKV Aachen, Hanser, Munich, New York, 1977
[330] *Bauer, E.:* Spritzgießen. VDI, Düsseldorf 1983 Documentation IKV Aachen, Hanser, Munich, New York, 1977
[331] *Schwab, W.:* Temperaturen messen und regeln. Series Ingenieurwissen, VDI, Düsseldorf, 1978
[332] *Allerdisse, W.:* Überprüfen und Angleichen von Thermofühlern, Thermofühlerleitungen und Reglern an gegebene Regelstrecken. Series Ingenieurwissen, VDI, Düsseldorf, 1978
[333] *Hengesbach, H. A.:* Plaste, Kautsch. 23 (1976) p. 38
[334] *Görmar, H., Pütz, H.:* Messen von Extrusionsdruck und Schmelzetemperatur unter besonderer Berücksichtigung der Messfühlerauswahl für den Produktionsbetrieb. Series Ingenieurwissen, VDI, Düsseldorf, 1978
[335] *Hardt, B.:* Plastverarbeiter 34 (1983) 3, p. 219
[336] *Scholl, K.:* Plastverarbeiter 33 (1982) p. 278
[337] Mehrfarbenspritzgießmaschine zur Verarbeitung von Thermoplasten. Print. information, Krauss-Maffei AG
[338] Kunststoff Magazin Prodoc (1990) 6, p. 36
[339] *Rothley, J.:* Rechner verkürzt Durchlaufzeit beim Spritzgießen. Special section in Hanser-Fachzeitschriften (1991) 12, p. CA304
[340] *Haupt, M.:* Diss. RWTH Aachen, 1989*

[341] *Burr, A.:* Sicherung der Qualität durch Prozeßüberwachung. 7. Kunststofforum, SKZ, Würzburg, 1987
[342] *Backhaus, J.:* Diss. RWTH Aachen, 1986*
[343] *Bourdon, K.:* Diss. Aachen, 1989*
[344] Reprints of papers, colloquia IKV Aachen, 1988
[345] *Eckardt, H.:* Plastverarbeiter 34 (1983) 4, p. 316
[346] *Steinbichler, G.:* Prozeßführung, Anlagenkonzept und Anwendungsbeispiele des Gasmeltverfahrens. Print. information Engel, Schwertberg, A, 11 (1990)
[347] *Rabe, J.:* Kunststoffe - German Plastics 83 (1993) 4, p. 291
[348] Infrarot Schmelze Temperatursensor MTT 900, Print. information Dynisco Geräte GmbH, Heilbronn / Germany
[349] *Hüttner, H.J.:* Kunststoffe - German Plastics 82 (1992) p. 967
[350] Plastverarbeiter 44 (1993) 3, p. 114
[351] *Thienel, P., Jehn, P.:* Plastverarbeiter 44 (1993) 3, p. 123
[352] *Gierth, M.M.:* Aachener Beiträge zur Kunststoffverarbeitung. Vol. 5, IKV Aachen, Verlag der Augustinus Buchhandlung, Aachen, 1992
[353] *Johannaber, F.:* Mehrkomponententechnik beim Spritzgießen. Paper SKZ Convention, May 6, 1992
[354] *Couglin, J.:* Mod. Plast. Int. (1992) 4, p. 57
[355] *Eckardt, H.:* Im Innern muß nicht immer Schaum sein. Paper, Dec. 14, 1982, Meinerzhagen / Germany
[356] *Kirkland, C.:* Plastic World, (1991) 2, p. 37
[357] *Kreisher, K.R.:* Mod. Plast. Int. (1990) 3, p. 56
[358] *Maskus, P.:* Engineering Plastics 1 (1988) 4, p. 260
[359] *Zipp, Th.:* Diss. RWTH Aachen, 1992
[360] *Zipp, Th.:* Aachener Beiträge zur Kunststoffverarbeitung. Vol. 3, Verlag der Augustinus Buchhandlung, Aachen, 1993
[361] GTZS-Verfahren. Print. information, Klöckner-Ferromatik Desma GmbH, Malterdingen, 1989
[362] *Dümmel, R.:* Plastverarbeiter 40 (1989) 12, p. 48
[363] *Friesenbichler, W., Ebster, M., Langecker, G.:* Kunststoffe - German Plastics 83 (1993) 6, p. 445
[364] *Rathgeb, R.:* Plastverarbeiter 43 (1992) 1, p. 50
[365] DE 3932416 A1
[366] *Klamm, M., Feldmann, L.:* Kunststoffe - German Plastics 78 (1988) 9, p. 767
[367] *Eckardt, H., Ehritt, J.:* Plastverarbeiter 40 (1989) 1, p. 14 and 2, p. 130
[368] DE 3905177 A1
[369] *Coesfeld, W., Wegmann, J.:* Graduation paper, FH Lippe-Lemgo, 1978
[370] Print. information Bayer AG, Leverkusen
[371] *Jaroschek, Ch., Utescheny, R.:* Plastverarbeiter 42 (1991) 5, p. 55
[372] Gasinjection finally feed for growth. European Plastic News 5 (1991) p. 61
[373] Eine Möglichkeit lange Zykluszeiten zu verkürzen. Plastverarbeiter 42 (1991) 5, p. 26
[374] Reprint, colloquium IKV Aachen, 1992
[375] *Jaroschek, Ch.:* Kunststoffe - German Plastics 80 (1990) 8, p. 837
[376] US patent for gas injection. European Plastics News (1990) 9, p. 58
[377] Orientierung prägt innere Werte. Kunststoff Magazin Prodoc 12 (1990) p. 44
[378] Mechanism of the Structural Web Process. Paper at the 2nd World Congress of Chemical Engineering, Montreal, Oct. 1981
[379] Reprint, colloquium IKV Aachen, 1969

[380] *Allan, P.S., Bevis, M.J.:* Plastics and Rubber Processing and Applications 7 (1987) p.3
[381] *Becker, H., Gutjahr, L.M.:* Neue Werkstoffe, Magazin, Berlin, No.3, Sept. 29, 1989
[382] *Gutjahr, L.M., Becker, H.:* Kunststoffe - German Plastics 79 (1989) 11, p.1108
[383] *Plaetschke, R.:* Internal report, Bayer, Leverkusen 1993
[384] *Thienel, P., Hoster, B., Schröder, K., Schröder, Th., Kretschmer, J., Ludwig, R.:* Kunststoffe - Synthetics (1993) 2, p.12
[385] *Michaeli, W., Lanvers, A.:* Plaste und Kautschuk 39 (1992) 7, p.241
[386] *Henschen, S., Brinkmann, T.:* Kunststoffe - German Plastics 82 (1992) 12, p.1274
[387] *Eckardt, H.:* Kunststoffe - German Plastics 82 (1992) 10, p.881
[388] *Sauer, R.:* Paper at seminar „Gasinnendruckverfahren beim Spritzgießen". SKZ, Würzburg, 1992
[389] Größere Flexibilität und Vielseitigkeit durch Heißgasspritzgießen. Plastverarbeiter 42 (1991) 6, p.54
[390] *Becker, H., Fischer, G., Müller, U.:* Kunststoffe - German Plastics 83 (1993) 8, p.54
[391] Mehrkomponenten- und Gasinnendruck-Spritzgießverfahren. Convention and handbook, SKZ, Würzburg, Sept. 1990
[392] *Anders, S., Johannaber, F., Steinbüchel, R., Sauer, R.:* Kunststoffe Europe (1991) 4, p.68
[393] *Trepte, H.:* Plaste und Kautschuk 11 (1964) 2, p.109
[394] *Trepte, H.:* Plaste und Kautschuk 12 (1965) 3, p.158
[395] *Michaeli, W., Eversheim, W.:* CIM im Spritzgießbetrieb, Hanser, Munich, New York, 1993
[396] Startzeichen für den Motor des Fortschritts. Print. information Montaplast GmbH, Morsbach and Bayer AG, Leverkusen
[397] *Anders, S., Littek, W., Schneider, W.:* Kunststoffe - German Plastics 80 (1990) p.997
[398] *Michaeli, W., Galuschka, S.:* Plaste und Kautschuk 39 (1992) 8, p.281
[399] *Bürkle, E., Rehm, G., Zweig, K.:* Kunststoffe - German Plastics 82 (1992) 3, p.102
[400] *Michaeli, W., Galuschka, S.:* Plastverarbeiter 44 (1993) 3, p.102
[401] *Nething, J.-P.:* Handbook, Convention „Der neue Trend im Automobilbau", SKZ, Würzburg, 1990
[402] *Roth-Walraf, H. A.:* Handbook, Convention „Der neue Trend im Automobilbau", SKZ, Würzburg, 1990
[403] *Mischke, J., Bagusche, G.:* Kunststoffe - German Plastics 81 (1991) 3, p.199
[404] Print. information Krauss-Maffei Kunststofftechnik GmbH, Munich, 1991
[405] *Jäger, A., Fischbach, G.:* Kunststoffe - German Plastics 81 (1991) 10, p.869
[406] *Eyerer, P., Bürkle, E.:* Kunststoffe - German Plastics 81 (1991) 10, p.851
[407] *Eyerer, P., Märtins, P., Bürkle, E.:* Kunststoffe - German Plastics 81 (1993) p.505
[408] Print. information Ikegai Corp., Tokyo, 1990
[409] Print. information Sumitomo Chemical Corp., Tokyo, 1990
[410] *Likar, M.:* Graduation paper FH Rosenheim / Germany, 1992
[411] *Schmid, G.:* Graduation paper IKP, University Stuttgart, 1992
[412] Neues Patent für Gasinjektion. Kunststoff und Plastic Zeitschrift, Aug. 23, 1990, p.7
[413] *Klotz, B., Bürkle, E.:* Kunststoffe - German Plastics 79 (1989) 11, p.1102
[414] DE 3913109 C2, Jan. 27, 1991, patent application, Kloeckner Ferromatic Desma GmbH
[415] *Eckardt, H.:* Kunststoffe - German Plastics 75 (1985) 3, p.145
[416] *Jaroscheck, C.:* Kunststoffe - German Plastics 80 (1990) 8, p.873
[417] *Steinbichler, G.:* Kunststoffe - German Plastics 77 (1987) 10, p.931
[418] *Knür, E.:* Technische Daten Spritzgießmaschinen 3 (1990)
[419] *Sakai, T., Nakamura, K., Morli, A.:* Paper at SPE-Antec, New York, 1989
[420] *Wübken, G.:* Plastverarbeiter 26 (1975) p.17

[421] Netstal News 1980
[422] *Menges, G., Bangert, H.:* Kunststoffe - German Plastics 71 (1981) p. 552
[423] *Bangert, H.:* Diss. RWTH Aachen, 1981
[424] *Zöllner, O.:* Hanser Fachzeitschriften (1991) p. CA 228
[425] *Haak, W., Schmitz, J.:* Rechnergestütztes Konstruieren von Spritzgußformteilen. Vogel, Würzburg, 1985
[426] *Menges., G., Mohren, P.:* How to Make Injection Molds. Hanser, Munich, New York, 1992
[427] *Gastrow, H.:* Der Spritzgieß-Werzeugbau. 4th ed. Hanser, Munich, New York, 1990+
[428] *Schürmann, E.:* Reprint of paper, 35th SPE-Antec, 23 (1977) p. 104
[429] *Backhoff, W., Lemmen, E.:* Stammwerkzeug für Probekörper. Print. information Bayer AG, Leverkusen, 1979
[430] *Hamer, B.:* Plaste und Kautschuk 39 (1992) 2, p. 56
[431] *Oberbach, K., Rupp, L.:* Kunststoffe - German Plastics 77 (1987) 8, p. 783
[432] *Oberbach, K.:* Qualitätssicherung bei Polymerwerkstoffen. VDI-Berichte No. 600.5, 1989
[433] *Breuer, H., Dupp, G., Schmitz, J., Tüllmann, R.:* Kunststoffe - German Plastics 80 (1990) 11, p. 1289 and Plastverarbeiter (1988) 4, p. 50
[434] *Michaeli, W., Lessenich-Henkys, V., Bücher, R.:* Plastverarbeiter 42 (1991) 9, p. 68
[435] *Menges, G., Schulze-Harling, H.:* Kunststoffe - German Plastics 65 (1976) p. 732
[436] *Schulze, P., Funke, A.:* Plaste und Kautschuk 21 (1974) p. 683
[437] Cand: Plast. Rech. 27 (1989) p. 74
[438] *Meyer, D.:* Handhabungsgeräte und Industrieroboter. Kunststoffmaschinenführer, Hanser, Munich, New York, (1992)
[439] *Harreis, J.:* Kunstst. Ber. 4 (1982) 5, p. 43
[440] *Hotz, A.:* Netstal News 6 (1991)
[441] *Johannaber, F., Schwittay, D.:* Reprint of paper 38th SPE-Antec, (1980) p. 146
[442] *Schwittay, D., Sajben, J.:* Plastverarbeiter 33 (1982) 11, p. 1366 and 12, p. 1477
[443] *Friel, P.:* Leistungsfähige Temperiergeräte. Spritzgießen, VDI, Düsseldorf, 1983
[444] *Gorbach, P.:* Handbuch der Temperaturregelung mittels flüssiger Medien. Print. information Regloplas, St. Gallen/Switz., 3rd ed., 1986
[445] *Johannaber, F.:* Die Trocknung von Kunststoffen. Print. information Bayer AG, Leverkusen, 1978
[446] Statistisches Bundesamt 1992
[447] *Potente, H.:* Entwicklungen bei der Auslegung von Plastifizierschnecken, Plastverarbeiter 43 (1992) 11, p. 118
[448] *Dietzel, H., Schumann, J., Müller, K.:* Kunststoffe - German Plastics 81 (1991) 12, p. 1138
[449] *Müller, H., Plessmann, N.:* Die Zukunft der deutschen Kunststoffindustrie. Opinion sampling by IKV Aachen, 1983
[450] *Potente, H.:* Auslegung von Schnecken-Maschinenbaureihen, C. Hanser, Munich, 1981
[451] Neues Patent für Gasinjektion. Kunststoff und Plastic Zeitschrift, Aug. 1990 p. 7
[452] *Lanvers, A.:* Analyse und Simulation des Kunststoff-Formteilbildungsprozesses bei der Gasinjectionstechnik. Aachener Beiträge zur Kunststoffverarbeitung, volume 11, 1993
[453] *Steinbichler, G.:* Print. information, Engel, Schwertberg/Austria, 9 (1990) D
[454] *Steger, R.:* Die Formgebung pulvermetallurgischer Werkstoffe auf der Spritzgießmaschine. Print. information Kloeckner-Ferromatic-Desma GmbH, Malterdingen, Mai 1993
[455] *Domininghaus, H.:* Die Kunststoffe und ihre Eigenschaften. VDI, Düsseldorf, 4th ed., 1992
[456] *Saechling, H.:* Kunststofftaschenbuch, Hanser, Munich, New York, 25th ed., 1992

[457] Several new technologies make thermosets more appealing for high volume markets. Mod. Plast. Int. (1990) 6, p. 34
[458] *El Sayed, A.:* Kunststoffe - German Plastics 83 (1993) 3, p. 213
[459] *Chung, C.I., Ree, B.O., Cao, M.Y., Lin, C.X.:* Advances in Powder Metallurgy. Vol 3, Metal Powder Industries Fed., American Powder Metallurgy Institute 1989
[460] *Graf, W., Pötschke, J.:* Das Krupp-Verfahren des Metallpulverspritzgießens, Pulvermetallurgie in Wissenschaft und Praxis. VDI, Düsseldorf 1993
[461] Gemischte Abfälle rezyklieren, Kunststoffe - German Plastics 83 (1993) 4, p. 264
[462] *Eyerer, P., Schuckert, M.:* Recycling im Spritzgußbetrieb, VDI, Düsseldorf 1993
[463] *Schuckert, M., Dekorsky, Th., Pfleiderer, I., Eyerer, P.:* Kunststoffe - German Plastics 83 (1993) 3, p. 195
[464] *Becker, H., Fischer, G., Müller, U.:* Kunststoffe - German Plastics 83 (1993) 3, p. 165
[465] *Bürkle, E.:* Hinterpressen und Hinterprägen - eine neue Oberflächentechnik. paper at convention SKZ, Würzburg, Dec. 1993
[466] *Jaroschek, C.:* Integrierte Teilefertigung durch Direktspritzguß von festen oder beweglichen Werkstoff- und Teileverbunden. paper at convention SKZ, Würzburg, Dec. 1993
[467] *Bourdon, K., Robers, T.:* Vorteile und Einsatzgebiete alternativer Arbeitstechniken für Spritzgießmaschinen. paper at convention SKZ, Würzburg, Dec. 1993
[468] *Wright, R.E.:* Molded thermosets. A Handbook for Plastics Engineers, Molders and Designers, Hanser, Munich, 1991
[469] *Hunold, R.P.D.:* Analyse der Verarbeitungseigenschaften Duromerer Formmassen und Ansätze zur Prozeßoptimierung beim Spritzgießen. Report IKV Aachen, 1992
[470] *Lauterbach, M.:* Ein Steuerungskonzept zur Flexibilisierung des Thermplast-Spritzgießprozesses. Report IKV Aachen, 1998
[471] Das Mono-Sandwich-Verfahren. Plastverarbeiter 46 (1993) 7, p. 14
[472] *Schwab, E.:* Plastverarbeiter 47 (1994) 2, p. 26
[473] *Bourdon, K.H., Jaroschek, Ch.:* Recycling im Spritzgießbetrieb, VDI, Düsseldorf 1993+
[474] *Robers, Th.:* Analyse der Übertragbarkeit von Maschineneinstelldaten beim Spritzgießen von Thermoplasten. Final report of a AIF research project, IKV Aachen, 1993
[475] *Menges, G.:* Werkstoffkunde Kunststoffe. 3rd ed., Hanser, Munich, 1990*
[476] *Knappe, W., Lampl, A., Heuel, O.:* Kunststoffverarbeitung und Werkzeugbau. Hanser, Munich, 1992
[477] *Zipp, Th.:* Fließverhalten beim 2-Komponenten-Spritzgießen. Aachener Beiträge zur Kunststoffverarbeitung, IKV Aachen, 1992
[478] *Obendrauf, W., Kukla, C., Langecker, G.R.:* Kunststoffe - German Plastics 83 (1993) 12, p. 971
[479] *Konejung, K.:* Kunststoffe - German Plastics 83 (1993) 12, p. 956
[480] *Findeisen, H., Galuschke, S., Herren, S., Smets, H.:* Neue Produkte, verbesserte Qualität, höhere Wirtschaftlichkeit. 17th Plastics Colloquium, IKV Aachen, March 1994*
[481] *Bluhm, R., Hohenauer, K., van Oepen, R., Robers, Th., Vakulik, R.:* Prozeßführung beim Spritzgießen, 17th Plastics Colloquium, IKV Aachen, March 1994*
[482] *Burr, A.:* SPIRIT - PC-Software für Spritzgießbetriebe zur Steigerung der Wirtschaftlichkeit und Qualität. Summary, PIK, FH-Heilbronn/Germany, 1992
[483] *Burkhardt, D.:* Form + Werkzeug (1993) 10, p. 505
[484] *Eyerer, P., Märtens, R., Bürkle, E.:* Kunststoffe - German Plastics 83 (1993) 7, p. 505
[485] *Jaroschek, C.:* Kunststoffe - German Plastics 83 (1993) 7, p. 519

* This item contains an extensive list of references for further information on the subject.
+ A detailed description of the subject matter is presented with this item.

Index